AF553850

ENVIRONMENTAL BIOTECHNOLOGY

ENVIRONMENTAL BIOTECHNOLOGY

S. K. AGARWAL
Head, P.G. Department of Botany
Government College, Kota

A P H PUBLISHING CORPORATION
4435-36/7, ANSARI ROAD, DARYA GANJ
NEW DELHI-110 002

Published by

S.B. Nangia
for **APH PUBLISHING CORPORATION**
4435-36/7, Ansari Road,
Darya Ganj,
New Delhi-110 002
Tel. 23274050
23285807
Fax 91-011-23274050
Email: aphbooks@vsnl.net

2026

© Author

Printed at
Balaji Offset
Navin Shahdara, Delhi-32

Contents

Preface

Biotechnology as an art has been practised for a long time in traditional activities such as fermentation and brewing. It has now emerged as a separate discipline, pointing to future revolutionary developments for the good of mankind. Spectacular advances that have taken place in the field of biotechnology in the last two decades have a profound impact on the socio-cultural scene, life-style, health care system, availability of food, employment opportunities, energy and clean environment. The promise of biotechnology offers hope to the most destitute people on the earth. Through its miracle we can conceive of winning the eternal struggle against hunger and disease.

Biotechnology stands on the understanding of molecular basis of biological cell functions and the ability of mankind to alter cell functions to make it produce products required by society. New techniques available with biotechnology holds potentials for developing products and processes in various sectors of agriculture, horticulture, floriculture, forestry, animal husbandry, health care, energy generation and environmental protection.

Virtually all types of human activity generate wastes and this places a heavy burden on the environment. So far we have relied more upon the physical and chemical methods of pollution control. However, micro-organisms are likely to prove as more suitable tools for pollution control due to their versatility and adaptability. In the field of environmental management biotechnology have helped in environmental monitoring, degradation of wastes, substitution of non-renewable resource base with a renewable resource base, and production of eco-friendly products and processes.

The application of biotechnologies with its risks that are difficult to define, much less to assess. The introduction of genetically engineered microbes can create new ecological niches and bring about large scale transformations in the structure and function of the ecosystem as a whole. There are many unknown risks and hazards associated with the accidental or deliberate release of self-propagating, genetically engineered novel forms of life into the biosphere. In the event of disastrous consequences, it is likely that people will be harmed who had little to gain from the use of biotechnology.

The first chapter deals with introduction with an emphasis on Biotechnology an agenda for development, biotechnological boom and ecological doom, Biohazards and Biosafety. This is followed by conventional fuels and their environmental impacts, Non-conventional fuels and their environmental impacts, Transgenic plants with an emphasis on improved fibre quality, Transgenic animals with an emphasis on improved wool production, Biotechnology for pollution abatement, Bioremediation, Biomineralization, Thuringiensis toxin as a natural pesticide, Biological control of insects swarming agriculture fields, Biofertilizers, and Biotechnology of food.

The book in no way wants to blindly support biotechnology as the solution, but definitely intends to impress upon the reader its importance as feasible for environmental problem solving.

It was interesting experience to hunt for literature on this subject. I would like to express my appreciation to Dr. (Mrs.) C.M. Nagalakshmi of NMKRV College for Women, Bangalore for helping me to get acquainted with people working in the field of Biotechnology, while attending UGC Sponsored Workshop on Biotechnology from 25th July to 10th August 1996. Thanks are also due to Prof. R.R. Das, Vice Chancellor, Jiwaji University, Gwalior, Prof. P.S. Dubey, Head, School of Studies in Botany, Vikram University, Ujjain; Prof. M.L.H. Kaul, Kurukshetra University, Kurukshetra; Prof. K. Dharmalingam, Madurai Kamraj University, Madurai, Prof. Gopinathan, Institute of Basic Medical Science, Madras; Dr. G.A.

Ravishankar, CFTRI, Bangalore; Dr. Prasad Rao, Pasteur Institute of India, Coonoor; Dr. S.L. Kothari, Rajasthan University, Jaipur; Dr. S.D. Purohit, M.L. Sukhadia University, Udaipur for their help in literature survey.

Two more persons who helped me also deserve my appreciation. Among them DaSilva for his article on Biotechnology an agenda for development, and Prof. R.R. Das for his article Biotechnological boom and Ecological doom deserve special mention. The author wishes to acknowledge with great pleasure all the scientists for their diagrams and tables that has been incorporated in this book, although the author has not been able to contact the authors and publishers in every case for permission. I would like to express my gratitude to my wife Sudha for her encouragement during the preparation of the manuscript.

The book is profusely illustrated with simple line drawings and data based with tables, readily reproducible by students. It is hoped that this book would be of reference value, so as to serve the needs of Teachers, Students, Entrepreneurs, Engineers, Technologists and Planners, as it covers the most important aspects of biotechnology.

S.K. Agarwal

Chapter 1

Biotechnology

An Agenda for Development

Biotechnology as an art has been practised for a long time in traditional activities such as fermentation and brewing. Many other biological processes have been in use in various industries such as production of antibiotics and enzymes. With so many recent advances in cell and molecular biology including virology, biotechnology has now emerged as a separate discipline. The real impact of this latest science that is, the application of biological processes has yet to show up, but a beginning is already pointing to future revolutionary developments for the good of mankind. It is another example of how basic biological research, besides providing an understanding of the life processes, has come to a stage where applications are coming out in many directions. It is this application of biological systems including various processes involving different organisms associated with them, this constitutes what is now known as "biotechnology". The potential is really infinite in agricultural, industrial and medical sciences. Following are some of the available definitions of biotechnology, and one of them may be really adequate:

(1) Biotechnology is "the application of biological organisms, system or processes to manufacturing and service industries."

(2) Biotechnology is the integrated use of biochemistry, microbiology and engineering sciences in order to achieve technological application of the capabilities of

micro-organisms, cultured tissue cells and parts thereof.

(3) Biotechnology is a technology using biological phenomenon for copying and manufacturing various kinds of useful substances.

(4) Biotechnology is the controlled use of biological agents, such as micro-organisms or cellular components, for beneficial use.

The recent discoveries which have led to the development of techniques of genetic engineering, cell fusion amongst plants and animals, and micro-organisms, immobilized enzymes etc. have opened up new possibilities for the production of fuel from renewable resources, fertilizing of crops, better yielding varieties of plants, cheaper products, better methods of early diagnosis of diseases, more selective methods of pest control, stabilization of waste water etc. Biotechnology, therefore involves several basic disciplines such as biology, biochemistry, pathology, physics, mathematics, chemistry and engineering. Scientists belonging to these disciplines have all been an essential part of this new and exciting concept and are responsible for this latest breakthrough in science.

A higher plant has between one and ten million genes, each of which holds information that specifies the structure of a product RNA and often a protein. Each has a nucleotide sequences involved in recognising on-off signals. Pieces of DNA carrying one or a few genes can now be removed, replicated in micro-organisms and studied at the level of nucleotide sequence. A gene serves the organism not only by encoding a product but also as an integral part of the organism as a whole.

Biotechnology, essentially involves the manipulation of cells and micro-organisms including viruses, bacteria, yeast, fungi, algae and cells of higher plants and animals. This tast in turn requires genetic manipulation involving transfer or alterations in genes or genetic material from one organism to another. This process is then made use of in achieving a desired objective, such as production of an enzyme(s), hormone or some biological material by the recipient which was not normally possible. Such a process has also been termed as 'genetic engineering' or recombinant DNA research. Obviously, handling such a basic process with so many diverse repercussions involving different living cells and organisms including human beings may have certain hazards. And these have been argued and discussed time and again. However, these problems can be controlled or eliminated by careful planning and judicious use.

One of the common techniques used now is that the DNA isolated from cells of a micro-organism, plant or animal which is termed as donor, is inserted into another piece of DNA through what is known as a vector and then transferred to a host micro-organism which acquires some novel genetic properties. Some of the steps included are:

1. Extraction of DNA and then cutting it in two small pieces of known sequences by means of restriction enzymes (endonucleases).
2. Splicing of DNA into vectors which are either plasmids or viruses; small sequences of DNA are joined or spliced into the vector DNA molecule by DNA-ligase thus creating an artificial DNA molecule.
3. Introduction of vectors into host cells, where they act as replicons and exist as extra chromosomal elements. They are normally transferred by transduction or transformation.
4. Selection and characterization is finally made of the recombinant clones with newly acquired DNA.

Human ingenuity has responded magnificently through the use of the green and gene revolution, not withstanding some disadvantages. In addition to these strategies, account must be taken of the beneficial activities of the invisible microbe, in fixing atmospheric nitrogen. A variety of market-products developed in the non-industrialized countries are now available to conserve soil fertility and increasing crop yield (Sasson, 1994).

The potential of the micro-organism for industrial uses and food production is legion and legend. Solvents, chemicals and bioenergy, such as acetone, butanol and methane result from the process of anaerobic fermentation. Microbial action has been used to produce food and feed from waste cellulosic materials. Textured mycelial food products, such as "Phycomeat" and "Quorn", that stimulate veal and chicken, alongwith microbial production of pharmaceuticals and immuno-active components, and the synthesis of interferons, rennet and growth promoters by genetically engineered microbes are all realities.

Microbial cultures or their enzyme components are the basic elements of industrial microbiology and biotechnological processes. Many microbes of economic significance, have over the years, been lost through lack of resources which are financial, technical or human in nature. In the MIRCEN (Microbial Resource Centre) network, emphasis has been given to the practical rather than to the spectacular aspects in the development and management of culture collections since they are safeguarded against extinction. Moreover, as a result of the Budapest Treaty pioneered by another U.N. Agency—the World Intellectual Property Organisation, culture collections also function as depositories of patented microbial germplasm. Culture collections are also considered as equivalent to seed banks, botanical gardens and zoos, because they conserve representatives of the world's micro-organisms, and make them available to other users, in the same way, for example, as the International Centre for Tropical Agriculture does for seeds.

Central to these issues is the uneven distribution and sharing of biotechnological resources amongst nations. In the technically advanced societies such resources spring from a research base in the life-sciences that sparkles with acknowledged talent, laboratories of excellence, adequate support funds, private sector capital, governmental commitment, and commercial incentives. Individual countries have reached different levels of activity. The United States, Japan and a number of European countries are often described with identifiable profiles in biotechnology. Hence, if one is to measure success of microbiology and biotechnology in terms of successful commercial products, then without doubt these are the countries which at present lead the world. The other side of the coin, that is the developing countries, are beset with the problems embodied in the "Trinity of Despair" that is hunger, resource depletion and disease.

This despair is further compounded by the lack of cornerstone infrastructure of basic and applied research and of required qualified personnel in different numbers in many of the developing countries of the Southern hemisphere and now since 1989, in East Europe and the newly-emerged countries of Central Europe. Again, in the least developed countries that constitute the planet's belt of the malnourished and the hungry, advanced microbiology and biotechnology, at best, is limited to the academic level since facilities for postgraduate industrial research are virtually non-existent.

DaSilva and Sasson, (1989) made an attempt to answer the question. Are biotechnologies going to be a panacea for the various problems the developing countries are facing, or will they merely add to the disparities which exist between these countries and the industrialized technologically advanced world? In response it was noted that:

> "there is not one singles trategy which may be adopted with a view to using biotechnologies and reaping their benefits. For instance, India, a country with a large budget for research, and numerous scientists and technicians, cannot adopt the same strategy as a small African country of the Sahel-Sudan zone. Nevertheless, even the poorest and least technologically and scientifically advanced countries can reap some benefits from the progress of biotechnologies and participate in the *biotechnological revolution*, thanks to international and regional cooperation networks. This participation does not necessarily concern the most advanced basic research but should rather aim at adopting local capabilities and capacities to development problems. It would be a sad mistake to try and imitate the policies adopted by industrialized and technologically advanced countries".

It was also noted that:

> "biotechnology unquestionably generate benefits or gains, they are also bringers of certain dangers or potential threats. Their impact on societies will be considerable and there will be winners and losers. But no country, and no community, is condemned to become a loser. This will depend on which strategy are adopted by the community, country, or group of countries in order to reap their legitimate portion of the benefits of biotechnologies. In view of the fact that the "*biotechnological revolution*" will affect even the most isolated societies, it is neither wise or justified not to participate in this

revolution and not to fight for gaining some of its expected advantages".

It is in this context that the microbial world enriched by the techniques of molecular biology can help the developing world, ranging from the newly industrialized countries to the least developed countries, to narrow the gap in the North-South dialogue on biotechnology.

In less developed countries, traditional farming is based on photosynthesis—the natural biosolar energy conversion mechanism that sustains life from the primitive to the most advanced. It is complemented by additional energy sources such as the use of organic waste and the deployment of human and animal power for cultivating crops. In more developed countries, agriculture is productive with costly inputs coming from the use of fertilizers, pesticides, improved irrigation and drainage methods, and the use of fossil fuels in highly mechanised farm management practices. Thus, agricultural practice in the developing countries is already at a disadvantage and it is characterized by sweat and lack of subsidy in comparison to that in the industrialized societies where agriculture enjoys subsidy and sophisticated machinery.

The emphasis on the use of beneficial nitrogen-fixing micro-organisms is justified by the low-cost production of rhizobial and algal inoculents that are easily accessible to farmers, particularly in less developed countries. Such farmers do not have the means to buy the more expensive petroleum derived fertilizers, which increase production but pollute the environment in varying ways. Today, in these times of economic recession, many developing countries (Brazil, Kenya, Thailand, Senegal) are turning to the production and use of biofertilizers to reduce, either partially or completely, dependency on the chemically-derived products.

The role of micro-organisms as biotechnological tools for rural processing of organic residues has been emphasized in several national programmes, for example, in China, Brazil, Mexico and India, for the production of fuel, fertilizers, feed and food.

The bioconversion of organic and agricultural residues by the microbial labour force is attractive because it is multi-purpose and has been proven by scientific and technological experience. A good example is the well-known Xinbu System in China. Non-polluting in its technology and feeding upon renewable resources, bio-conversion of bio-degradable resources is appealing since it contributes to energy production, counteracts the deterioration of the environment, and facilitates waste management. The use of renewable resources, irrespective of the approach, that is, classical or modern through the use of genetically-engineered microbes, has been further endorsed by the G-7 countries in Article-20 of their Economic Declaration of the Summit Meeting of 17th July 1991 which states:

> "the commercial development of renewable energy sources and their integration with general energy sources should also be encouraged, because of the advantages these sources offer for environmental production and energy security".

It is thus possible to integrate food, fodder, fuel, fibre and chemical production into a new type of highly economic, multi-purpose agriculture. They produce more food for the planet's increasing population and to maintain technological growth, it is evident that deep inroads will continue to be made into nature's energy bank. But to change lifestyles and programmed industrial growth patterns in midstream, though possible, is neither easy nor attractive. The impact of such a move on the labour and market, sectors would be devastating.

In the field of microbial insecticides, various viral, bacterial and fungal preparations, have been developed for use against a variety of pests encountered in agricultural sector and some of these are already on sale. Compared with chemical pesticides, microbial products are not harmful to non-target organisms and do not contaminate the environment.

Furthermore, research in the use of microbes for the extraction and recovery of metals indicates the development of a new microbial technology-bio-metallurgy. Potentially profitable, bio-metallurgy is expected to contribute to the growing supply of a number of critical minerals. Bio-metallurgy involving the extraction or leaching of insoluble minerals and the recovery of metals from solution by microbes is popular since it allows for economic extraction of valuable metals from low-grade ores, involves low-capital costs and energy inputs rather than the brute forces of temperature and pressure of pyrometallurgy, and uses simple technology that does not pollute the atmosphere. The reporting of the constant hiking of *tin* prices in Malaysia and the concerns of the *Copper* producing nations are serious indicators of the turmoil surrounding the availability of minerals. Like oil and food, the acquisition of strategic non-fuel minerals is being enmeshed into socio-economic strategies. One of the most industrialized countries appears to be dependent upon 24 of the 32 strategic and critical minerals which originate in exports from Zaire, Zambia and Zimbabwe. And, in response to growing competition from low-cost producers in the developing world, the U.S. Bureau of Mines and CANMET of Canada have encouraged the practice of microbial *in-situ* leaching of copper in their countries.

The cultural heritage of virtually every civilization includes one or more fermented foods made by the souring action of microbes, for example *Leben* (Egypt and Syria), *taettemjolk* (Scandinavia), *matzoon* (Armenia), *dahi* (India), *piner* (Lapland), *chak* and *mopwo* (Kenya), *yakult* (Japan), *kefir* (Bulgaria), and the *blue-veined cheese* produced by the fungus *Pencillium roquefortii* (France) are all well-known examples. Men and women thus found ways to deploy a mix of microbes and traditional domestic skills to make new protein-rich foods, which tested better and kept longer. Another example of the role of microbes in food science and technology is to be found in the high-tech *nouvellie cuisine* of "fish chewies" and golden "meat nuggets" derived essentially from myco or microbial-protein. Indigenous fermented milks and foods are very important as they are socio-culturally bound in village household and community traditions. They also contribute towards providing a sense of national identity.

Rural societies, in feeding themselves, have resorted to fermenting corn, rice and milk to produce low-cost nutritional value foods for local consumption. For example, the bean curd factory at Zhengzhou, China is one such example that relies on microbial fermentation to reduce waste through stockage to concentrate the protein content of the curd to help sustain human health.

Another example of microbial technology, using gene-manipulated strains of *Zymomonas mobilisce* to bioconvert organic agricultural residues is that of alcohol. Widely recognised as an echo of the oldest form of biotechnology, this commodity is now being used as an environmental-friendly fuel in the transportation sector of some national economies. Indeed, one of the best-known biotechnological approaches to import substitution is to be found in the introduction of ethanol fuels. Brazil, in providing the insights into the tapping of the natural bioresources available, has the world's largest alcohol fuels programme dating back to 1902. Other examples are to be found in the national bioenergy programmes initiated by China and India to meet needs of electrification, and in Brazil and Thailand for the transportation sector, which programmes benefit from a rather important reduction in expenditure of valuable foreign exchange.

Virtually all types of human activity generate waste and this place a heavy burden on the environment. For centuries, soil micro-organisms have been involved in the degradation processes at landfills, agricultural waste sites, and sewage treatment facilities. Today, the use of the modern techniques of genetic engineering offer possibilities of developing organisms with a number of promising applications such as heavy metal sequestration, degradation of persistent toxic chemicals such as chlorophenols, oil spill dispersion, biomass conversion, and volume reduction of municipal wastes.

However, the most promising areas of application of these novel organisms is in clean technology or better still, preventive technology, wherein they are applied directly to potential waste streams at the point of origin in manufacturing process. Such waste minimization procedures have the advantage that the organisms used can be confined and therefore present less of a burden in terms of the assessment of possible risks particularly to the environment. Examples include deamination of residual amino-napthalenes in the dye industry; denitrification of cynide-producing industrial processes; partial denitrification of certain hazardous chlorinated by-products of pesticide manufacture; digestion of lignins in the pulp and paper industry; the development of cleaner technologies in distillery and tannery related processes and the concentration of heavy metals in waste streams either at the point of origin or at sampling stations.

In addition to bacterial components of the world of microbes, other microscopic entities have contributed to the welfare and progress of humanity. For centuries microbes have been consumed as gastronomic delicacies. Today, mushroom cultivation is a lucrative practice and holds significant economic potential for developing countries particularly in the fields of economic growth and nutrition. The mushroom used, *Volvariella, Pleurotus, Lentinus, Morchella* and *Agaricus* contain 30 to 40 per cent protein on a dry weight basis and are of significance in vegetarian diets.

Furthermore, mushroom cultivation is an excellent example of a low-capital technology through which unattractive waste such as vegetable and agro-industrial residues are converted into proteinaceous palatable food that is within the purchasing power of the middle-class consumer. It is indeed as effective weapon in counteracting malnutrition and poverty of food intake.

There are many areas of scientific interest and commercial potential in biotechnology that do not involve the latest research findings in molecular genetics or immunology. It is important to stress the wealth of untapped natural genetic variation in the microbial fauna of the tropics which has immense potential for production of specialty chemicals from inexpensive natural resources.

Specialty chemicals are a developing area of microbial technology. As yet there has been only limited interest in using the vast array of biotechnological diversity of micro-organisms to produce such chemicals. One example is the use of *Xanthomonas* for the production of extracellular polysaccharides, that is the Xanthon gums for use in the food industry. It is clear, however, that many micro-organisms have the ability to produce polysaccharides, which have many uses other than in processed food, for example, in crude oil recovery, the paint and textile industries, pharmaceuticals and cosmetics. A recent development has been the use of Methylotrophs, organisms which can use inexpensive methanol as their carbon and energy source, to produce an entirely new range of biopolyment gums.

A further example of micro-biological diversity used for biotechnology is the production of Poly Hydroxy Butyrate (PHB) from *Alcaligens eutrophus* (Imperial Chemical Industries, UK). This organism can be grown using a variety of carbon sources including natural sugars and ethanol. *A. eutrophus* accumulates PHB as discrete intracellular granules of polymer. Thus, bacteria are capable of producing a range of polymers with commercially interesting properties. One such plastic in *Biopol* which is biocompostible, biodegradable, inexpensive to manufacture and thermoplastic. Applications include biomedical applications as well as pesticide and herbicide controlled released systems for use in agriculture (Holmes, 1985).

Advances in recombinant DNA technology, protein engineering and process development have begun to influence the industrial enzyme market in a positive manner (Arbridge and Pitcher, 1989). These advances indicate that the developing countries are progressively losing out in the export market. This competitive setback is accompanied by loss of valuable foreign-exchange earnings and subsequent massive unemployment on the home front. Several instances of drops in sugar prices that have adversely affected the trade balances of sugar producing countries like Haiti, the Dominican Republic and Mexico in the markets of North American (Session, 1987). Likewise, other sugar producing countries like Argentina, Brazil, Colombia, China, India, Indonesia, Pakistan, Philippines and Thailand are also experiencing the changing geo-economic influence of biotechnology.

Moreover, research on the biotechnological production of cocoa and coffee flavours is already beginning to have an impact on the exchange-earning market products of Cameroon, Cote d'lvoire and Ghana, these countries benefited immensely in export in the "non-biotechnologically engineered" period of yesteryear. A similar situation is to be found with tea production in Kenya, Malawi and Tanzania that have not kept abreast with the introduction of new cultivars arising from genetic improvement.

Biotechnological Boom-Ecological Doom

Modern man endowed with the power of science and technology has not yet acquired the capacity to veto his evolution but also to alter the evolutionary path and pace of a host of organisms constituting the biosphere. The intervention of man in a large scale through rapid strides in the field of biotechnology resulting in the biopiracy of plant materials has brought about such changes in the functioning of biosphere that they warrant immediate attention. On turning over the pages of human history and going back in time, one finds that 'Man' has been practising and harnessing the fruits of biotechnology since the very beginning of his existence. Much before man learnt to raise fire at his command to cook his food, he learnt to soften the meat and other edibles through the activities of micro-organisms. The story of cultural evolution of man from the nomadic primitive society to the present technological civilization is the success story of acquisition of biotechnological knowledge and their application in every day life.

Selection, breeding and improvement of crop plants and domesticated animals by the nomadic clan might have transformed the food gatherers and hunters into food growers. Whereas, selection of the best genotypes and their subsequent cultivation led gradually to the plant and animal improvement. The early breeders remained principally interested in improving the genetic potential of plants and animals in order to maximise economic gain per unit of input and land. Later on, the breeders goal was to improve plants and animals to yield better, to grow faster, and to stay disease-, pest-, stress-, and growth-resistant, to cater to the needs of changing socio-cultural scenario. The outcome has been the "green" and "white" revolutions. Ignacimuthu (1995) has shown that the prehistoric sewerage system of Akkadian City of Eshnunna close to the Baghdad, Mohanjodaro, Kalibunga had inherent treatment systems. The evidence are also available for the practice of anaerobic waste treatment. Clear indication of the use of flushing toilets by Aegean civilization of Minos on Island of Crete are available. They were also using byproducts generated as a result of biotransformation of wastes in agriculture, forestry and food industries.

One may brand the story associated with the birth of 'Kauravas' as merely fantasy but it clearly indicates the rearing of human embroids in well defined, carefully managed and protected *in-vitro* rearing technique. The whole process has been described as follows:

वितथं नोक्तपूर्पं मे स्वैरेष्वपि कुतो अन्यथा।
घृतपूणा कुण्डशतं क्षिप्रयमेव विधियताम् ॥18॥
सुसुप्तेषु च देशेषु रक्षा चैव विधियताम्।
शीताभिरद्रिरष्ठी लाभिमां व परिषेवय ॥19॥
सासिच्यमाना त्वष्ठीला बभूव शतधा तदा।
अंगुष्ठपर्वमात्राणां गर्भाणां पृथगेव तु ॥20॥
एकधिक शतं पूंर्ण यथा योगं विशाम्पते।
मांरापेश्यारतदा राजन् क्रमशः काल पर्ययात् ॥21॥

The rishi averres that his blessings will come true and the hard lump aborted out will yield hundred sons. Then through a technique this lump splitted up into hundred and one small parts, they being reared *in-vitro* yielded hundred sons and one daughter. But this clearly indicates that by shattering the ideals enshrined in immemorial traditions results in bringing chaos into the world through the disturbance of social equilibrium. It was perhaps the loss of this social equilibrium that the battle of Mahabharat was fought to eradicate the creation of man and once again supermact and balance of nature was restored.

The Biotechnological Boom

Spectacular advances that have taken place in the field of biotechnology in the last two decades have had a profound impact on the socio-cultural scene, life-style, health care system, availability of food, employment opportunities, energy and clean environment.

The immunisation against the dreadliest diseases like small pox, polio, tuberculosis, cholera, typhoid etc. has served the humanity by combating the pain and death caused by these diseases. In the present context huge amount of wastes are being generated and many new chemicals are being introduced through industrial development, the world is facing the problem of their degradation and recycling. As per estimates of United Nations Environmental Programme (UNEP) about 1700 million tonnes of straw are generated per year. Sasson (1984) pointed out that advances in biotechnology leading to development of more efficacious strains of microbes has helped in effective and rapid conversion of wastes and agricultural and industrial byproducts. These microbes are now being used in pollution control, energy production from biomass and recycling of matter.

The rapid advances in biotechnology has resulted into a new industrial revolution, the ramification of which extend far beyond the realm of industries directly affected by it, like industries based on fermentation technology. It extends to provide potential benefits in better health, production of more food and as a very effective source of renewable energy, besides providing a cheaper and efficient industrial process, it reduces pollution (Prentis, 1985). Rehm and Reed (1986) have dealt in detail the refinement that have been achieved in aerobic activated sludge process, aerobic fixed process, biomethanation process, biofiltration of drinking water, combating the problem of environmental pollution etc., through the biotechnological and bioengineering process. They have also exhaustively dealt the role played by biotechnological and bioengineering process in meeting the challenges offered by environmental problems.

Brown et al (1987) pointed out that the anaerobic processes evolved through this technique have helped in effective biogas generation under diverse conditions. As per estimate of United States Environmental Protection Agency about 60000 chemicals are being currently marketed within United States and these are let loose in the environment to cause deleterious effects of diverse kinds and dimensions in the environment. It has also been estimated that about 1200 new chemicals are added to this list annually. Nature does not posses microbial scavengers to scavenge these chemicals from the environment. But for the helping hand offered by biotechnological and bioengineering processes the biodegradation of these xenobiotics could not have been achieved. They offer the development of a very effective synergistic microbial community for the degradation of a host of man-made chemicals (Figure 1).

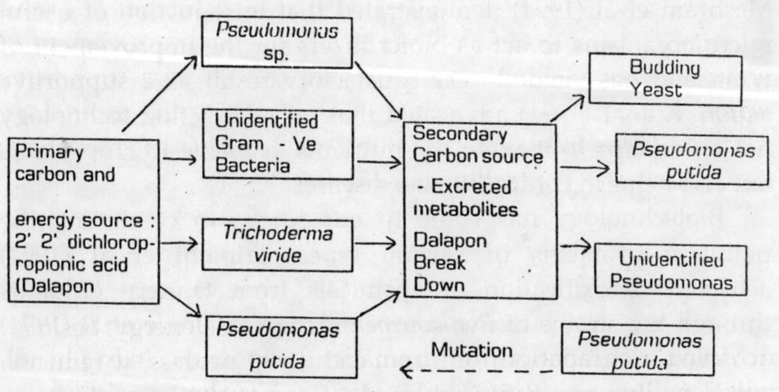

Figure 1
Microbial degradation model for man-made chemicals (Brown et al, 1987).

Bhawalkar and Bhawalkar (1992) have discussed in detail the vermiculture biotechnology and its use in environmental protection, sustainable agriculture, waste water treatment, garbage treatment, BERI vermiprocessing toilet etc. The use of this biotechnology in wasteland development has also been emphasized. The process is eco-friendly and cost effective.

Mookerjee (1990) has pointed out that a cell is constantly subjected to the environmental stress through diverse stress agents and a number of varieties are produced in response to these stresses and best suited amongst them are selected by nature.

This natural process is allowed to occur faster through biotechnological processes. Newer stresses have been generated in the form of xenobiotics and thus there is a need to unravel the general plasticity of adaptive nature of the cell and they meet the challenges in a variety of ways, thus producing the strains better adapted to meet the vagaries created by man. The monoclonal antibodies have yielded a very quick and sure method of disease diagnosis and thus helped in scientific treatment of these diseases (Webb, 1993).

Bhattacharya and Mishra (1994) have suggested the use of biofertilizers as integrated plant nutrient supply system. Meshram et al (1994) demonstrated that introduction of useful micro-organisms to act as biofertilizers for the improvement of wasteland has yielded very satisfactory result as a supportive action. Kumar (1994) advocated that microbe acting technology not only helps in making the nutrients available to crop plants but also helps in controlling the diseases.

Biotechnology has come to our rescue in combating the pollution problems of several types. Tripathi et al (1994) achieved detoxification of chemicals from tannery effluents through the strains of *Pseudomonas aeruginosa.* Ingle et al (1994) achieved degradation of aromatic compounds like phenol, cresol, aniline, resorcinol etc. by *Micrococcus, Alicaligens* etc.

Discussing the role of biotechnology in environmental management, Khanna (1994) indicated that it has played very vital role. Some of the most important ones are as follows:

1. It has very effectively contributed and helped in environmental monitoring.
2. It goes a long way in restoration of environmental quality through degradation of wastes and their recycling.
3. It has played a very vital role in substituting the non-renewable resources base with a renewable resource base.
4. It has provided several methods and alternatives of re-source/residue/waste recovery/utilization/treatment.

5. It helps in producing environmentally benign products through highly efficient biotechnological transformations.

The foregoing averments are but a few examples of the assistance offered by the biotechnological, and bioengineering processes which have been responsible for the cultural evolution. It is the biotechnological achievements which have been responsible for providing the food to escalating human population even though the food producing land has been shrinking due to the loss of croplands in a variety of ways. It has provided a host of new drugs to treat the ailing patients. It has also provided several kits for quick and dependable diagnosis of diseases. Allured by the achievements of biotechnology one gets an expression that answer to all the human maladies and problems of this technological civilization rests with the progress in biotechnology.

Modern civilization characterised by consumerism exceeding the limits of production; desire more than the need leading to wastage and pollution. The fossil fuel based industries and commerce have led us to a situation where exploitative technology overshoots the restorative technology. The result has been the fast depletion of fossil fuel, the harbinger of the present way of life, which has necessitated the search of alternative source of energy especially to fuel the automobiles which are likely to be converted into junks as soon as petroleum gets exhausted. World looks with great expectation towards biotechnology to come to its rescue by providing alternate renewable and sustained supply of vehicular fuel in place of present non-renewable and exhaustible fuel, through the development of appropriate fermentation biotechnology.

The Ecological Doom

A naturalist lives in an ocean of ideas sailing in a little boat on the voyage of discovery of the secrets of nature. Sharples (1983) has warned of the hazards that could be in offing through the introduction of a new strain in the ecosystem. The total ecosystem is thus thrown in a state of stress and the outcome is extremely unpredictable. Stressing this aspect Alexander (1983) also points out that with such changes the possibility of maintenance and survival of ecosystem is very small but potential consequences may be significant. The ramifications of these processes are likely to put social, economic and political axions in disarray (Prentis, 1985). Discussing the effects of human exposure by genetically modified bacteria Levy (1986) pointed out that it affects the survival and if the survival takes place, the fighting mechanism of host goes into jeopardy. Similar observations have been made by Hullman (1986) regarding the exposure of human beings to biotechnologically altered viruses.

Rehm and Reed (1986) outlined the legal, ethical and international questions originating out of biotechnological transformations in the biosystems. Elaborating the environmental risks associated through the application of biotechnology. Fiksel et al (1986) indicated that many alterations become imminent in the ecosystem when genetically engineered microbes and other organisms are introduced in the ecosystem. These create new ecological niches and bring about large scale transformations in the structure and function of the ecosystem as a whole. Misra (1971) discussing these aspects with reference to human beings observed that it can not be assumed that with the cultural and social evolution the unwanted genes have been weeded out and in that regard man can perhaps be assumed as a disease of the ecosystem and hence he can explode the whole ecosystem (Meshram et al, 1994).

Genetically modified organisms produced by evolving biotechnologies are alien to the ecosystem in which they are released and can pose uncharted risks for the ecosystem. There is much debate and discussion aimed at ensuring that they do not harm human beings and the ecosystem with a demand for a Biosafety Protocol.

Several studies have shown all biotechnological innovations in the field of Agriculture or medicine are based on the biodiversity and traditional knowledge resources in the South which are being stolen without any compensation or even acknowledgement.

Ignacimuthu (1995) indicated that life on this planet is finally turned delicate balance of all forms of life, a balance that has been achieved as a result of millions of years of evolution and selection. Technical innovations and their possible repercussions require ethical reflections so that the limits of actions are not transgressed.

A new vista has recently been opened in the realm of biotechnology through the consolidation of Uruguay round of multilateral trade negotiations approved by 118 signatory countries in 1994. This GATT agreement besides other aspects of trades includes Intellectual Property Rights (IPRS), and using its clout and influence, the United States managed to include in the GATT agenda trade related intellectual property system (TRIPS). Thus, a surge started towards patentisation of new genetic materials. To understand the debate on patenting of life forms it will be useful to refer to the developments in the USA and the noted case of Diamond vs Anand Mohan Chakravarthy, granting patent to *Pseudomonas* in which plasmids of 3 bacteria were transplanted in the fourth one, on the ground that the engineered bacterium was not a product of nature. The verdict pronounced by United States Supreme Court in 1980 held that though the genes have not been created *de-novo* yet their shifting is the work of man, hence are patentable. This precedent opened the gates for similar claims, first to patent plants and in 1988 to the first animal patent, for the Onocomouse—a mouse engineered to develop cancer. The claim that patenting of life forms is based upon someone's invention is itself dubious. What the genetic engineers do is merely to exchange genes between different organisms and not create anything new. Traditionally, living organisms have been viewed as non-patentable, since they are products of nature and not inventions. Since micro-organisms are living organisms their patenting has been called the slippery slope that could ultimately lead to patenting of all life forms.

Control over genetic resources by a few large companies will also lead to ecological problems from ill effects on humans and animals that consume engineered crops to disruption of natural ecosystems. The most challenging risks of biotechnology is the loss of biodiversity—loss of wild plants and wild relatives of land races via pollen transfer. Crop genetic diversity is already diminishing at a rapid rate, as farmers around the world are persuaded to abandon the numerous land races of past in favour of a relatively few modern crop varieties, with the resultant risk of famines when large sectors of crops can be wiped out by a new pest.

The Human Genome Organisation encourages research of human genes. In 1992 Gaig Center of National Institute of Health applied for obtaining a patent thousands of gene sequences from human brain. United States applied to obtain a patent on cell lines of a 26 year old Guaymi Indian Women from Panama (Wo 9208784). The cell line from the women was found to provide cure against blood cancer (Leukemia). Any American pharmaceutical company that buys the patent right can develop an anti-leukemia drug from the women's cell lines and a market of billions of dollars is readily available. The patent application has now been stalled due to persistent protest by the activists but still anyone who wants to use an "immortalised" sample of the cell lines for research will have to buy it for $ 126.

Dr. George Annas of Boston University asks—"Since cloned human embroys are not persons protected by the constitution and theoretically atleast could be as immortal as cloned lines, could a particularly "noval" and useful human embryo be patened, cloned and sold?

Plato cautioned that man cannot follow his scientific desires without destroying himself. We are at the cross-road, just like with the use of nuclear energy. The profits from the use of genetic knowledge must not remain in the hands of biotechnologists alone. Indigenous people whose genes contributed to the gain in knowledge should also reap benefits. We have not to be swayed away with immediate benefits *vis-a-vis* long range effects on man and the biosphere as a whole. Science has to interact with value systems and we have to advance with caution so that indiscriminate and hasty adoption of technology does not land us in environmental deterioration. The indiscriminate and irrational application of biotechnology can put us face to face with the total annihilation of human species from the biosphere. Thus, there is need to preserve human dignity and philosophical comprehensions far beyond impulses and stunts so that human life, dignity and the nature is preserved for all times to come.

Biotechnology and Biohazards

(a) The Scientists Call for Safety

Technological innovation and scientific change do not merely bring benefits. They also carry social, ecological and economic costs.

It was the scientists closest to genetic engineering who first expressed concerns relating to the emergence of the new technology. In 1973, a group of prominent scientists called for a moratorium of certain types of research due to unknown risks and hazards associated with the possible escape and proliferation of noval forms of life. In 1975, at the Asilomer Conference, part of the scientific community led by Paul Berg, a molecular biologist from Berkely, attempted to agree on the need for regulation of biotechnological research (Kriwisky, 1982).

Later on many scientists got involved in the commercial application of the new technologies—what Congressman Al Gore has called the selling of the tree of knowledge to Wall Street—the self-criticism and self-restraint of the scientific community faded away.

The sustaining of the social impact analysis of the new technologies then became the responsibility of individual scientists and activists. The most persistent theme of the criticism has been the fear of adverse ecological and epidemiological consequences that might stem from the accidental or deliberate release of self-propagating genetically engineered organisms into the biosphere. Prominent scientists like Licbe Cavaliert, George Wald and David Suzuki have argued that the very power of the new technology outstrips our capacity to use it in safety, that neither nature's resilience nor our own social institutions are adequate protection against the unanticipated impacts of genetic engineering.

(b) Public Outcry Against Testing and Deliberate Release

(i) The "Ice Minus" Story

Since frost damage is a major threat in the colder climate of the North, and runs upto US $ 14 billion annually worldwide, biotechnologists are trying to make plants more tolerant to frost. They have isolated a gene which triggers ice nucleation in plant cells, and have deleted it from a certain bacterium called *Pseudomonas syringae.* The idea is that when this ice-minus bacterium is sprayed on a crop, such as Californian Strawberries, it displaces the naturally occurring ice-forming bacteria, and the plants do not freeze when they normally would.

In 1983, Steven Lindow of Berkely, and Advanced Genetic Sciences, a firm which was funding his work, were permitted by the National Institute of Health (NIH), Recombinant DNA Advisory Committee to run a field test. However, on September 4, a group of Citizens and the environmental interest groups based in Washington DC—including Jeremy Rifkin, and the Foundation on Economic Trends, the Environmental Task Force, Environmental Action and the Humane Society—field a suit against the NIH for approving the project. Among other things the suit charged that the NIH had not conducted an adequate assessment of the potential environmental risks of Lindow's field test and had 'been grossly negligent in its decision to authorise the deliberate release of the first genetically engineered life-forms'.

Among the risks that the public interest suit against NIH pointed to was the dramatic possibility that the frost-preventing bacteria might be swept into the upper atmosphere, disrupting the natural formation of ice-crystals, ultimately affecting local weather pattern and possibly altering the global climate. Eminent scientists like Eugene Odum, and Peter Raven pointed to the ecological hazards of deliberate release of micro-organisms since they reproduce rapidly, and their inter-relations with higher plants are not known.

(ii) The BST Story

Bovine growth harmone, BST (Bovine Somatotropin) is the first harmone of the new biotechnological generation. Natural

BST is a protein hormone that cows produce in sufficient quantities. In young animals it regulates muscle formation and growth, whereas in adults it controls milk production.

Genetically engineered BST is produced not by cows but by genetically engineered bacteria. Administered regularly to cows daily it increases milk yield by 7-14 per cent.

Among the undesirable and negative side effects of biotechnological BST are severe deterioration of the health of the cow and higher surplus in regions where milk surpluses are already driving dairy farmers out of business. One estimate shows that if BST were licensed in the United Kingdom, by 1994-95 there would be 10 per cent more dairy farmers going out of production then if it were not licensed. It is also not known whether hormone fragments will have side effects on the human body. There is no test for checking whether the growth hormone in cow's milk is natural or genetically engineered. There is no test for finding out what the recombinant version can do to the hormonal balance of people consuming BST-treated milk (Ram's Horn, 1991). In addition, the reduced immunity of the cow to disease will imply increased use of drugs and decreased quality of milk.

Animal rights activists, farmers and consumers in the North have achieved a ban on BST, such as in Wisconsin and Vermont in the USA. Three Canadian provinces have banned the selling of BST milk, and a national 'Pure Milk Campaign' has been launched to block the licensing of BST. The European Parliament supported a resolution calling for a worldwide ban on BST. BST has been banned in Denmark, Sweden and Norway.

(iii) Fluoroacetate Story

Large number of cattle and sheep in northern Austria die from fluoroacetate poisoning after eating native plant foliage. Keith Gregg of the Institute of Biotechnology, Armidate and his team produced a new bacterium by inserting a gene from a soil bacterium into the bacterium that lives in the rumen of sheep and cattle. The gene helps the new bacterium to produce an enzyme which detaches the fluorine atom from fluoroacetate, turning it into a harmless compound. However, it is feared that if the organism was released, it might spread to mammalian pests, making them immune to the poison and encouraging their spread and thus be capable of wide irreversible dissemination. Hence, it has been banned in Australia.

(c) Export of Hazards to the Third World

As bans and regulations delay tests and marketing in the North, biotechnology products will increasingly be tested in the South to bypass regulation and public control.

The public, the scientists, and the official agencies of countries where these technologies are being developed, are aware of these hazards. Genetic engineering companies therefore face regulatory constraints, public protests and Court injunctions domestically, and have started to conduct their release experiments involving recombinant organisms in countries where obstacles appear to be fewer due to laxer legislation and lower public awareness. As Dr. Alan Goldhammer of the Industrial Biotechnology Association of the United States had stated, 'the pathway may be clearer in foreign nations to getting approval. The Indian government has welcomed the biotech bandwagon of foreign companies by diluting the regulations and eroding the democratic structures that have existed within the country. VAP (Vaccine Application Program) is clearly designated to by-pass safety regulations prevalent in the United States because the memorandum of understanding states that all genetic engineering research 'will be carried out in accordance with the laws and regulations of the country in which research is conducted'. Since, India has no laws regulating genetic engineering, testing vaccines in India amounts to totally unregulated deliberate release.

The VAP was initiated in 1985 as part of the Reagan-Gandhi Science and Technology initiative, and the agreement was signed in Delhi on July 9, 1987. The project document states that 'the announcement of the VAP is an important recognition that vaccines are among the most cost-effective of health technologies, and their widespread use in both countries is key to controlling the burden of vaccine-preventible diseases'. The primary purpose of the project is to allow an extended range of trails of bio-engineered vaccines on animals and human subjects. The priority areas have been identified as Cholera, Typhoid fever, Roravirus, Hepatitis, Dysentry, Rabies, Pertusis, Pneumonia and Malaria, but these could change in succeeding years of the project as other areas of research opportunity are identified.

In 1986, the WISTAR institute based in Philadelphia hit the headlines for testing bio-engineered rabies vaccine on cattle in Argentina without the consent of the government or people of Argentina. When the Argentinian government became aware of the bovine rabies vaccine experiment in September 1986, it was immediately terminated. The Argentinian Ministry of Health alleged that farm hands who cared for the vaccinated cattle had been infected with the live vaccine.

WISTAR was driven out of the Argentinian government, but has been welcomed by the Indian government for participation in VAP. In fact, the project paper for VAP prepared by the US government applauds WISTAR for its achievements in the field of vaccine development, and specifically mentions the bovine rabies vaccine for field trails and other research.

The US government is evidently dictating the terms and conditions for these experiments, under VAP. The programme is financed by USAID (United States Agency for International Development) and the US Public Health Service. The total project cost is US $ 9.6 m, of which the US and Indian components are 7.6 m $ and 2 m $ respectively. Through the financial input the US government controls the agreement. Thus, all documents, plans, specifications, contracts, schedules and other arrangements with any modifications therein, must be approved by the USAID. On the other hand, scientists and scientific agencies in India directly concerned with the subject have been excluded from discussion on the programme.

Argentina and India are not the only countries to which biohazards are being exported. USAID officials have been pressing African countries to allow field trials of genetically altered organisms that might not be allowed in the regulatory system in the North.

(d) Biohazards and Biosafety

Ignorance about the ecological and health impacts of new technologies has far outweighed the knowledge needed for their production.

It took 200 years of production based on fossil fuel before scientists realised that the burning of fossil fuel has unanticipated side effects—the destabilisation of the climate, the pollution of the atmosphere, and the creating of the green house effect.

DDT was celebrated as an ultimate tool for ensuring public health. A Nobel Prize was awarded for its discovery. Today, DDT and other toxic pesticides are known to carry very high ecological and health costs, and many have been banned in the industrialised countries.

Hazardous substances and processes have been manufactured faster than the structures of regulation and public control have evolved. We do not yet have full ecological criteria of testing for environmentally safe management of fossil fuel technologies of the mechanical engineering revolution. The test for environmentally safe management of the chemical engineering revolution are still in their infancy, leading to the marketing of the products, processes and wastes which are proving to be ecologically unmanageable. Tests for safety in the genetic engineering revolution are yet to be conceived, since how the genetically modified life-forms interacts with other organisms is totally unknown.

Further, unlike hazardous chemicals such as pesticides and chlorofluorocarbons, the products of genetic engineering cannot be removed from the market. As George Wald has said in 'The case against genetic engineering', 'The results will be essentially new organisms, self-perpetuating and hence permanent. Once created they cannot be recalled'.

(e) Technology Transfer and Technology Choice

In biotechnology more than in any other area, lack of knowledge of hazards cannot be treated as safety. Restraints and caution is therefore considered the only wise strategy for unleashing powerful technologies with potentially serious risks in a context of near total ignorance.

For the Third World countries, a special danger exists for being used as a testing ground. In their haste to get access to the new biotechnologies, the Southern governments could unwillingly place themselves and their people and environment in this role of testing ground.

Therefore, to increase the benefits from the new technologies and to reduce their negative impacts, the Third World needs to evolve a framework of assessment of biotechnology on the basis of ecological, social and economic impacts. Transfer of technology, an important issue for the South needs to be negotiated within such an assessment framework, so that socially desirable transfer of technology can take place while undesirable and hazardous transfer can be prevented.

Chapter 2

Natural Resources

Natural resources have been variously defined. For working purposes we can be satisfied with the crude definition that a natural resource is something that can be altered in quality but not diminished in quantity (Dorfman, 1991). Natural resources are the materials which are required for the survival and prosperity of human beings (Seetharam et al, 1995). Five basic ecological variables—energy, matter, space, time and diversity—are sometimes combinedly called natural resources (Saxena, 1990). A resource is a form of energy and/or matter, which is essential for the functioning of organisms, populations, and ecosystems (Ramade, 1984). Natural resources are those components of the atmosphere, hydrosphere and lithosphere which can be drawn upon for supporting life. These include energy, atmosphere (air), water, land (soil), minerals, plants and animals.

The *atmosphere* comprises the gaseous envelope of the planet. It is made up of a series of invisible shells of air covering the earth. Each new shell is just a little bigger and completely covers the previous shell of protecting air. The gaseous mantle forming the atmosphere extends into the outer space some 100 kilometers or so, above the earth's surface. It maintains its contact with all the major types of environment of the earth, interacting with them and greatly affecting their ability to support life. The atmosphere is a reservoir of several elements essential to life and it serves many functions including the filteration of radiant energy from the sun, insulation from heat loss at the earth's surface, and stabilization of weather and climate owing to the heat capacity of the air.

There are five concentric layers within the atmosphere which can be distinguished on the basis of temperature (Figure 1). These are: Troposphere, Stratosphere, Mesosphere, Thermosphere, and Exosphere.

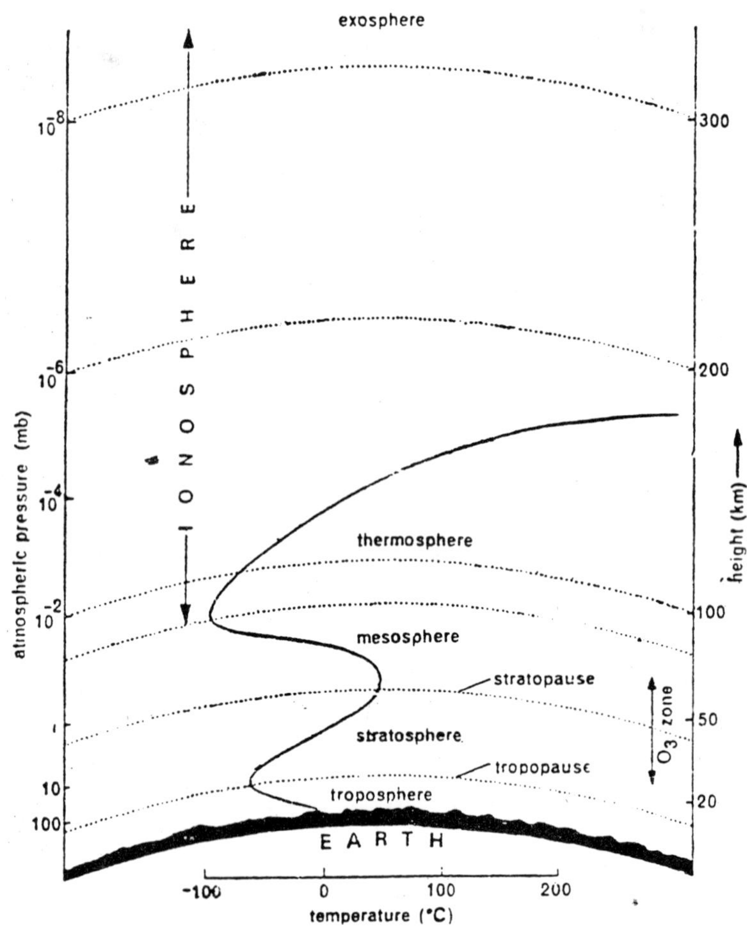

Figure 2: Diagrammatic sketch showing the principal zones of atmosphere along with various temperature

The earth's atmosphere from the surface upto 20 km (8 km on poles) is called *troposphere.* It concerns us most because it is the only part that supports plants and animals. The important weather events, such as cloud formation, lightning, thundering etc. take place in the troposphere. Air temperature in this zone gradually decreases with height at the rate of about 6.5° C per km. Towards the upper layers, the temperature might decrease upto –60°C. Water vapour in the atmosphere is confined to the troposphere. Air in the troposphere moves side-ways (laterally) and up-down (vertically) keeping it well mixed. The troposphere gradually merges into the next zone, which is known as tropopause.

Stratosphere is the second zone of about 30 km, where temperature is about 90°C. The increase of temperature in this zone is due to ozone formation under the influence of ultraviolet rays of solar radiation. Ozone is formed from oxygen by a photochemical reaction, in which solar energy splits the oxygen molecules to form atomic origin, which then combines with oxygen molecules to form ozone. Stratosphere is the region of horizontal motion with no up and down wind motion. In the lower layers of the Stratosphere, the horizontal winds reach the highest speeds of any wind in the entire atmosphere. These high speed winds are known as *jet-streams.* The stratosphere is above clouds, violent storms and precipitation. The upper layers of the stratosphere form the stratopause.

Mesosphere is the third layer of the atmosphere, about 40 km in height. In this zone, temperature shows a decrease upto –80°C. The upper limit of the mesosphere is termed mesopause.

Thermosphere is the fourth layer of the atmosphere, which extends upto 500 km above the earth's surface, and is characterised by steady increase in temperature with height. In this region, ultraviolet radiation and cosmic radiation cause ionisation of oxygen and nitric oxides. Hence, this region is also called as *ionosphere.*

Exosphere is the fifth layer of the atmosphere. It lacks atoms except that of hydrogen and helium, and extends upto 32190 km from the earth. The earth's magnetic field becomes more important, than gravity in the distribution of atomic particles in the exosphere. Exosphere has a very high temperature due to solar radiation.

All these layers of the atmosphere are of interest, since together, they form the total blanket of air, which moderates the solar radiation reaching the earth, and also serves as a blanket and regulates the earth's radiation escaping into the space.

The *hydrosphere* is made up of oceans, lakes, underground aquifers, rivers, ice-caps etc. The amount of water which is present on the earth is vast. More than two-third of the earth's surface is covered with water. The water which is present today, was formed centuries ago, during a time when the earth was changing from a molten mass to essentially the form in which we know it today. During the cooling period, changes occurred, gases were released, and from these gases, water was formed. The water has appeared at different times in different ways and has covered varying amount of the earth. However, overall of these changes and times and places, the amount of water has remained the same.

Out of the total amount of water present on the earth 97 per cent of it is salty, which exists in seas and oceans. Of the remaining 3 per cent, 4/5 is blocked in the form of polar ice and glaciers, and gets eventually lost to the sea. Another major fraction remains in the deep crevices of the land as underground water, leaving about 0.3 per cent for use of plants, animals and human beings.

For practical purposes we can classify our hydrosphere as existing of generally four conditions—Salt water, atmospheric water, fresh water and ice-and-snow. The most abundant is salt water. Second is the water vapour which exists in the atmosphere and the soil. Third is the fresh water which exists in a liquid form in lakes, rivers and underground aquifers. Fourth is the ice and snow, which is a solid state found on high mountains and poles.

The *lithosphere* includes the earth's solid crust down to an average depth of about 60 km into the interior of the earth.

Classification of Natural Resources

The classical sub-divisions of natural resources includes *renewable* and *non-renewable* resources.

Renewable natural resources are also called replaceable natural resources. They are replaced by natural processes like quick recycling, reproduction etc. They are all interlinked and not likely to be exhausted. However, over consumption of these natural resources may cause a depletion and make them non-renewable, for example, water, soil, plants and animals.

Non-renewable natural resources are the resources which cannot be replaced or recycled after use. They are not reproducible and are obtained from the finite non-living reserves, such as metals (iron, copper, zinc, silver, gold etc.), other minerals and their salts (phosphate, nitrate, carbonate etc). The minerals are often called stock or non-renewable resources, because, their new materials can only be extracted from the earth's crust once. But even in the transformed state in which they are used, they are not lost to the planet and so are ideally available for reuse. Solar energy although having a finite life, a special case, is considered as a renewable resource in as much as solar stocks are inexhaustible on the human scale.

Some authors prefer to classify natural resources into *biotic* (or living) and *abiotic* (or non-living) resources. Crops, forest, fish and wildlife are biotic resources, whereas air, water, soil, minerals etc. are abiotic resources.

Owen (1971) proposed a detailed classification of resources (Figure 3). Quantity, mutability and reusability are the attributes on which this classification is based. *Inexhaustible* and *exhaustible* are the main categories of resources based on their abundance and availability.

There is a real possibility that the world's stock of inexhaustible resources will be depleted permanently and irretrievably. This awareness has given rise to the doctrine of "maximum sustainable yields". The doctrine is based on the "logistic law of population growth", which holds that any environment or habitat has a maximum carrying capacity for each species that inhabits it. When the population of any species is small in relation to the carrying capacity for it, the population grows rapidly, simply because the number of individuals growing and reproducing is small. As the population increases, the growth rate declines. The maximum sustainable yield is the absolute growth rate at that critical point at which population is growing most rapidly. For, if the harvest taken each year is equal to the natural annual amount of growth, the population will remain stationary, and that size of harvest can be sustained indefinitely.

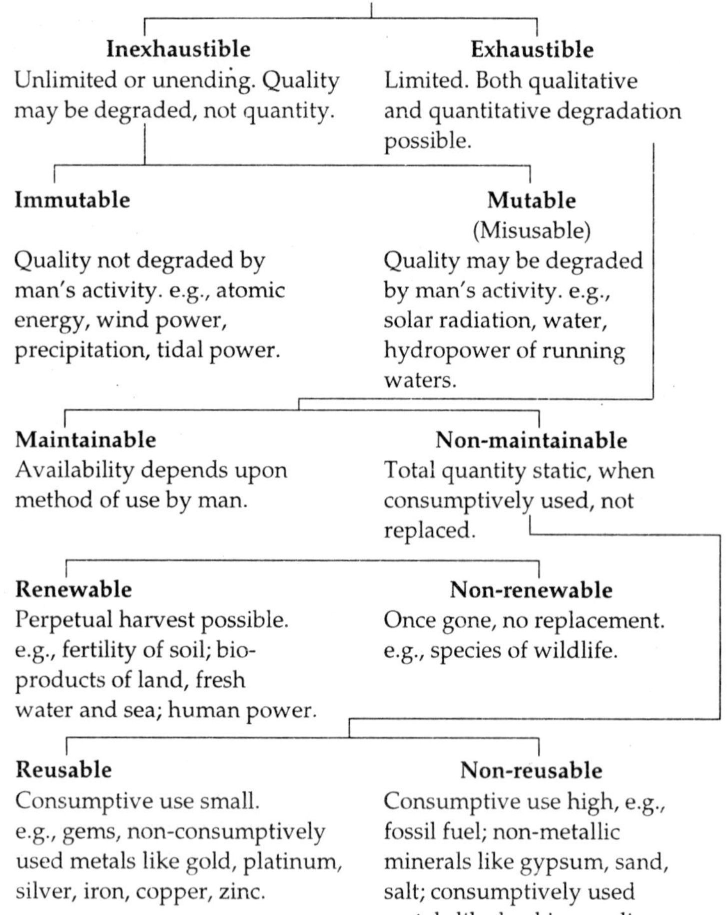

Figure 3
Schematic representation of classification of natural resources

The exhaustible resources are available in finite quantity and likely to be exhausted. The market for any exhaustible resource will generate a path of gradually increasing prices. Increase in price will be inconvenience to future generations. It will induce them to use the resource more sparingly, to extract and process it more carefully to reclaim and recycle it more extensively, and to substitute other, somewhat less suitable resources for it if they are cheaper. Most of the critical exhaustible resources are generally imported in industrial countries as raw materials for their industries. The developing countries depend on exhaustible resources as a way of paying for current imports and storing wealth with which to finance their own development.

The inexhaustible resources are further divided into two groups immutable and *mutable* resources, depending upon the possibility of their quantitative degradation as a result of human activity. On the other hand, exhaustible resources, being limited in occurrence are vulnerable to quantitative and qualitative degradation. Thus, their availability depends upon the method of use, so these are classed into *maintainable* and *non-maintainable* resources. Supply of maintainable resources could be made to last for long through wise use. The maintainable resources are of two types *renewable* and *non-renewable.* Resources which have the inherent capacity to reappear or replenish themselves by (i) quick recycling, (ii) reproducing, and (iii) replacement within a reasonable time and maintain themselves are called renewable resources. Soil, water and living organisms constitute the main renewable resources. Resources which do not undergo recycling and replacement are called non-renewable resources. Substances having a very long recycling time are also regarded as non-renewable resources. For example fossil fuels, (like coal, and petroleum, and minerals) are non-renewable resources.

The non-maintainable resources have static supply and when destroyed or over-consumed are irreplaceable, yet some of them have potentiality for reuse and are called *reusable* resources, while others whose permanent exhaustion is certain are called *non-reusable* resources.

Demand for Resources

Human history has revolved around demand for, pursuit of, and access to resources. Wars have been fought for fertile valleys, good grazing grounds, or desirable habitations on protected hills. These wars were actually fought for resources: food, minerals and energy.

Food

The demand for food have been concentrated on few key crops like wheat, rice and maize, followed by rye and barley, potatoes and sugarcane. The development in agriculture have led to varietal improvement, and enhancement in cultivation, distribution and yield. With increased distribution, people in Europe now eat a great deal of rice and potatoes, crops that in origin are Oriental and American respectively. America produces acres of wheat, which was Middle Eastern initially. Maize, the staple crop of Pre Columbian America, is now also that of Africa.

A second characteristic of the world food scene has been the decrease in variety and growth in consistancy. Many countries enjoy a relative abundance of food in comparison with the past. This is possible due to increase in yields, increase in the amount of land under cultivation, and efficient distribution system by rail-roads and trucks. Access to food has been a direct cause of nearly every significant upheaval, the French and Russian revolutions included.

Minerals

The demand for minerals, like that of food, has centered on just a few kinds. Precious metals (gold and silver) remain in top demand, even though tin, copper and iron are also in great demand. Precious stones have been equally sought after. Wood for housing, furniture and ships, has also been transported from one place to another from the earliest times.

Iron achieved critical importance, and has been in constant use since its usefulness was discovered in Asia Minor about 2500 BC. The acquisition of iron weapons and agricultural tools was probably behind the southward and eastward thrust of the

Negro peoples of West Africa between the second and first centuries AD. Medieval Europe was reconstructed by ironsmiths able to make iron axes to level the great forests and iron stirrups for knights in iron armour bearing iron weapons. The wars in the Renaissance further increased European demand for iron not only for weapons, but also for cooking pots, porks, and knives for new foods; agricultural implements for new crops; and anchors for trans-atlantic ships.

Energy

In the remote past, energy was produced at home by people, animals or water. The agricultural revolution (1000-8000 BC) was characterised by the shift from human power to animal power with the domestication of reindeer, sheep, dogs and goats. The harnessing of water followed a few thousand years later. Subsequent useful innovations such as castration of bulls to create oxen, the taming of the horse and subsequent breeding of the cart force, and the fashioning of water mills were variations on ancient themes rather than essentially new developments.

Wind power was first used by Chinese, and the use of coal for furnaces were later innovations. Coal was used in blast furnaces and in houses in China atleast since about 2000 BC. The widespread use of coal was hampered by its weight and by the lack of good transportation system. Access to indigenous coal helped some countries readily for industrialization (Britain, Belgium, United States, France and Germany).

The use of petroleum as a source of energy began by 1900. It required great self-confidence for European nations to turn to oil from coal as the main source of energy, just after the first World War. The leaders who made the decisions believed that they could control the places from which oil came indefinitely (as the British had controlled India), and that their navies could keep the sea lanes open for international trade, come what may. The liquid fuel showed advantages—speed, refueling at sea, and benefits for sailors. Wars had been won by those who had controlled the waves for oil. The two superpowers that emerged in the late 1940's (USSR and USA) were rich in this natural resource (Thomas, 1985).

Population Resource Conflict

Over the past decade it has become obvious that the population-resource conflict cannot be resolved without addressing the problem of unsustainable pattern of consumption.

The total resource consumption of 20 average families in the developing world is even less than the consumption of just one British family (having two children). An average Northern citizen consumes copper, for example, 17 times more than his counterpart in the South. In the case of Aluminium, the difference is a factor of 20 against the south. India, uses less paper for printing school text books than the Scandinavian countries use for producing poronographic literature. One fifth of the human population that occupies the North consumes 80 per cent of the world's resources and if we have to save the world from an irreversible resource crises, the world should focus its attention on the North.

There is an enormous difference in the economies of developing and developed countries in natural resources terms. The former continue to depend heavily on the exploitation of their natural capital to meet their current consumption need and generate the investment needs to built up a stock of human-made capital and a knowledge and skills resource base. The developed countries, on the other hand, have already gone through a prolonged phase of natural resource exploitation, both within their own countries and outside, to built up a massive base of human-made capital, knowledge and skills.

The control of powerful technological capacity and the systems of world trade and finance has enabled the industrial countries to suck out forest, minerals and metal resources from the Third World, and to make use of its land and labour resources to produce the raw materials that feed into the machinery of industrialism. The rich nations with around one-fifth of world population use up four-fifths of world resources, a large portion for making luxury products. The Third World gets to use only 20 per cent of resources. Since incomes are also unequally distributed within Third World nations, a large part of these resources are also used up in making or importing high tech products as are enjoyed in the rich nations. Thus, only a small part of world resources flows towards meeting the basic needs of the poor majority in the Third World.

Automobile is basic necessity. It is wrong to look upon them as a luxury. In recent years production of cars light commercial vehicles, tractors and two wheelers have shown a significant upward trend. Leaving economy and pollution, land use and highways aside, let's examine simply the raw materials (natural resources) necessary to produce automobiles, and to make things simpler still, consider the availability of metals.

In America, they have one automobile for every two people, giving them the highest per capita of automobiles in the world. The result of their love affair with the automobiles has been the spawning of new automobiles at an alarming rate. On the basis of number of vehicles as percentage of our population, only one out of 500 persons in India has a car and out of 100 persons one has a two wheeler. We have about 10 lakh trucks and 2.5 lakh buses, which are woefully inadequate for a country of our size. China has one automobile for every 13000 people, and the Chinese have almost as many bicycles as human beings.

Now consider what would happen if India and China produced automobiles at the same domestic level as the United States? Are there enough metals available worldwide for such production? What would happen if citizens of all other nations fall as madly in love with automobiles as the Americans are? There are, of course, simply not enough resources to produce an automobile for every two people on the face of the earth! Where will they get the resource base—where will they put the cars—what will happen to public transportation—what will be the impact on environmental pollution (Dill, 1986).

Modern civilization characterised by consumerism exceeding the limits of production; desire more than the need leading to wastage and pollution. The fossil fuel based industries and commerce have led us to a situation where exploitative technology overshoots the restorative technology. The result has been the fast depletion of fossil fuel, the harbinger of the present way of life which has necessitated the search of alternative source of energy especially to fuel the automobiles which are likely to be converted into junks as soon as petroleum gets exhausted. Natural resources hold a key to the prosperity of the people. The intensive use and abuse of natural resources with the greed for profit making, is ecocidal treatment of the environment and invariably results into homicidal consequences which bring chaos in human society.

Not only in consumption but also in discharging polluting waste, the industrial countries of the North play a devastating role. The per capita emission of greenhouse gases by the United States population, for example, is twelve times more than that of South Asia's. Thus, by depleting the resource base and impairing environmental space, Northern countries are offering us an irreversible civilisational crises (Faizi, 1995).

Conservation of Natural Resources

The word "conservation" has been derived from the Latin words—"con" (together) and "servare" (guard). It deals with practices and customs through which human beings tries to ensure a continuous yield of renewable resources and afford protection of non-renewable resources from wastage. A conservationist has two basic aims: (i) to ensure the preservation of a quality environment that considers aesthetic and recreational as well as product needs, and (ii) to ensure a continuous yield of useful plants, animals and materials by establishing a balanced cycle of harvest and renewal. Thus, conservation process remains chiefly concerned with the use, preservation and proper management of the natural resources of the earth and their protection from the destructive influences, misuse, decay, fire or waste. It deals with biogeochemical cycles so that the process of removal or dispersion of energy and materials are only as fast as they are synthesized. A good conservationist always aims at continuous yield of renewable resources through the inbuilt resources of the ecosystem, the external application of materials and methods for increasing the output of desired products, and the restriction of undesirable ones. Many dwindling resources or attractive animals need protection in sanctuaries, where the ecosystem is specially maintained to suit selectively the protected species. Preservation of rare species is certainly not all that conservation means. Conservation is in-fact never a "hoarding" or even "rational use" of a limited resource, but a constant effort to increase the resources and rapid resynthesis of the materials.

Conservation of some of the important resources have been discussed under the following headings:

Conservation of the atmosphere

The atmosphere is a thin gaseous envelope, surrounding the rock and water of the earth. In addition to approximately 78 per cent nitrogen and 21 per cent oxygen, the atmosphere contains carbon dioxide, rare gases, water vapours and suspended particulates (Table 1). Technological advancement and industrial revolution has added various gaseous and particulate matter to our atmosphere. As the advancement of technology is in progress, various brands of pollutant are added to the air by different types of industries.

Table 1
Composition of dry air by volume (Clapham, Jr. 1973)

Gases	*Per cent (By volume)*	*Gases*	*Percent (By volume)*
Nitrogen	78.0841	Crypton	0.00011
Oxygen	20.9486	Xenon	0.00009
Argon	0.9340	Hydrogen	0.00006
Carbon dioxide	0.0318	Methane	0.0002
Neon	0.00182	Nitrous oxide	0.00005
Helium	0.00052	Ozone	0.000004

To breath clean air is everybody's natural right, which is more important and sacrosanct than one's fundamental right to practise any profession, or to carry trade, or business, which may pollute the air. Man has forced a change in the composition of the atmosphere by using the atmosphere as a disposal site for the many wastes of industries.

The atmospheric contaminants create a nuisance by restricting visibility. They also affect plant and animal life and property. They too adversely affect human health through the air we breath. These atmospheric pollution problems have been getting worse with increasing industrialization.

The atmospheric pollution recognises no boundaries. The soil, plants, animals and man himself on this planet are being affected by massive pollution now taking place. It spreads in the air across nations for great distances and even around the world. Therefore we must act quickly, or the severe atmospheric pollution episodes will become universal. The deterioration of the air we breathe must be stopped and the process reversed, or the future will be one of an intolerably dirty and smelly existence for us all.

(i) Natural atmospheric cleansing

Natural oxidation of many pollutants aids in their removal. Carbon monoxide and methane are oxidised to carbon dioxide. Similarly, nitric oxide is oxidised to nitrogen dioxide, and sulphur dioxide is oxidised to sulphuric acid in the atmosphere, as part of the natural cleansing processes. All these gaseous substances together with abiotic and biotic particulate matter suspended into the atmosphere, are easily washed out of the atmosphere by rain or snow.

Vegetation offers an important sink for atmospheric contaminants (Figure 4). Plants growing in polluted environments act as an important sink of air pollutants and produce pockets of clean air. It has been estimated that global uptake of carbon dioxide by plants is 0.31×10^{14} gallons per year. The vast forests of the northern hemisphere, especially the marshy peat heads and huge bogs together absorb more carbon dioxide than other terrestrial plants. Air pollution abatement capacity of trees have been demonstrated in England and the former Soviet Union. The Hyde Park of one square mile in the centre of London, showed a 27 per cent reduction of smoke concentration in comparison to non-green areas. In Soviet Union, 400 poplars, which were the poorest dust collectors, spread over 2.5 acres, filtered out 0.375 tonnes of dust during the leaf bearing season. It has also been shown that a 500 meter wide green belt surrounding industries will reduce sulphur dioxide and nitrous oxide concentration by 70 and 67 per cent.

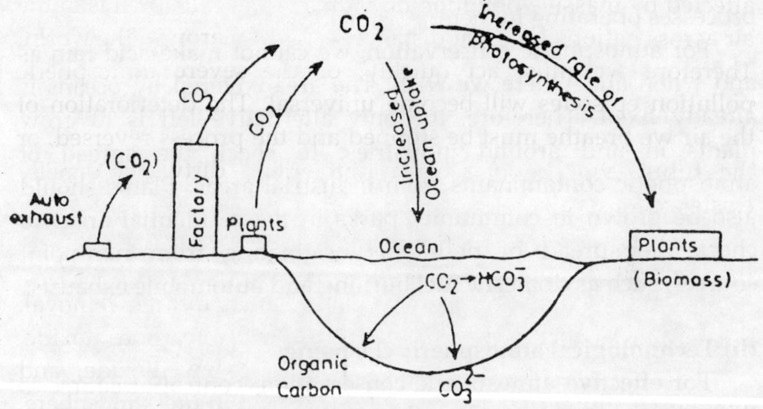

Figure 4
Sources and sinks of carbon dioxide

Gases penetrate vegetation canopies rapidly and are absorbed by plants by diffusion through stomata. They are absorbed in relation to their solubility in water. Particulate matter are absorbed/deposited on the leaf surfaces. The ability of plants to improve air quality by absorbing gaseous and particulate pollutants is proven by the fact that the polluting gases do enter plants and get detoxified into less injurious forms.

Thus, vegetation offers various self-cleansing mechanisms and maintains the purity of air through ages. However, the self-cleansing mechanism always have some definite threshold limits. After some level of accumulation of toxic chemicals in the plants, their self-cleansing power is exhausted.

Another natural sink for atmospheric pollutants is the ocean. About half of the carbon dioxide released into the atmosphere is utilised by plants or absorbed by water of the oceans. Part of the carbon dioxide becomes dissolved in the oceans, where it may be precipitated as carbonaceous earth, or incorporated in the marine organisms. In this respect, aquatic plants in the oceans plan a vital role in maintaining the carbon dioxide equilibrium between the atmosphere and the surface layers of the oceans. The capacity of the oceans for the uptake of carbon dioxide depends on their circulation dynamics and the biological processes operating in them.

For atmospheric conservation, we cannot make acid-rain as and when and where we wish. The area covered by oceans is already fixed. Therefore, the only alternative left is to grow plants in and around industries, to check the spread of atmospheric contaminants from industrial areas. Plants should also be grown in community parks in the residential areas to check the spread of pollutants originating from non-point sources, such as domestic fuel burning and automobile exhausts.

(ii) Technological atmospheric cleansing

For effective atmospheric conservation, control at source is the only viable option. Once the air contaminants are identified, efficient control equipments can be designed and used to control it. Some of the procedures are as follows:

(a) Absorption

This is a process in which a soluble gas is transferred from a gas stream into a liquid, in a gas-liquid contacting equipment. The gas may simply dissolve in or may react with the liquid. Usually, the gaseous components to be removed are present to an extent of one per cent or less. Water is the most suitable scrubbing medium. Acidic gases like chlorine and sulphur dioxide are absorbed in alkaline solutions to form non-volatile salts. Hydrogen sulphide is absorbed in ethanolamine.

(b) Adsorption

In this process, molecules from a gas or liquid stream attach themselves on the surface of a solid. The process is reversible, economic, and the same adsorbent can be used more than once. Activated carbon, silica gel, and diatomaceous earth are some important adsorbent. The adsorbent placed in a suitable container captures gas or liquid or odour molecules passed through it. The attractive forces between atoms, molecules and ions, which held together a solid are unsatisfied at the surface, and thus are always available for holding other materials. With an increase in the porosity of a solid, its adsorptive capacity also increases.

(c) Condensation

Gaseous vapours can be controlled by condensation. The method is most suitable for hydrocarbons and organic compounds having reasonably low vapour pressure at ambient temperature. They can be controlled satisfactorily in water or in air cooled condensers.

(d) Chemical reaction

The gaseous pollutants can be controlled by chemical reaction. For example, sulphur dioxide can be controlled by reacting it with copper oxide or limestone, to produce copper sulphate or calcium sulphate.

(e) Incineration

It is actually a thermal oxidation process. Fume burners operating under high temperature consume combustible material in a waste gas fed to the equipment. It converts hydrocarbons to carbon dioxide and water, haloginated material to halogen acid. Often auxiliary fuel is required to be burnt to heat the fumes to a sufficiently high temperature.

(f) Particulate control

Various types of settling chambers, electrostatic precipitators and bag filters are used for controlling particulate emissions. These devices are placed across the gas stream. The choice of the device varies with the efficiency required.

(g) Fuel change

A change to less polluting fuel, gives an ideal solution for atmospheric conservation. The use of fuels that are low in sulphur is not really practical because the world supply of these fuels is limited. Alternatively, the use of natural gas instead of coal will avoid sulphur dioxide and flyash emission from power plants.

(h) Process change

A process change can be either in operating procedures for an existing process or the substitution of a completely different process.

Water conservation

Water requirement and water withdrawal usually exceeds or tends to exceed, the natural water requirement. In addition to this, effluents and excessive water consumption impair water quality, restricting its further utilization (Table 2). The water problems of the future is not going to be one of a global water shortage on a planetary basis, but one of local crises due to the concentration of people, industry, agricultural expansion, and the vagaries of weather. The need to search for means of managing water economy usually arises as a result of an actual or expected deterioration in water resources caused by pollution of an area in question or of a whole country. An adequate utilization of water resource and a proper control of the use of water are impossible without an adequate utilisation of all available means—legal, institutional, technical, economic, personal and moral. These criteria provide effective tool for rational and integrated development and use and conservation of water resource. It is indispensable to use all the above tools for protection of human society before unuseful wastage, misuse and depreciation of water resources and before excessive water withdrawal, demands and arrangements threaten or have a negative impact on the environment, thereby restricting the future development or negatively influencing the living standards or life-styles of the society concerned.

Special emphasis has to be given to conservation of the quality of water from the sources to the point of consumption by making use of different techniques such as:

Table 2
Unintentional wastage of water

Activity	*Method adopted*	*Quantity used (1)*	*Method to be adopted*	*Quantity required (1)*	*Quantity saved (1)*
Brushing teeth	Running tap for 5 min	45	Tumbler or glass	0.5	44.5
Washing hands	Running tap for 2 min	18	Half filled wash basin	2.0	16.0
Shaving	Running tap for 2 min	18	Shaving mug	0.05	17.75
Shower	Letting shower run while soaping staying under shower too long	90 rinse off	Wet down, tap off, soap up,	20.0	70.0
Flushing	Using old fashioned large capacity cistern	13.5 or more	Dual system short flush liquid waste, full flush solid waste	45. 9.0	4.5 or more
Watering plants	Running house for 5 min	120	Water can	5.0	115.0
Washing fioor	Running hose for 5 min	200	Mop and bucket	18.0	182.0
Washing car	Running hose for 10 min	400	Buckets (two)	18.0	382.0

(i) Efficient use

One of the most important conservation activities is the use of fresh water in such a way that we get the very most for our efforts without depleting it. Efforts should be directed to increase the usability of low grade or polluted water. Sewage irrigation useful for raising valuable crops has long been known, but application of industrial wastes for crop raising or their hygienic disposal on land or in waterways is a comparatively new development.

(ii) Reuse

It has been estimated that only 30 per cent of domestic water supply is consumed and remaining 70 per cent is discharged as waste water. Industries return about 80 per cent waste water. Waste water from any source need not be considered as water wasted. Waste water can be looked upon as a source of additional water for its reuse for various beneficial purposes for which water quality does not have to be brought to a high level of purity.

The reuse of water reduces the demand on fresh water. Consequently, industries should be able to lower production costs and simultaneously improve the quality of water in downward reaches. The reuse of waste water in arid areas of western Rajasthan is practised because of the depletion of ground water and scarce availability of new water resources.

Water reuse has a special significance in mining, steel mills, pulp and paper mills, textile mills and oil refineries where the resource is scarce and the impurities can be separated by sedimentation, filteration, clarification, separation or any other suitable treatment. In many cases the cost of the treatment is modest as compared to the overall benefit. However, in some cases, each reuse cycle, is increasingly costly in terms of treatment, and each reuse cycle requires disinfections and other industrial chemicals. Reuse of water from domestic effluents needs to be given top priority in the allocation of water. Trade effluents also needs reuse to maintain water quality. Sometimes reuse of water is difficult as in the case of radioactive wastes, where the cost of reuse may prove prohibitive as compared to the cost of providing water from a new source.

(iii) Biofilteration

In biofilter process, the organic waste water—like sewage, food processing, paper making, tanneries, distilleries and dairy industry water is converted into reusable water. The process developed by the Bhawalkar Earthworm Research Institute (BERI) involves filtering waste water through a vermifilter formed by enclosing earthworms and worm-castings (earthworm excreta, which harbour cocoons and a wide spectrum of beneficial microflora) in a specially developed medium. The earthworms convert water impurities into worm-casts and after repeated filtering, clear water is obtained. The water so obtained is free of colour, odour and harmful bacteria. However, it is not potable, but can be used for gardening and toilet flushing. The worm-casts accumulate in the vermifilter and can be periodically harvested for use as fertilizer (Figure 5).

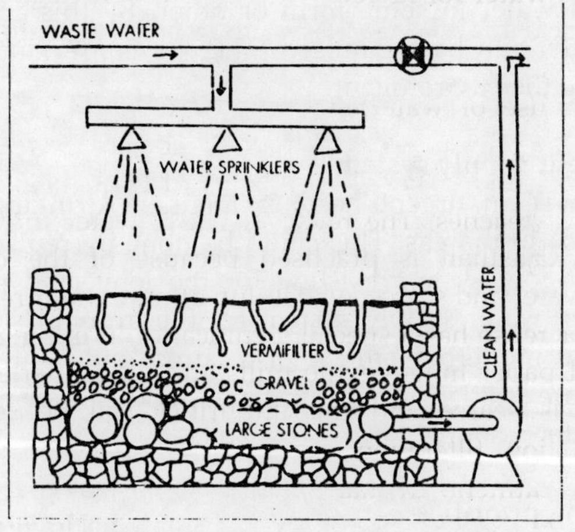

Figure 5
Demonstration of biofilteration process

(iv) Waste water treatment

Our fresh water rivers, lakes and wells; and salty oceans had self-correcting mechanisms, which used to check the excess of one thing over the other. Human interference was also taken care of by nature, until the modern technology created a situation where excessive exploitation and pollution caused a serious threat to the very existence of life. Naturally, the common approach the "Solution to pollution is dilution", is not adequate now.

Waste water from toilets, wash basins, bath rooms, kitchens etc. can be subjected to minimal treatment for making it fit for flushing toilets and watering lawns. Treated waste water can gainfully be used for irrigation. In fact, the nutrients present in treated waste water are an advantage for agricultural production. Industrial treated water can be used for cooling purposes, since cooling can tolerate low quality water. In Bombay, 5 tall buildings utilise the treated sewage recycled water for air-conditioning. ISRO Satellite centre, Bangalore, and Bajaj Auto Limited, Aurangabad are utilising their treated waste water for horticulture, gardening and other uses. About 45 industries around Bombay have adopted reuse of their processed water in some form or other. By this approach, the waste water can be minimised. Thereby conserving water and improving the environment.

(v) Efficient supply system

There is an urgent need to augment drinking water for urban and rural areas. This could be done by replacing rusted-choked supply system existing for too long, and provision of emergency water supply system for the drought stricken rural areas. The modern methods of assured and controlled water supplies are still lacking in rural areas and call for management of our water resource.

(vi) Setting priorities

While setting priorities for water, we must all start from the premise that water, irrespective of where and how it is available, is the gift of nature and therefore a national asset. No individual or family can be allowed to claim that a particular body of water belongs to that individual or family as a private property.

Whenever there is shortage of water, *first priority* should be given to provide drinking water for human beings and cattle. This use should have priority over any other use, irrespective of the consequences. *Second priority* should be given to agriculture which is the mainstay of the Indian economy and the very life of the people of India. *Third priority* calls for allocation of any surplus allocation of any surplus available water to industries. However, the industries catering directly to the needs of the common people should be given preference over those which cater to their needs somewhat indirectly.

Soil conservation

The main aim of soil conservation are (1) to protect the soil from erosion, and (2) to maintain the productive capacity of the soil. For soil conservation certain practical methods have been employed. These methods can be broadly grouped into (a) biological measures, and (b) mechanical or engineering measures.

(a) Biological measures

The following are the biological measures which are helpful in soil conservation:

1. Agronomic practices
2. Agrostological practices
3. Dry farming practices

1. Agronomic practices

The important agronomic practices which contribute to the conservation and productivity of cultivated lands are also known as conservational farming or advanced agronomical measures:

(i) Contour farming
(ii) Tillage and keeping land fallow
(iii) Crop rotation, sowing of leguminous crops and mixed cropping
(iv) Mulching
(v) Strip cropping

(i) Contour farming

It is practised in hilly regions or on the slopes. In such areas, the rain water is absorbed in very little amount because of its quick downward movement on the slopes, the heavy rainfall may cause gully development. Therefore, the sloppy areas are ploughed and seeded against the slopes i.e., in circular furrows around the slopes. The contour catches the downward moving water until it is absorbed in the soil. The ridges reduce the flow of water. The circular rows of plants across the slopes check the soil erosion. Thus, contour farming reduces run-off, saves more water for crops, reduces soil erosion and increased the yield of crops.

(ii) Tillage operation and keeping the land fallow

In dry areas, shallow ploughing gives comparatively good crop yield. Shallow ploughing removes the weeds and enables the soil to absorb water. Deep ploughing often leads to soil erosion, but in the areas where rainfall is sufficiently high, deep ploughing (upto 15-30 cm) is effective in removing weeds and increasing crop yield.

If the land is left uncultivated, sheeps, goats and cattle are allowed to graze over it for some time, the soil becomes fertile. Though this practice is useful, yet it is not possible to leave the land fallow because of the growing human demands for food.

(iii) Crop rotation, sowing of leguminous crops and mixed cropping

When the same crop is grown in the field every year, the soil becomes depleted in certain minerals. The soil looses its fertility even after the use of fertilizers and ultimately erosion begins. Rotation of crops is an important practice for checking soil erosion. Selection of crops for rotation should be made accordingly to the local climate, soil, slope, economy etc. Deep rooted crops should be rotated by shallow rooted crops. Deep rooted crops absorbs nutrients from the deeper strata of soil, while the shallow rooted crops absorb nutrients from the upper strata of the soil.

Rotation of cereals by legumes is useful in maintaining productive soil. Leguminous plants increase nitrogen content in the soil. The nitrifying bacteria present in root nodules of leguminous plants fixes atmospheric nitrogen into nitrates, nitrites, ammonium salts, amino acids and proteins. These nitrogenous compounds return to the soil on death and decay of root nodules.

In mixed cropping, one main crop and one or two subsidiary crops are grown together on the same field. This practice checks soil erosion and avoids the risk of crop failure.

(iv) Mulching

It involves covering the soil surface by straw. Mulches of different kinds check soil erosion, increase soil fertility and minimise moisture evaporation from the top soil. Various types of surface tillers and crop residues are helpful in obstructing soil erosion.

(v) Strip cropping

Strip cropping is employed to reduce the velocity of raining water. This involves growing of alternate rows of erosion permitting and erosion resisting crops. In wind strip cropping tall cereal plants strip is alternated with short leguminous or pulse crop across the direction of wind.

2. Agrostological practices

The following are the important agrostological soil conservation practices: (i) cultivation of grasses, (ii) afforestation, (iii) checking overgrazing.

(i) Cultivation of grasses

Growing grasses in rotation with crops, improves soil fertility and helps in binding the soil. Besides their role in soil conservation, the grasses provide a good fodder for cattle.

(ii) Afforestation

Afforestation measures involve growing forests at places where there were no forests before owing to lack of seeds or due to adverse factors such as unstable soil, aridity or swampness. Plantation of trees in short blocks is known as *wind-break,* and extensive plantation of tress is known as *shelter-belts.* The plantation is usually done in two or three belts. Small sized plants are planted on windward side and tall trees on leeward side. These plantations reduce the wind velocity and also check the transport of lifted sand and soil particles. Selection of species, and season for planting varies with the climate. Pure plantations are easily damaged by insects and fungus diseases. Hence, mixed plantations with more than one species should be grown.

Plants are the most effective means of conserving soil. The leaves and branches of trees, shrubs and grasses helps to break the force of the falling rain and together with the plant litter on the ground, keep the rain from eroding the soil particles.

Sand dunes are mostly barren land devoid of vegetation and most of them remain unexplored due to their inaccessibility. The drifting sand engulfs the fertile agricultural fields, canals, wells, railway lines, roads, buildings etc. To check the onslaught of shifting dunes and accumulation of sand dunes on the adjoining fertile and cultivable land, and to prevent the natural resources from over exploitation and to make the arid regions more productive with assured subsistence to the inhabitants, a viable programme of dune stabilization through vegetation cover hardly needs any emphasis (Muthana, 1977).

Suitable species for afforestation of the shifting dunes are trees such as *Prosopis cineraria, P. juliflora, Acacia tortilis, A. senegal* and *Albizzia lebbek.* Well developed seedlings when planted in pits 45 cm^3 or 60 cm^3 prove very successful. Such deep planting facilitates the root system to get in contact with the moist zone at a lower depth, which are generally met with at 50-60 cm below the surface.

Micro-wind-breaks can be erected using locally available bush species such as *Crotolaria burhia, Leptadenia spartium, Zizyphus nummularia, Aerva pseudotomentosa, Calligonum polygonoides, Panicum turgidum* etc. These plants should be collected at the site and buried vertically crown downwards in lines 2 to 5 meter apart to reduce the wind velocity at the dune surface. Closer spacing and cross barriers to form a checker board pattern may be adopted in localities subjected to high wind velocity or in places where disorderly local wind movement frequently occur. The afforestation should be completed before the commencement of the monsoon (Mann, 1981).

The road side trees protect from erosion in two ways. Firstly, the roots of the trees firmly bind the soil along roadside and thus save it from flying and flowing away with wind and water. Secondly, the roots of trees mechanically breaks the force of wind and rain, and protect the soil. In high mountains, roadside trees stop the shifting snow sheets and blocks from falling on the roads, they also protect landslides. They also divert the run-off water along the level of the road, thereby breaking the force of sudden gush of water.

(iii) Checking overgrazing

Since grazing in all the areas subjected to soil erosion cannot be stopped, a restricted and rotational grazing may be helpful in soil conservation.

3. Dry farming practices

In arid regions, an integrated rational system of cultivation of purely rainfed, drought enduring and early maturing crops are grown. Both organic and inorganic manure are necessary to improve the soil fertility.

(b) Mechanical measures

Although the biological practices offer the cheapest ways of soil conservation, it is not always possible to employ them. The mechanical measures of soil conservation aim at: (i) reducing the velocity of run-off water to retain it for long period so as to allow maximum water to be absorbed in soil, (ii) dividing a long slope into several small parts to reduce run-off, and (iii) protection against erosion by wind and water. The mechanical measures include:

(i) Basin leaching

In this practice, a number of small basins (water reservoirs) are made along the contour. Basins collect and retain water for long periods and also catch and stabilize downwardly moving soil on the slopes.

(ii) Pan breaking

In some areas, formation of hard sheet of clay occurs below the surface soil, which does not permit infiltration of water. By pan breaking, drainage and percolation of rain water is improved and the soil is saved from residual run-off and erosion.

(iii) Contour terracing

Drainage channel or properly spaced ridges or soil mounds are formed along a contour to check soil erosion. These are called terraces.

(iv) Contour trenching

A series of deep pits or trenches are made across the slopes at convenient distances. The soil excavated from the trenches is deposited along the lower edge in the form of bund. On the ridges trees seeds are sown. The main objective of contour trenching is to cut-off the force of water by a series of hurdles in the form of these trenches. Contour trenches may be continuous or staggered.

(v) Gully control

Gully formation along the slope can be checked by (a) making perimeter bunds around gullies to check flow of water through it, (b) growing suitable soil-binding vegetation on the gullies to check soil erosion, and (c) making diversion trenches around gullies.

The soil conservation measures discussed above can be employed either singly or in combination, according to the circumstances.

Heritage Conservation

The cultural heritage has been considered as one of the forms in which the characteristic spirit of a society, and people of a nation reveals itself most clearly. It is an expression of people's historical experience and its collective personality. It represents the very basis of cultural identity in the consciousness of individuals and the community.

The cultural heritage, and specially its architectural component, are at present exposed to serious risk of damage owing to the effects of urbanisation and industrialization, air pollution and some types of climate.

In the face of these dangers, which are rapidly becoming more serious, the whole world has become more aware of the need to preserve the heritage in all its forms, tangible and intangible. Both public and private action for their conservation and enhancement has been considerably stepped up. No country today can refuse to protect atleast one aspect of its architectural and artistic heritage. Inventories are being drawn up, legal and practical protective measures have been taken and restoration projects have been started or are continuing. The population at large now more willingly accepts the duty to preserve not only the outstanding and illustrious features of their heritage but also more modest relics of their past.

Furthermore, in the context of action to preserve the heritage, many countries have reintegrated ancient monuments into economic and social life, putting them to use again for modern activities. Historical monuments are no longer merely places where specialists carry out their researches but fulfill a cultural and educational mission, making the population aware of the various aspects of their culture.

As a result of changes in ideas and attitudes, a parallel has been established between concerns about protection of the cultural heritage on the one hand, and the natural environmental and genetic resources on the other. Cultural property and natural property have been incorporated into the single concept of "world heritage". Although they have different characteristics and often come under the responsibility of separate departments. Immovable cultural property, such as historical monuments and groups of buildings, and natural property, such as national parks and outstanding sites, raise a set of common problems regarding their preservation and enhancement, and call in the same way for international cooperation.

Some countries have already framed legislation defining the content and scope of action to protect the cultural heritage. UNESCO has launched a number of major appeals in support of international campaigns for the preservation and safeguard of sites of exceptional historical interest. As a result of international action, each society's heritage has began to be seen as forming part of the common heritage of mankind.

Preserving the natural heritage means, first and foremost, preserving the living resources for the purpose of genetic selection, on which ultimately depends the maintenance and expansion of agricultural production, as well as for the purpose of preservation of the diversity and beauty of nature.

Cultural heritage

As threats to the world's treasures grow, nations become increasingly aware of the need for international measures to ensure that the cultural heritage of mankind can be passed on to future generations. This led to the organisation of UNESCO's World Heritage Convention. The Convention established a list of 136 properties of outstanding universal value in 1972. The World Heritage Committee in 1983, added 29 more cultural monuments, thus bringing the total to 186. This list contains six sites from India—Ajanta caves, Elora caves, Agra fort, Taj Mahal, Sun Temple (Konark), and groups of monuments at Mahabalipuram.

Natural heritage

Conservation of natural heritage has long been supported by UNESCO, when in 1948, it sponsored the creation of the International Union for Conservation of Nature and Natural Resources (IUCN). Twenty years later UNESCO organised an inter-governmental conference on the scientific basis for the "Rational use and conservation of the resources of the biosphere". It was this conference, which gave rise in 1970 to the "Man and Biosphere" programme, which was launched in November 1971. It was the central feature of the UNESCO's activities concerning conservation of natural resources. Efforts have been made to provide extra protection for a number of specially selected biosphere reserves and world heritage sites, in order to make them more permanent and stable, not only as conservation units but also as field laboratories for research, training and demonstration.

The *biosphere reserves* are intended to preserve genetic diversity in representative ecosystems and provide for *in-situ* conservation of plants, animals and micro-organisms. They are in a sense, the world's heritage sites of rich flora and fauna for posterity. They belong to all mankind. The thematic roles envisaged for the biosphere reserves have universal appeal. They may be summarised as follows:

(1) The biosphere reserves are to be considered as forming an extensive network of natural laboratories spanning continents and as world's heritage sites of humanity for conservation and management. The emphasis on environmental protection and resource conservation will be from a holistic viewpoint. Stress will be on the ecosystems as a whole, minimising if not eliminating altogether, the "human syndrome".

(2) A nexus between natural and social sciences in the management is conceived. In the tropics, at many of these "heritage sites", the tribal population live as co-evolutionary parts of the natural ecosystem. They live in ecological harmony with the ecosystem. The people who live in urban areas and cities are the real migrant tribals.

(3) Management for a wise husbandry of resources and restoration, as far as possible, to its climax status.

(4) The limits to ecosystem resilience have to be determined. The degree of human impact prescribes the limitations to ecosystem functions and the balance of inputs and outputs for its healthy functioning.

(5) Wise human investment for protection, preservation and conservation for ecosystem diversity, sustainability and productivity.

(6) Organisation of research into pilot-scale projects, short-term, medium-term, long-term projects. Comparative studies on regional and national scales will be necessary to improve the efficiency of management plan of action.

The Indian subcontinent has been divided into 8 bio-geographical regions (Blandford, 1901). The modern classification of Udvard (1975) on a joint consideration of biogeography and biome types divides India into 12 biographic provinces viz. (1) Himalayan highlands, (2) Thar desert, (3) Malabar rain forest, (4) Indus-Ganges monsoon forest, (5) Deccan thorn forest, (6) Coromandel, (7) Mahanandian, (8) Bangalian rain forest, (9) Burma monsoon forest, (10) Laccadives islands, (11) Maldives and Cagos islands, and (12) Andaman and Nicobar islands.

In India 13 potential sites for setting Biosphere Reserves were identified in 1979. Later on Great Nicobar island was also included, making a total of 14 Biosphere Reserves (Table 3).

Table 3
Biosphere Reserves in India

Site		State
1.	Nilgiri	Tamil Nadu, Karnataka, Kerala
2.	Namdapha	Arunachal Pradesh
3.	Nanda Devi	Uttar Pradesh
4.	Uttarakhand (Valley of flowers)	Uttar Pradesh
5.	North island of Andamans	Andaman & Nicobar
6.	Gulf of Mannar	Tamil Nadu
7.	Kaziranga	Assam
8.	Sundarbans	West Bangal
9.	Thar desert	Rajasthan
10.	Manas	Assam
11.	Kanha	Madhya Pradesh
12.	Nokrek (Tura range)	Meghalaya
13.	Little Rann of Kutch	Gujarat
14.	Great Nicobar island	Andaman & Nicobar

The biosphere reserves are considered as an extensive network of laboratories spanning continents, and as world's heritage sites of humanity for conservation and management. These world heritage, which include both man-made and natural ones, will always require careful protection. The emphasis in them, has to be given on environmental protection and resource conservation from a holistic view. Protection has to be afforded on a day-to-day basis. In any event, the need for it will become increasingly frequent, and the work will continuously increase.

National parks and sanctuaries are our other natural heritage which have been established to conserve the resources contained in them.

A *national park* is an area dedicated to conserve the scenery, natural objects and the wildlife therein. In these areas, all private rights are non-existent, forestry operations and grazing of domestic animals are prohibited. Certain parts of the park are developed for tourism, enjoyment and study in such a way that it will not disturb or scare the animals. Some of the important national parks are: Kaziranga (Assam), Hazaribagh (Bihar), Bandipur, Nagarhole and Silent Valley (Karnataka), Kanha, Shivpuri, Bandhavgarh, and Panna (Madhya Pradesh), Tadoda (Maharashtra), Keibul Lamjao (Manipur), Simlipal (Orissa), Ranthombore (Rajasthan), Kanchenjuga (Sikkim), Corbet, Rajaji, and Dudhwa (Uttar Pradesh).

A *wildlife sanctuary* is dedicated to protect the wildlife, but it considers the conservation of species only and also the boundary of it is not limited by state legislation. In a sanctuary, killing, hunting or capturing of any species of birds and mammals is prohibited except by, or under the control of the highest authority in the department responsible for management of a sanctuary. Private ownership may be allowed to continue in a sanctuary. Forestry and other usages are permitted to the extent that they do not adversely affect the wildlife.

Today, India has 421 wildlife sanctuaries covering an area of about 1.4 lakh square kilometers constituting more than 4 per cent of the total geographical area of the country and about one-fifth of all the forest areas. Some of the important wildlife sanctuaries are—Pakhal, Pocharam, Kawal (Andhra Pradesh), Namidapha (Arunachal Pradesh), Manas (Assam), Mollen (Goa), Wild Ass and Nal Sarovar bird sanctuary (Gujarat), Sultanpur lake bird sanctuary (Haryana), Sechu-tun-nallah sanctuary (Himachal Pradesh), Dechigam Wildlife sanctuary (Jammu & Kashmir), Ranganthitoo bird sanctuary (Karnataka), Periyar wildlife sanctuary and Wynad wildlife sanctuary (Kerala), Yawal wildlife sanctuary (Maharashtra), Dampa wildlife sanctuary

(Mizoram), Intangki wildlife sanctuary (Nagaland), Chilka lake bird sanctuary (Orissa), Abohar wildlife sanctuary (Punjab), Sariska Wildlife sanctuary, Ghana bird sanctuary, Darrah game sanctuary (Rajasthan), Madumalai wildlife sanctuary and Vedanthangal water bird sanctuary (Tamil Nadu), Jaldapara wildlife sanctuary (West Bengal).

Chapter 3

Conventional Fuels and their Environmental Impacts

Fuels has been used for comfort heating and cooling (including cooking). The demand for fuel has been gradually increasing with the increase of human population. We are quite sure that it is not possible to meet this demand by conventional sources like—fire wood, plant and animal waste, coal, crude oil and natural gas, because these sources are limited and they are going to finish sooner or later. The very high consumption of fuel in advanced countries is due to ready access for heating, cooking, lightening, domestic work etc. In the developing countries, the fuel consumption is concentrated in the urban areas for industrial, commercial and even domestic work.

The fuel consumption for lighting and cooking in rural areas and urban areas shows that in the rural areas 84 per cent lighting is through the use of kerosene, while 94.5 per cent cooking is through fire wood. The corresponding figures in urban areas are lighting 45.2 per cent from kerosene and cooking 58.1 per cent through conventional fuels and 26.5 per cent through kerosene (ABE, 1985).

Different epochs of history gave preference to different forms of fuel use. Ever since man has discovered fire, about 150 thousand years ago, he felt warmth against cold and protection from predators, and could get more palatable food and higher survival rate. The fuelwood, plant and animal waste soon became the principal sources of energy and even today man gets about 50 per cent of his fuel requirements from wood. During the industrial revolution, coal was the cost favoured form which brought about far reaching changes in the commercialization of the conventional sources. This was followed by the use of crude oil and natural gas, which helped to usher in many significant transformations, particularly in the transportation and industrial sectors. The fuelwood is a renewable resource. However, fuelwood resource is dwindling fast. There is a big gap between demand and supply of the fuelwood, resulting in the depletion of the forest cover. Thus, this renewable resource is tending to become a non-renewable resource due to over exploitation. The reserves of coal, crude oil and natural gas are limited in quantity. These conventional fuels are also dwindling fast due to over exploitation. Here we shall take up the description in details about each one of the above mentioned conventional fuels separately.

Fire Wood

More than one-third of the world's population depends on wood for cooking and heating. Around 86% of all the wood consumed annually in the developing countries is used for fuel, and of this total half is used for cooking. In the arid tropics the woodlands have been axed down at the rate of 4 million hectare per year. The outcome is a ruinous destruction (Oza, 1984).

The households in India and much of the rest of Asia have relied principally on the dried wood for cooking. Such fuel remains in plentiful supply to villagers in all but the arid parts of India. As urbanisation has proceeded, prices for fuelwood have increased, providing an increasing lucrative source of income to villagers. Thus, the total energy requirements of the rural and much of the urban population is covered by wood from the nearest forests. Usually small trees are felled or the limbs of large trees are lopped off until only the crown twigs are left. Wood cutting parties from the cities work their way in the forests even though the ban on cutting is strictly supervised, people go out in the forests in night, trees are felled and cut and the next day they are offered for sale as firewood in the city market.

The demand for firewood is likely to increase with a corresponding increase in population, and until we are able to search an alternative energy source in place of wood. As per the Energy Survey Committee of India (1965) fuelwood consumption per day in rural and urban areas of our country is 0.58 kg/head/day and 0.41 kg/head/day respectively. On the basis of this study, annual per capita consumption in rural and urban areas works out to be about 212 and 100 kg respectively. The projected demands for the years 1981, 1991 and 2001 have been shown in Table 4.

Table 4
Projected demand for fuelwood in India

Year	*Firewood requirement (Lakh tonnes)*		
	Rural	*Urban*	*Total*
1976	50.95	07.89	58.86
1981	56.07	08.97	65.64
1986	62.94	10.17	73.11
1991	69.91	11.85	81.76
1996	76.91	11.92	89.93
2001	84.60	14.49	99.03

According to ABE (1985) the firewood demand is going to be of the order of 300-330 MT against the present level of 120-130 MT (Table 5) lists such demand statewise.

Table 5
Estimated fuelwood consumption in 2004/06 in Mt.

Region/State/UT	*Fuelwood Consumption*		
Norther Region	Rural	Urban	Total
1. U.P.	37	10	47
2. Rajasthan	17	4	21
3. Punjab	3.5	0.5	4
4. Haryana	2.0	0.5	2.5
5. Himachal Pradesh	3.0	0.5	3.5
6. J & K	2.5	0.5	3.0
Eastern & NE Region			
7. West Bangal	9.5	0.5	10

Contd.

Region/State/UT	Fuelwood Consumption		
8. Bihar	19.5	1.5	21.0
9. Orissa	13.0	3.5	16.5
10. Assam	11.5	1.0	12.5
Southern Region			
11. Andhra Pradesh	26.5	8.5	35.0
12. Karnataka	17.5	7.5	25.0
13. Tamil Nadu	20.0	10.5	30.5
14. Kerala	11.5	4.5	16.0
Western Region			
15. Madhya Pradesh	20.5	6.5	27.0
16. Maharashtra	22.5	5.5	28.0
17. Gujarat	10.0	3.0	13.0
Total	247	69	316

Source: ABE, 1985

It is clear from the table that nearly 70 per cent of firewood demand pertains to the rural area. All the demand for firewood has to be met as firewood, atleast in the rural sector. If firewood can be substituted, it would be so in the urban sector to start with. Only 50 MT of firewood may become available from the natural forests. So far most of the removals have been unrecorded and according to National Commission on Agriculture (IX), hardly 10 per cent are recorded. Furthermore, in Gujarat in 1970-71 nearly 15 per cent of firewood was derived from forests, while the remaining 85 per cent came from tree lands (ABE, 1985). According to NCA for the next 20 years if so the average fuelwood contribution from natural forests would be 0.75 t/ha/year and the remaining part will have to be met from plantations. An idea about this have been shown in Table 6.

Table 6

Estimated fuelwood requirements in 2004/05 to be met through plantations

State	*Quantity (Mt)*
Northern Region	
1. U.P.	66.5
2. Rajasthan	22.5
3. Punjab	7.0
4. Haryana	6.0
5. Himachal Pradesh	2.5
6. J & K	2.0
Eastern & NE Region	
7. West Bengal	13.5
8. Bihar	20.0
9. Orissa	19.0
10. Assam	11.0
Southern Region	
11. Andhra Pradesh	35.0
12. Karnataka	27.5
13. Tamil Nadu	32.0
14. Kerala	15.5
Western Region	
15. M.P.	28.0
16. Maharashtra	30.5
17. Gujarat	14.5

Source: ABE, 1985

Whole of this requirement has to be grown on non-agricultural land, degraded forest land, culturable wasteland, barren/unculturable land, permanent pasture and grazing lands and even private perimeter land. Though the estimates of such land vary considerably, there is reasonable amount available throughout the country. Roughly 60 M hectare of such land are available but it might be difficult to bring more than 50 per cent of this land under plantation. On the basis of present day 'Chulah' efficiency of 8 per cent, the scenario for different states have been shown in Table 7, together with the possible strategy for bridging the gap.

Table 7
Fuelwood scenario 2004/05

Firewood demands	*States*	*Strategy for self-sufficiency*
Likely to be met	Rajasthan, HP, J & K MP, Gujarat and Maharashtra (?)	—
Likely to be met after:		
— all available land planted	WB, Bihar, Orissa and AP	—
— fuelwood yields increased from existing forests	Karnataka	—
Not likely to be met	UP and Tamil Nadu	Alternatives to need to be developed fuelwood
Acute shortage	Punjab, Haryana and Kerala	Large scale use of agricultural residues as briquetts

Source: ABE, 1985

Energy Plantations

Energy plantation is the practice of planting trees, purely for their use as fuel. Mitchell (1979) estimated that the need for wood for cooking purposes in India is 0.8 kg per capita per day. But this value does not hold good for cold areas where firewood is also required for keeping houses warm. In recent years, to meet the demand of energy, plantation of energy plants has been re-emphasized. If firewood plants were not raised rapidly, by 2000 AD more than 250 million people would not be able to manage fuels for cooking purpose (Anonymous, 1980).

In India, conditions of hills are quite different from that of plains. The villagers hardly get firewood plants, as they have to go to interior of forest and collect wood-falls. Even they have very limited right to fell trees and, therefore, they depend on wood-falls and loppings of minor tree branches (Singh et al, 1988).

Recently, energy plantation has got much boost in our country. Government has started many plans, for example social forestry, silviculture, agri-horticulture and afforestation on wastelands. In view of getting maximum biomass, afforestation and forest management systems will have to be developed.

Social Forestry

Social forestry will certainly decrease the gradually increasing pressure on the forests. This includes planting trees along roadsides, canals, railway lines and on wasteland in villages. Some important species are—*Acacia nilotica, Albizia labback, A. procera, Anthocephalus chinensis, Azadirachta indica, Bauhinia variegata, Butea monosperma, Cassia fistula, Dalbergia sissoo, Eucalyptus globulus, E. citriodora, Ficus glomerata, Lagerstroemia speciosa, Madhuca indica, Morus alba, Populus ciliata, P. nigra, Terminalia arjuna, Toona ciliata, Salix alba, S. terasperma* etc.

The choice of tree species for fuelwood social forestry varies with the climate.

Fuelwood Species for Arid and Semiarid Regions

Acacia brachystaschya, A. cambagei, A. cyclops, A. nilotica, A. saligna, A. senegal, A. seyal, A. tortilis, Albizia lebbeck, Anogeissus latifolia, Azadirachta indica, Cajanus cajan, Cassia siamea, Colophospermum mopane, Emblica officinalis, Eucalyptus camaldulensis, E. citriodora, E. gomphocephala, E. microtheca, E. occidentalis, Haloxylon aphyllum, H. persicum, Parkinsonia aculeata, Pinus helepensis, Pithocellobium dulce, Prosopis alba, P. chilensis, P. cineraria, P. juliflora, P. pallida, P. tamarugo, Tamarix aphylla, Zizyphus mauritiana, Z. spina-christi.

Fuelwood Species for Humid Tropical Regions

Acacia auriculiformis, Calliandra calothyrsus, Casurina equisitifolia, Derris indica, Gliricidia sepium, Gmelia arorea, Guazuma ulmifolia, Leucaenia leucocephala, Mabgroves spp, *Mimosa scabrella, Muntingia calabura, Sesbania bispinosa, S. grandiflora, Syzyium cumini, Terminalia clappa* etc.

Fuelwood Species for Tropical Highlands

Aciacia mearnsii, Ailanthus altissima, Alnus acuminata, A. nepalensis, A. rubra, Eucalyptus globulus, E. grandia, Grevillea robusta, Inga vera etc.

Silviculture Energy Farms (Short Rotation Forestry)

Firewood demand can be met by growing energy crops involving short rotation forestry plantations. It involves growth of tree species which give rapid juvenile growth and can be regrown from harvested stumps. This type of fuelwood production is suited to a variety of climates and gives a very high yield. Feasible ways of using wood for energy will vary from country to country. Agroforestry involves inter-cropping trees and field crops, so that both wood and fuel can be produced from the same land. The crops are chosen carefully so that they compliment rather than compete with the trees. *Eucalyptus* for example, is not a good choice for agro-forestry because it soaks up a lot of moisture and tends to produce toxic substances. *Leucaena leucocephala* on the other hand, helps to fertilize the whole area around it because it is a leguma. It also has an open leaf structure which does not shade out ground crops as much as some other trees (Hall et al, 1982). Many of the tree planting schemes based on fast-growing trees have several complementary objectives and are aimed at providing jobs, improving the environment, and bringing social benefits to the local people as well as growing renewable energy.

Leucaena leucephala, known in India as subabul, is one of the fastest growing tree species. It grows luxuriently even on marginal or wastelands. It can be cut every five years after initial planting, and thereafter once every three years, for an indefinite period. It yields approximately 75 tonnes of fuelwood per hectare, per each cutting, under rainfed conditions. Subabul readily grows again from the cut down stumps. The calorific value of the dry wood is between 3895-4640 K Cal per kg.

Clonal Propagation or Micropropagation

It has been demonstrated that a variety of plant species can be convincingly propagated through the techniques of cell, tissue or organ culture. This is popularly described as clonal propagation or micro-propagation. The major benefits of this method include the following: (i) rapid multiplication of superior clones and maintenance of uniformity, (ii) multiplication of disease free plants, and (iii) multiplication of sexually derived sterile hybrids. In most cases, clonal propagation is achieved by placing sterilized shoot tips or auxiliary buds onto a culture medium that is sufficient to induce formation of multiple buds. This method has already been used propagate a large number of marketable ornamentals. Following steps are involved in the method of clonal propagation: (i) Stage-I involves establishment of tissue *in-vitro,* (ii) Stage-II involves multiplication of shoots (often media are not changed between stage I and stage II, (iii) Stage-III concerns root formation and conditioning of propagules prior to transfer to the green houses; this stage requires high intensity and alteration of media for promotion of root formation, (iv) Stage-IV involves growth in pots followed by field trails. The number of steps in micropropagation may sometimes be reduced to three or two as shown in Figure 6.

A wide range of plants have now been regenerated through technique of tissue culture. This has been found particularly useful for propagules of trees species, so that a large number of plant species have been successfully grown by tissue culture (Table 8).

Burning Fuelwood for Power Generation

In Ballnao, Philippines, there exists a series of small (3 megawatt) power generation station operating on subabul fuelwood alone. This project aims in the long term, to provide rural development and to settle shifting cultivators.

Today there is a trend towards a planned reversal of the destruction of the earth's forest cover, based on large fuelwood energy plantations raised specifically to feed industrial projects. They also point the way towards reversal of the trend for concentration of industries in big cities, and the continuous flight of economic refugees from the villages pouring into big cities in search of employment. Since fuelwood energy plantations can best be established in rural areas, new industry could also be attracted to locate on these cheaper energy resources, and rural people could get employment in their own home territory.

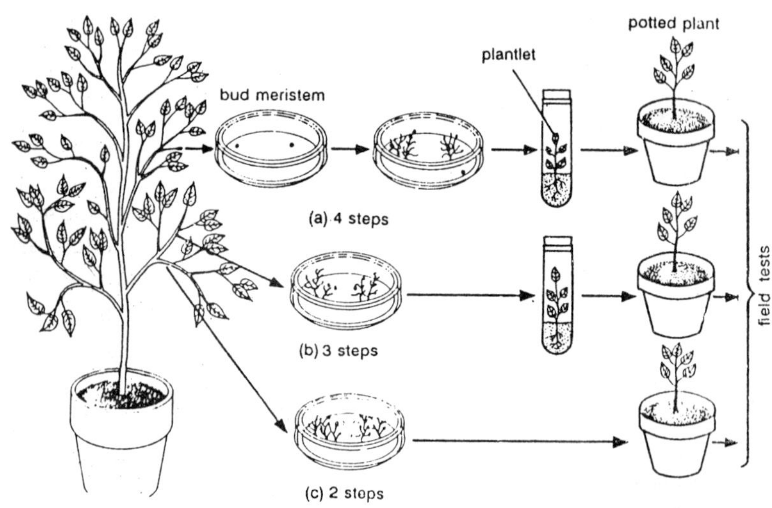

Figure 6
Three different methods of micropropagation of plant material

Table 8
Plants which can be propagated by tissue culture (Modified from Swaminathan, 1991).

1.	*Alpinia*	25.	Cardamom	49.	*Gypsophila paniculata*
2.	*Alstroemeria*	26.	Cattleya	50.	*Hydrangea*
3.	*Anthurium*	27.	Crocusmia	51.	*Hamamelis*
4.	*Asparagus*	28.	*Cucumis sativus*	52.	*Iris*
5.	*Actinidia*	29.	*Capsicum*	53.	Kalanchoe
6.	*Aglaonema*	30.	*Coffee*	54.	Lily
7.	*Agapanthus*	31.	*Dianthus*	55.	Maranta
8.	*Amaryllis*	32.	*Dieffenbachia*	56.	Monstera
9.	*Amibias*	33.	*Dendrobium*	57.	Malus
10.	*Artemesia*	34.	*Eryngium*	58.	*Mimosa*
11.	*Aster*	35.	*Epidendrum*	59.	*Narcissus*
12.	*Arillata*	36.	*Eucalyptus*	60.	*Petunia*
13.	*Allium*	37.	*Ficus benjamina*	61.	*Philodendrons*
14.	*Alocasia*	38.	*Ficus elastica*	62.	Pineapple
15.	*Banana*	39.	*Ficus lyrata*	63.	Papaya
16.	*Begonia*	40.	*Ficus robusta*	64.	*Prunus*
17.	*Betula*	41.	*Ficus mini*	65.	Potato
18.	*Babaco*	42.	*Ficus compacta*	66.	Poplar
19.	*Bamboo*	43.	*Ficus foliole*	67.	Rose
20.	*Calathea*	44.	*Fragaria*	68.	*Spathiphyllum*
21.	*Cordyline australis*	45.	*Gerbera*	69.	*Syngonium*
22.	*Chrysanthemum*	46.	Ginger	70.	Turmeric
23	*Cymbidium*	47.	Grape vine	71.	Tomato
				72.	Vanilla

Plant and Animal Wastes

These conventional fuels primarily originate from plants and animals and they include various crop residues, wild growing biomass, forest and wood processing wastes, fruit and vegetable processing wastes, animal wastes etc. Depending upon their origin, they may be further classified as crop wastes, forestry wastes, animal wastes, fish and marine wastes, and agro-industrial wastes. Paddy and wheat straw, sugarcane tresh, peels and vines of vegetables and fruits, stalks, stems or sticks of maize, jute and cotton etc. are crop wastes. Saw dust, bark, clippings, under utilized or non-edible plants constitute the forestry wastes. Dung, urine, skin, skeleton etc. are animal wastes. Heads and bones of fishes, prawn wastes, squilla, frog wastes etc. constitute fisheries and marine wastes. Rice husk, baggasse, molasses, cotton linters, bran etc. are examples of agro-industrial wastes.

The exact quantities, nature and disposal of plant and animal wastes are difficult to establish because of their varied and scattered availability. However, based on some all India Surveys conducted by DST, NPC and ICAR, the estimated quantities of different types of plant and animal wastes and their seasonal availability have been shown in Table 9.

Table 9
Quantity of agricultural residue/waste/byproducts

Sl. No.	*Agricultural residue/ waste/by-product*	*Estimated quantity/ year, million tonnes*
1.	2	3
I. Crop Residue		
1.	Wheat straw	31.80
2.	Paddy straw	88.00
3.	Maize stalk	14.25
4.	Maize cobs	6.85
5.	Sorghum sticks	32.80
6.	Pearl millet straw	8.80
7.	Finger millet straw	10.50
8.	Barley straw	4.40
9.	Gram straw	10.80
10.	Pigeon pea stubble	3.64
11.	Groundnut shell	5.75
12.	Sugarcane trash & dry leaves	1.77
13.	Jute sticks	2.05
14.	Jute bark & leaves	2.05
15.	Cotton sticks	10.05
16.	Mango peel & kemel	1.25
17.	Pine apple wastes	0.02
18.	Citrus fruits peel, pimace, sed etc	0.03
19.	Banana pseudostems	0.2

Contd.

1.	2	3
20.	Pea shells & vines	0.01
21.	Tomato seeds and pomace	0.08
22.	Coconut shell	0.88
23.	Coconut water	0.50
24.	Coconut husk	0.06
25.	Arecanut husk	0.10
26.	Arecanut leaf sheaths & leaves	0.90
27.	Cashew apple	30.00
28.	Cashew testa	0.04
29.	Cashewnut shell liquid	0.14
30.	Tea fluffs, stalks & sweepings	0.01
31.	Coffee residue	0.45
32.	Tobacco leaf scrap & stalks	0.62
33.	Tobacco seeds	0.15
34.	Rubber wood	0.25
35.	Bark from industrial wood	0.30
II. Agro Industrial Residues/Waste/By products		
36.	Rice husk	18.0
37.	Rice bran	3.0
38.	Deoiled rice bran	2.1
39.	Jute mill waste	0.5
40.	Cotton dust	0.3
41.	Groundnut deoiled cakes	2.5
42.	Deoiled cakes of other major oil seeds such as rape seed and mustrad, linseed, seasmum and castor	2.0
43.	Deoiled cake of non-edible oil-seeds, such as, Kusum, Neem, Karanza etc.	0.17
44.	Sugarcane bagasse	5.25
45.	Molasses	1.70
46.	Press mud	0.20
47.	Saw dust	2.00
III. Animal & Poultry Wastes		
48.	Wet cow dung	9.00
49.	Cattle urine	37.00
50.	Carcasses	12.50
51.	Blood	0.50

Contd.

1.	2	3
52.	Sheep/Goat wastes	45.00
53.	Poultry wastes	1.00
IV. Fisheries Wastes		
54.	Prawn wastes	0.40
55.	Squilla	1.00
56.	Frog wastes	0.05

Source: Srivastava PK and RC Maheshwari, 1981. Position Paper on Utilization of Agricultural Wastes and By-products in India. Paper presented in XVIII Annual Convention of ISAE held at CSSRI Karnal on Feb 26-28, 1981.

Plant and animal wastes are generally organic materials, traditionally, they are burned in the field. However, they are now being used for energy production.

Traditional use of plant and animal wastes 'domestic fuel' still consumes maximum portion of the available residue and it has been estimated as 137 million tonne (35%) in 1989-90. This application has efficiency less than 10 per cent. Many crop wastes as well as other plant and animal wastes have very good energy content (Table 10). In fact some of such wastes as stalks and stubbles of pigeon pea, jute, cotton, dung cakes etc have been traditionally used as cheap domestic fuel in rural areas of India, while baggasse, rice husk, groundnut shell etc. are now extensively being used as boiler fuel. The total energy released by firewood and plant and animal wastes including dung consumed in our country amounts to 942.83 million MWh (Table 11).

Table 10
Heat potential of agricultural wastes

Sl. No.	Agricultural Waste	M.C. % (a, b)	Ash content %	Calorific value kcal/kg	Estimate quantity mt	Energy potential kcal
1.	Paddy straw	10.6	21.1	3000	88.00	264.00
2.	Rice husk	9.6	15.5	3440	18.00	61.92
3.	Mango leaves	9.8	18.0	3390	0.08	0.27
4.	Groundnut straw	12.1	1.3	4200	—	—
5.	Groundnut Shell	12.0	1.3	4200	5.75	—
6.	Cowdung	8.5	21.7	3290	900.00	2961.0
7.	Cowdung	4.3	33.2	3240	...	...
8.	Sugarcane bagasse (fermented)	15.0	1.0	3800	5.25	—
9.	Wheat straw	18.0	18.0	3800	31.80	120.84
10.	Cotton sticks	12.0	13.5	3300	—	—
11.	Maize stalks	11.5	14.2	4700	14.25	66.980
12.	Maize cob	8.0	13.8	3500	6.85	23.975
13.	Gram straw	9.2	13.2	3950	—	—
14.	Mash straw	7.8	13.4	3920	—	—
15.	Bajra stalks	11.2	17.5	3850	8.80	33.88
16.	Massor straw	10.1	12.8	3810	—	—
17.	Moong straw	10.3	12.6	3820	—	—
18.	Coconut shell	6.0	—	4350	0.88	3.89

Contd.

Sl. No.	Agricultural Waste	M.C. % (a, b)	Ash content %	Calorific value kcal/kg	Estimate quantity mt	Energy potential kcal
19.	Oilseeds	—	—	4775	—	—
20.	Bamboo Cane	10.5	—	3925	—	—
21.	Buckwheat husk	10.0	3925	—	—	—
22.	Oak wood	13.0	—	38.0	—	—
23.	Oak bark	7.0	—	4310	—	—
24.	Fire wood	—	—	5000	—	—

Source: Srivastava PK and RC Maheshwari, 1981. Position Paper on Utilization of Agricultural Wasts and By-products in India.

Table 11
Non-Commercial Consumption During 1975-76

Sl. No.	*Source*	*Consumption million tonnes**	*Calorific value Fcal/kg*	*Total energy million MWh*	*Percentage of total energy*
1.	Firewood	133.1	4000	619.07	65.65
2.	Agricultural waste	41.0	3000	143.02	15.17
3.	Animal dung	73.0	2130	180.80	19.18
	Total	247.1		942.89	100.00

Source: Report of Working Group on Energy Policy 1979, Government of India.

Even if the efficiency of burning these wastes for cooking purpose alone is taken as 10 per cent, the total energy used is 94.29 million MWh, which is approximately 1.36 times the energy used for crop production and post harvest operation (Randhawa and Maheshwari, 1981).

Paddy husk can be converted into smokeless solid fuel briquettes suitable for use in domestic cooking, hotels, kilns and boilers. In India, this is being done at Briquetted Fuel Plant installed at Alternate Hydro Energy Centre, Roorkee. This process is based on principle of pyrolysis and involves heating the raw material at 250-300°C in absence of air. The fuel briquettes are sold at Rs. 0.60 per kg which is cheaper than firewood and coal.

The mixture of wood and bark waste burnt directly is termed as *'hog fuel'*. Hog fuel combustion technology is used for large boilers sizes from 15000 lb/h to 500000 lb/h. In United States, a cogeneration technology has been developed to generate electricity from high-fuel, and to use the exhaust heat in the form of process steam for manufacturing operations. This is done by burning the hogfuel to make high pressure in the hogfuel boilers (600-1200 lb/inch^2), passing this steam through an extraction turbine driving a generator, and then using the steam exhausted from the turbine at low pressure (50-300 lb/inch^2) for process heat. When wood containing 55 per cent moisture is used, the power generation has about 60 per cent efficiency compared to that of 39 per cent from fossil fuels and 33 per cent from nuclear fuels (Ignacimuthu, 1996).

Coal

Coal is a fossilized material with a complex chemical structure (mostly carbon and hydrogen) derived from plant organic matter. Variations in coal consumption are attributed to the amount and distribution of the original plant material, major elements like carbon, oxygen, hydrogen, nitrogen, sulphur, and trace elements such as sodium, chlorine and mercury. The geological conditions during the formation of coal influences the extent of impurities and their subsequent treatment or removal. Besides organic material, coal contains mineral matter such as clays, sulfides, carbonates, silica with other trace elements. The organic material and the minerals present in or associated with coal determine the levels of emissions during processing/ utilization of coal.

Coal is formed from the remains of trees. Most of the world's coal beds appear to represent accumulations of plant material in swamps near the shores of ancient seas that alternately transgressed and receded from the land. Such alterations probably caused by the instability of the continental margins under crystal stress, allowed, first, the creation of swamps in which the fallen trees rotted slowly, if at all and, second, the burial of the malled remains of beach sands laid down during the ensuring transgression. In most coal fields there are many coal layers and seams, indicating repeated alternations of living swamp and sedimentary burial. Individual coal seams commonly lie on what are called underclays, the very fine material deposited by the sluggish streams of a subsiding coastal area immediately before the creating of a coastal swamp. Because of the rather unusual geological conditions required for their formation, coal fields are localised both in space and in geological age.

There is a continuous series from peat through lignite and bituminous coal to anthracite, reflecting a progressive metamorphism (change in form and composition) of the organic material. Peat is brown and porous; it contains visible plant remains and has a high moisture content. The weight of overlying rock and the earth's heat combine to drive moisture out of burial peat and change it into lignite. As the metamorphosis of peat into lignite, and lignite into bituminous coal proceeds, carbohydrates are converted to hydrocarbons and then to elemental carbon. The higher the percentage of elemental carbon and the lower the percentage of moisture and volatile hydrocarbons, the greater is the potential-heat content and the higher the rank of the fuel. The term bituminous is term of rank, based on heating value. In addition to rank, these fuels have grade. The grade of a coal or lignite is determined by its content of waste materials, notably ash and sulphur. A low-grade bituminous coal is one that contains much ash, sulphur or both.

About 6000 billion tonnes of coal lies under the earth and now over 200 billion tonnes have been used. Coal reserves are beds at a depth as little as 35 cm to 1220 meters, or in a few cases at a depth as much as 1830 meters. At present about 50 per cent of coal reserves are mined economically. The total coal production in the world has increased from 273 crore metric tonnes in 1980 to 323 crore metric tonnes in 1986, registering an increase of 18.4 per cent. Table 12 shows coal production in major countries of the world. Though our coal reserves are limited, yet we have more energy in the form of coal than any other fossil fuel. At the consumption of 625 million tonnes per annum, it may not run more than 20 years, even if we are able to extract all.

Table 12
Coal producing countries of the world

1980		*1986*	
Country	*% of world production*	*Country*	*% of world production*
USA	26.04	USA	22.38
China	21.85	China	27.78
USSR	18.10	USSR	15.88
Poland	7.07	Poland	5.95
U.K.	4.77	U.K.	3.62
South Africa	4.25	South Africa	5.49
India	3.99	India	5.06
F.R. Germany	3.46	F.R. Germany	2.70
Australia	2.70	Australia	4.94
Czechoslovakia	1.04	Czechoslovakia	0.78
Total world	273 crores metric ton		323 crore metric ton

Coal accounts for 60 per cent of India's commercial power requirements. Major coal-fields in India are Raniganj, Jharia, East Bokaro and West Bokaro, Panchkankam (Tawa Valley), Singrauli, Talcher, Canda-Wardha, and Godavari valley. The major states known for coal reserves are Bihar, Orissa, West Bengal, Madhya Pradesh, Andhra Pradesh and Maharashtra. By and large, the quality of Indian coal is rather poor in terms of heat capacity. Coal is used for generation of electricity. Many of our thermal and super-thermal power stations are located on the coal fields to produce electric power to feed regional grids.

Coal mining till power generation, is a dirty job, that makes the whole area barren, scarred and leaves tonnes of ash at the power station in the form of flyash and residual wastes. Hence, if at all coal is to be used, it should be done sparingly and only in emergency. A major problem with coal based thermal power generation is that of air pollution by stack exhausts. It is necessary to lay down and enforce air quality standards. Japan has raised from 0.02 ppm to a range of 0.04—0.06 ppm nitrogen oxide because of a feeling that these levels may not be harmful to health, whereas the cost of attaining the lower levels would be extremely high. Similarly, in United Stated of America, nitrogen oxide standards have been raised from 0.08 ppm to 0.12 ppm (8 hours average).

There has been considerable advancement in the use of alternate energy sources but fossil fuels still continue to be a more popular choice. Inspite of its known shortcomings, coal still reigns as the major source of energy in several heavy industries and electric utilities. With its vast reserves worldwide, coal will remain the single largest source of primary fossil fuel energy.

Crude Oil and Natural Gas

Petroleum (rock oil) and natural gas are found in similar geological environments and often together; indeed some authorities classify natural gas as gaseous petroleum. Commercial accumulations of crude oil and natural gas are almost entirely limited to sedimentary rocks. Such accumulations are often called *pools.* A oil field consists of one or more pools geographically isolated. Most oil and gas fields are a few square kilometers in extent; only a few fields underlie more than a hundred square kilometers on the earth's surface.

The geological requirements for an oil pool are a source rock, a sufficient permeable reservoir rock, a sufficiently impermeable cap rock, a favourable trapping structure, and water in the formation. Because most oil and gas is found in sedimentary rocks, because most accumulations are close to thick deposits of marine or deltaic sediments that contain large volumes of shales having an appreciable organic content, and because this organic content contains compounds of carbon commonly manufactured by plants or animals, geologists believe that the *source rock,* for most, if not all, petroleum is organic shale. As with coal, accumulation took place under conditions that prevented oxidation and promoted concentration of the organic remains and was followed by burial and subsidence. Subsequent metamorphism was physically and chemically unlike than that produces coal, which is marked by a progressive diminution of fluid and volatile constituents that ultimately yields pure carbon. The metamorphism of petroleum appears to yield increasing fluidity and volatility and produces no elemental carbon. Crude oil consists almost entirely of liquid hydrocarbons with some gaseous hydrocarbons dissolved in them. Tar and asphalt are semisolid or solid forms of crude oil.

Natural gas consists of hydrocarbons simpler and lighter than those of crude oil; the most abundant hydrocarbon in natural gas is methane. In addition, natural gas may contain water and gases that are not hydrocarbons.

Crude oil is lighter but viscous than water. Natural gas is much lighter than crude oil and much less viscous than crude oil or water.

Some traps contain oil but no gas, possibly because no gas was generated in the source rock, but more likely because gas, having a much lower viscosity than oil, can escape upward through rocks that petroleum cannot penetrate.

Some traps contain gas but no oil. This may be due to natural distillation, under heat and pressure, of the light hydrocarbons of natural gas from the heavy hydrocarbons of crude oil (Cook, 1976).

The sedimentary rocks containing plant and animal remains—about 10 to 20 crore years old are the source of mineral oil. Mineral oil is very unevenly distributed over space like any other mineral. There are six regions in the world which are rich in mineral oil. USA, Mexico, former USSR and the West Asian Region (Iraq, Saudi Arabia, Kuwait, Iran, United Arab Emirates, Qatar and Bahrain) are the major oil producing countries of the world.

Oil production by major oil producing countries of the world. It is clear that the oil production has declined from 300 crore metric tonnes in 1980 to 275 crore metric tonnes in 1986.

India has a large proportion of tertiary rocks and alluvial deposits particularly in the extra-peninsular India. Such potential oil bearing area is estimated to be over a million square kilometer, one third of the total area. It covers the northern plains in the Ganga Brahmaputra valley, the coastal strips together with the off-shore continental shelf (Bombay high), the plains of Gujarat, the Thar desert and the area around Andaman and Nicobar Islands.

Till independence Assam was the only state where mineral oil was drilled. In India oil was first found at Makum (north-east Assam) but drilling was started at Digboi in Lakhimpur district. After independence Gujarat plains and the major reserves were found off the Bombay coast—the richest oil field of the country, known as Bombay High (115 km from the shore). The latest oil deposits have been found off shore areas of the deltaic coasts of Godavari, Krishna, Kaveri and Mahanadi. The gas reserves are generally found in association with oil fields. However, exclusive natural gas reserves have been located in Tripura, Rajasthan and almost in all the off-shore oil fields of Gujarat, Maharashtra, Tamil Nadu, Andhra Pradesh and Orissa.

The natural gas is used both as energy source and also as an industrial raw material in petro-chemical industry. It is taken from Gujarat gas field to Madhya Pradesh, Rajasthan and Uttar Pradesh. Hazira-Bizapur-Jagdishpur (HBJ) gas pipeline is 1730 km long and carries 18 million cubic meters of gas every day. It feeds six fertilizer industries and three power stations. The liquefied petroleum gas (LPG) also called the cooking gas is now a very common domestic fuel in our country.

Oil reserve are insignificant in quantity. It has been estimated that after 30 years, nothing will be available even in the world market. The consumption of oil in our country has increased from 31 lakh tonnes per annum in 1951 to 25 million tonnes in 1971 and 33 million tonnes in 1981. Thus there has been an eleven fold increase in our oil consumption. With the present trend of increase in our energy demand, we shall need more than 60 million tonnes of oil per year by 2000 AD.

Natural gas is presently the fuel of choice to reduce problems associated with environmental problems. Experience in a number of countries including Germany, Netherlands, United Kingdom and Italy, which have high penetration of natural gas in residential and commercial areas clearly point out the desirability of natural gas a non-polluting fuel source.

Compared to other fossil fuel, natural gas contains certain inherent qualities which lack pollutants such as sulphur and also provide greater efficiency in combustion. The absence of sulphur in natural gas reduces appreciably the emission of sulphur dioxide in the flue gas from appliances and other gas fired-equipments.

Natural gas has currently become established as a major world energy source. With the annual demand touching 1570 million cubic meters. Natural gas accounted for around one-fifth of the world's primary energy consumption in 1984. With a global proven resource base of some 50 years relative to current rate of production and consumption and exploration activities continually adding to this resource base.

Natural gas is emerging as an important energy source in the commercial energy scene of India because of the large reserves of gas. This country has about 479 billion cubic meters of natural gas reserves and the average annual production was estimated to go up from 1986 level of 8 to 19 billion cubic meters by the year 1990. Nearly three-fourth of these reserves are in the off-shore areas. The Oil and Natural Gas Commission (ONGC) has indicated that availability of 120 million cubic meters of gas per day could become a reality by the turn of the century.

In India, natural gas is now making significant contribution to the household sector by way of liquefied petroleum gas (LPG), extracted from associated gas at the two extraction plants at Uran near Bombay and Duliajan in Assam. The government has recently decided to make use of natural gas on a selective basis for power generation. However, the first priority for utilization of gas continues to be as feedback for the production of fertilizers, petrochemicals and extraction of LPG.

The country's first gas based power project came into operation on January 20, 1989, near Anta, 60 km away from Kota district in Rajasthan. The first 100 MW unit of this project started power production. Anta project has the target of 410 MW, out of which there shall be three units of 100 MW and one unit of 110 MW. National Thermal Power Corporation is also establishing similar gas based power projects at Orriya (Uttar Pradesh) and Kawas (Gujarat). Both these later projects shall be of 600 MW capacity each.

Nevertheless, natural gas industry has certain environmental problems. Exploration, production, processing, storage, distribution and transportation of natural gas are an integral part of the gas industry, and each of these operation can contribute to a certain degree of environmental degradation, such as the disposition of drilling fluids and cutting, gas leakages from pipelines, production of gas with a high concentration of sulphur.

Pollution Caused by Conventional Fuels

The rapid growth of industrialization and urbanisation has resulted into new way of thinking. Now the advancement of a society is gauged by its per capita energy consumption. India stands 113th with an estimated per capita annual energy consumption of 100 KWH as compared to 5000 KWH in USA, which is at the top. The energy used for industrial and domestic purposes usually comes from the conventional sources viz., coal, firewood, petrol and diesel. All these fuel use cause environmental disturbances and disruptions. The production, conversion and use of energy from these fuels have important environmental implications. These implications can be particularly acute in areas of concentrated population and activity, such as a large metropolitan city. For example, in Delhi, where these conventional fuels are used (Table 13), the environment remains highly polluted. Firewood produces appreciably higher amounts of carbon monoxide during combustion per unit mass than other conventional fuels. The composition of petrol and diesel exhaust is characterised by large amounts of carbon monoxide, hydrocarbons, oxides of nitrogen, particulates as well as relatively small amounts of sulphur oxides. There are no comprehensive studies on conventional fuel combustion and their contribution to pollutants in the city. Using the data on annual consumption statistics and assumed emission factor, an estimation (Table 14) of major air pollutants have been made by Pandeya (1995). Apart from polluting the atmosphere, water and soil, some of the above mentioned pollutants mix directly with the products of human consumption.

Table 13
Conventional fuel consumption in Delhi (Pandeya, 1985)

Fuel	*Quantity*
Firewood	0.2 million metric tons
Coal	1.0 million metric tons
Cowdung cakes, twigs	3000 metric tons
Diesal	449000 metric tons
Petrol	171000 metric tons

Table 14
Estimation of major air

Fuel	*CO*	*TSP*	*NO_x*	*SO_2*	*HC*
Fuelwood	12000	700	170	75	1000
Coal	500	3000	3000	8000	150
Petrol	60700	—	2000	—	24700
Diesal	6700	—	5400	2500	1150

Pollutants Generated by Burning Fuelwood

Over five hundred compounds have been identified in wood smoke arising from domestic heating associated with cooking based on fuelwood (Hubble et al, 1982). Undoubtedly, there are many more compounds as well that remain to be identified. However, scientists have focussed on 3 or 4 of these classes of compounds. These are carbon monoxide, particulate, polycyclic organic matter and formaldehyde (Figure 7).

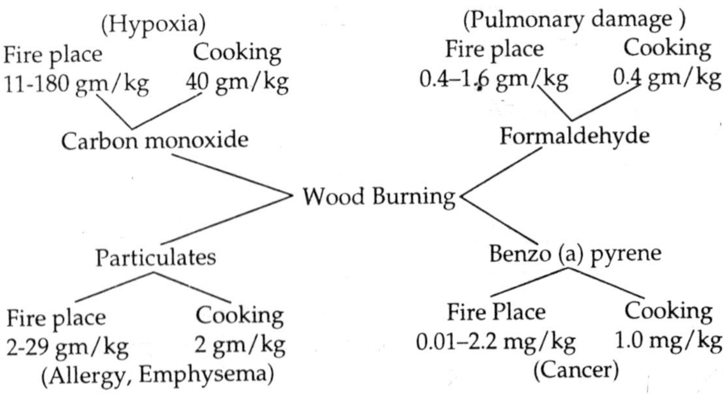

Figure 7: Major air pollutants emitted per unit mass of firewood consumed (After Smith, 1985)

Carbon Monoxide

It occurs naturally in small amounts. The background concentration are between 0.04—0.2 mg/m^3. Combustion of biomass contributes 11-180 gm/kg from fire places and 40 gm/kg from domestic cooking. The WHO recommended limit for carbon monoxide 1 hour exposure is 35 ppm. The concentrations measured in Guatemala and New Guinea are in the same range. Hence, carbon monoxide exposure should not be too much of a problem.

Particulates

Total suspended particulates emitted from wood burning range from 2-29 gm/kg from fire places and 2 gm/kg from domestic cooking. 8 hour Indian Standard for industrial area is 0.5 mg/m^3 and for sensitive areas it is 0.1 mg/m^3. The concentration in village kitchen are greater than the higher standards (Table 15).

Benzo (a) pyrene

Fire places generate 0.01–2–2 mg/kg and cooking fires generate 1.0 mg/kg Benzo (a) pyrene. Only Russia has an annual average standard of 1.0 mg/m^3. The concentrations resulting from biomass combustion in rural areas of developing countries are much greater. It is generated by incomplete combustion of firewood. High temperature burning in an open fire, effectively destroys the carcinogenic compounds produced by destructive distillation of the wood. However, in a burning stove with the minimum intake of oxygen, these compounds leak out in the kitchen.

Formaldehyde

Natural background level seems to be a few parts per billion (1 ppb = 1.25 ng/m^3). Fire places generate 0.4-1.6 gm/kg formaldehyde, while cooking fires generate 0.4 gm/kg formaldehyde. The standard for formaldehyde in Western countries vary between 0.1 and 0.5 ppm. The little data that exist indicates that these levels are exceeded.

Table 15
Indoor air pollution concenrations from biomass combustion (Smith et al, 1983)

Location	No. House-holds	Averaging Time	TSP mg/m_3	BaP ng/m_3	CO ppm	HCHO ppm
Papua New Guinea (PNG)		All				
Western highlands	6	night	.36	—	11	0.67
PNG Eastern highlands	3	"	.84	—	31	1.2
PNG Eastern highlands	6	"	1.3	—	—	—
Kenya, highlands	5	?	4.0	145	—	—
Kenya, Sea level	3	?	0.8	12	—	—
Guatemala						
Poorly ventilated	180	—	—	—	26-50	
Well ventilated	—	—	—	—	15-31	
India, Ahmedabad						
Wood	5	15 min	7.2	1270	—	—
Cattle dung	4	"	16.0	8250	—	—
Dung+Wood	7	"	21.2	9320	—	—
India, Gujarat*	10	45 min	6.5	3550	—	—
Boria						
Denapura	11	"	3.5	2720	—	—

Contd.

Location	*No. House-holds*	*Averag-ing Time*	*TSP*	*BaP*	*CO*	*HCHO*
			mg/m3	*ng/m3*	*ppm*	*ppm*
Rampura	5		5.9	4230	—	—
Meghva	10	"	7.5	3170	—	—
Meghva two mouth chul	1	"	14.0	4270	—	—
Meghva rainy season	1	"	56.6	19300	—	—
Standards			0.5	1	35	< 0.1
Averaging times			8 hrs	annual	1 hr	
Country			India	USSR	WHO	USA

Smith et al (1984) observed that the air inside the kitchen changes every 13 minutes at the lower end and about once every minutes at the upper end. There are two times in the year when the concentration are even higher than at other times. One is during the winter when near ground level inversions are common in India and the other during rains when whatever openings that might exist in the ceilings of the dwellings are closed. The ventilation rates and the diffusion of pollutants, consequently are both reduced. In addition, there is the phenomenon of stratification of smoke at any given time within a room with lower concentration being found at levels closer to the ground.

The change from an open combustion stove to one with flue/chimney should result in lower pollutant concentration inside the house. It may however, result in some increase.

The principal health effects of carbon monoxide inhalation is hypoxia induced by reduction of oxygen availability to body tissues. The haemoglobin in red-blood cells has about 200 times the affinity for carbon monoxide as for oxygen. Hence, relatively small amounts of carbon monoxide can lead to a significant reduction in the blood's ability to carry oxygen. The amount of carboxy-haemoglobin (COHb) in the blood is a function not only of the concentration but also of the duration of exposure.

The environmental fate and possible health effects of particulates are determined by its size and composition. Total suspended particulates are defined to be those less than 100 microns in diameter; inhalable particles are less than 15 microns in diameter and the most significant from the health point of view are the respirable particulates which are less than 3 microns. The respirable particles are deposited in the gas exchange areas of the lung; the inhalable particles are removed by the upper respiratory tract.

The polycyclic organic vapours containing Polycyclic aromatic hydrocarbons (PAH) and Benzo (a) pyrene (BaP) have been shown to be cancerous and mutagenic.

Formaldehyde (HCHO) in small amounts is harmless. At large exposures, irritation (0.1-25 ppm), pulmonary damage (50-100 ppm) and death (100 ppm) can occur. Chronic exposures at levels below 5 ppm have been associated with respiratory diseases (Ahuja, 1985).

Environmental Impact of Burning Coal

Coal is used in the domestic sector for cooking, and in the industrial sector for power generation. The composition of coal with respect to moisture, ash, volatile matter, carbon, sulphur, hydrogen and nitrogen is dependent upon the source from where it has been taken out. However, the proximate values have been shown in table 16.

Table 16
Proximate analysis of coal (Deshpande, 1985)

Particulars	*Unit*	*Proximate value*
Ash	%	21.8-35.2
Moisture	%	5.8-8.2
Volatile matter	%	25-28
Carbon	%	34-53.2
Sulphur	%	0.5-0.8
Hydrogen	%	2.8-3
Nitrogen	%	1.1-1.2
Calorific value	K Cal	4000-5000

The increasing use of coal as a fuel for thermal power stations, certainly increases the amount of particulate matter in the ambient air (Dubey et al, 1982). Particulate matter comes mainly from the inorganic compounds in the coal, as the unburnt carbonaceous matter constitutes a minor source. The burning of coal releases ash as fine dust particles that escape, atleast partially, with the combustion gas. The amount of ash in the flue gas varies with the ash content in coal. Combustion of inferior grade of coal increases coal consumption in the power stations, which subsequently increases particulate emissions (Deshpande, 1985). The average ash content of Indian coal is very high (50%). About 20 per cent of the non-combustible components are converted into bottom ash and 80 per cent are converted into flyash particles. Nationwide estimates of particulate emission for the coal-fired power stations in India indicate that these emissions are sharply increasing from a minimum (0.54 million tons) in 1947 to a maximum (8.06 million tons) in 1980.

The major air pollutants emitted from coal burning are particulates, sulphur dioxide, nitrogen oxides, carbon monoxide, hydrocarbons and trace elements (Table 17 and 18).

Table 17

Estimated gaseous and particulate emissions from stacks of coal fired power stations without control in India (Kumar and Upadhyay, 1983)

Year	*Particulate (Million Tonne)*	*SO_x 10^3 tonne*	*NO_x 10^3 tonne*	*CO 10^3 Tonne*	*HC 10^3 Tonne*
1947	0.54	32.40	19.79	1.07	0.32
1981	0.75	44.68	26.45	1.47	0.44
1956	0.96	57.23	33.88	1.88	0.56
1961	2.06	123.00	72.83	4.05	1.21
1966	2.57	163.68	96.62	5.38	1.62
1969	3.74	222.97	132.02	7.33	2.20
1970	4.21	250.97	148.60	1.26	2.48
1971	4.49	267.38	158.31	8.80	2.64
1972	4.89	291.64	172.68	9.59	2.88
1973	5.05	301.16	178.32	9.91	2.97
1974	5.27	313.85	185.83	10.32	3.10
1975	5.86	349.28	206.81	11.49	3.45
1976	6.90	411.20	243.47	13.53	4.06
1977	7.28	433.68	256.78	14.27	4.28
1978	7.58	451.60	267.40	14.86	4.46
1979	7.81	465.67	274.72	15.32	4.60
1980	8.06	480.70	284.62	15.81	4.74

Table 18
Estimated Trace elements emissions from stacks of coal-fired power stations without control, (in Tonne) in India (Kumar and Upadhyay, 1983)

Year	*Hg*	*Cr*	*Fe*	*Zn*	*Co*
1947	23.27	36.90	4906.8	316.66	7.85
1951	32.08	50.87	6765.1	436.59	10.82
1956	41.09	65.16	8665.2	559.21	13.86
1961	88.31	140.04	18624.9	1201.97	29.79
1966	117.51	186.36	24784.5	1599.48	39.65
1969	160.08	253.87	33764.9	2178.91	54.01
1970	180.18	285.74	38001.6	2452.45	60.79
1971	191.96	304.42	40486.3	2612.80	64.76
1972	209.39	332.06	44164.2	2849.96	70.64
1973	216.22	342.89	45601.9	2942.94	72.95
1974	225.32	357.33	47522.8	3066.91	76.02
1975	250.77	397.68	52889.0	3413.22	84.60
1976	295.22	468.17	62264.1	4018.25	99.60
1977	311.36	493.77	65667.6	4237.89	105.04
1978	324.23	514.18	68382.0	4413.06	109.39
1979	334.32	530.19	70511.7	4550.51	112.79
1980	345.11	547.30	72787.6	4697.39	116.43

Most of the sulphur present in coal leaves as sulphur dioxide on burning. Sulphur trioxide usually accounts for only a small proportion (0.5-1%). Fortunately Indian coal have relatively low sulphur content. However, large scale combustion of coal in super thermal power stations greatly offsets this inherent advantage. It has been estimated that in India, sulphur dioxide emissions from coal combustion has increased from 0.6 million tonnes in 1964 to 1.1 million tonnes in 1978 (Varshney and Garg, 1978). Sulphur dioxide combines with water vapour in presence of sunlight to form *sulphuric acid mist*. This acidic mist condenses on unburnt carbon particles and forms *'acid smut'* (Agarwal, 1980).

Apart from sulphur dioxide, combustion of coal produces large amount of oxides of nitrogen. Nitrogen oxides formation in the flame zone involves two mechanisms. The first mechanism, termed as *'thermal fixation'* involves the reaction of atmospheric oxygen and nitrogen through a series of free radical reaction steps. The second mechanism, termed as *'fuel-nitrogen conversion'* involves the oxidation of nitrogen contained in the fuel. Generally, more than 90 per cent of the nitrogen oxides emitted to the atmosphere is NO, with the remainder being NO_2. Subsequently, most of the NO oxidises further to form NO_2 which in turn can combine with water to form nitric acid aerosol.

The amount of carbon dioxide evolved from coal burning depends entirely on the carbon content of coal. Carbon dioxide is regarded as an air pollutant because of its ability to transmit heat from sun through the earth's atmosphere.

Incomplete oxidation of coal (specially during cooking) liberates a lot of carbon monoxide, which is a poisonous gas.

The hydrocarbons present in coal are volatilized at high temperature. However, they are emitted at low temperature of combustion.

Almost all varieties of coal have been found to contain significant quantities of $Krypton^{40}$, $Uranium^{238}$, $Thorium^{232}$, and other decay products. Indian bituminous coal has 180-870 Bq/kg $Krypton^{40}$, 12-666 Bq/kg $Uranium^{238}$, and 19-93 Bq/kg $Thorium^{232}$. Excessive coal combustion can increase environmental radioactivity (Ramachandran, 1985).

It is important to note that coal is used as a domestic fuel in many parts of India. The conditions under which coal is put to such a use, especially in poorly ventilated homes, suggests that inhalation exposure could be significant.

Environmental Impact of Crude Oil and Natural Gas

The major pollutants emitted on combustion of petrol and diesel fuel are carbon monoxide, hydrocarbons, nitrogen oxides, sulphur dioxide, particulates, zinc and lead. Minor pollutants are aldehydes. Air pollution from petrol and diesel combustion is a serious problem especially in metropolitan areas (Table 19). These observations reveal that carbon monoxide is the chief pollutant followed by hydrocarbons, sulphur dioxide, nitrogen oxides and so on.

Table 19
Pollutants emitted from petrol and diesel combustion
(Prakash et al, 1980)

Pollutant	*India* *Metric ton/yr*	*World* *Million metric ton/yr*
Carbon monoxide	755.0	65
Sulphur dioxide	101.6	23
Nitrogen oxides	73.0	8
Hydrocarbons	238.0	15
Hydrogen sulphide	11.6	—
Ammonia	11.6	—
Hydrogn chloride	2.6	—
Particulate	—	12
Other gases and vapours	—	2

Carbon monoxide is an intermediate product of the petrol combustion. If there is not enough air in the air/fuel mixture to completely burn the fuel, or if there is insufficient time in the cycle for complete combustion, all the carbon in the fuel can not be burnt to carbon dioxide and some of its stops midway to form carbon monoxide. Even if enough air is present, the rapid cooling of gases during expansion allows the combustion process to retain a small amount of carbon monoxide.

Nitrogen oxides are formed in the combustion chamber where the temperature is as high as 2500°C. This high temperature combines atmospheric oxygen and nitrogen. Nitric Oxide (NO) formation increases with temperature and oxygen availability. Nitric oxide gets oxidised to nitrogen dioxide in the atmosphere.

Sulphur content of crude oil varies widely. Only about 10 per cent of the global crude oil contains less than 1 per cent sulphur. This sulphur gets oxidised to sulphur dioxide during combustion process. The diesel oil has a higher sulphur content which could be as high as twenty times that in petrol.

Hydrocarbons generate during combustion of petrol and diesel. The mechanism of its formation has been attributed to the destruction of flame propagation radicals due to quenching in the combustion chamber.

Lead contamination arises from combustion of petrol to which alkyl lead compound in the form of tetraethyl lead and tetramethyl lead have been added during refining as an antiknock agent, in order to prevent the fuel from spontaneously exploding before its ignition. Lead in emitted as particles varying from 0.001-10 microns.

Soot emissions from combustion of petrol and diesel results from decomposition of hydrocarbon fuel, nucleation of carbon particles in the flame, growth of soot nuclei, agglomeration of particles and finally soot oxidation.

Besides these chief pollutants aldehydes, primarily in the forms of formaldehyde (70%) and acetaldehyda (30%), acrolein and bensaldehyde are also generated due to partial oxidation of fuel in the preflame reaction zone.

Effect of Pollutants Generated from Conventional Fuels on Human Health

The concentration of air pollutants and the duration of the exposure to the pollutants have been shown to play significant role in the body-fluid and tissue toxicity. These pollutants cause health hazards from short-term and long-term exposure (Table 20).

Table 20
Relationship between air pollutants from conventional fuel combustion and resulting health hazards

Pollutant	*Health effects*	
	Short term	*Long term*
Unburnt hydrocarbon	Difficulty in breathing	Impaired lung function
Nitrogen oxides	Soreness & coughing	Lung cancer
Sulphur dioxide	Asthma	Lung cancer
Particulates	Asthma	Silicosis
Carbon monoxide	Asphyxiation	Fatal at COHb level of 2-5%
Inorganic lead	Effects kidney, liver, gastro-intestinal function	Mental impairement

Carbon monoxide inhalation and absorption in the lungs, cause hypoxia induced by reduction of oxygen availability to body tissue. This gas combines with the haemoglobin in the blood and some extra vascular haemoglobin. The oxygen of the haemoglobin is displaced to form carboxyhaemoglobin.

$$O_2Hb + CO \Leftrightarrow COHb + O_2$$

The carboxyhaemoglobin is a stronger complex, so that the net result is reduction in the oxygen carrying capacity of the blood, and also interference with the release of oxygen that it carries to the tissues. Carbon monoxide has an affinity for haemoglobin 240 times greater than that of oxygen. Carboxyhaemoglobin is therefore a more stable compound than oxyhaemoglobin (WHO, 1972).

Carbon monoxide may be toxic to human beings at concentration of 30 ppm for 4 hours as it causes dizziness, headache, convulsions, loss of consciousness and asphyxiation. Concentration of 200 ppm may produce symptoms of poisoning if inhaled for few hours, while 500 ppm can probably be inhaled for little more than one hour without serious effects. The gas kills quickly at concentration of about 1000 ppm. This concentration can cause unconsciousness in about one hour and death in 4 hours. Formation of carboxyhaemoglobin decreases the overall capacity of the blood to carry oxygen to cells resulting into oxygen deficiency (hypoxia). The amount of carboxyhaemoglobin in the blood is a function not only of the concentration but also of the duration of exposure. Carboxyhaemoglobin hence produce a biological end point for relating concentration, exposure and doses. However, carbon monoxide is not a cumulative poison, but is excreted, depending on its partial pressure in the ambient air and the saturation of the haemoglobin.

Sulphur dioxide is intensely irritating to the eyes and the respiratory tract. It is highly soluble and consequently is absorbed in the moist passage of the upper respiratory system, leading to airway resistance (swelling) and stimulated mucous secretion. It appears that most individuals suffer irritation at a concentration of 5 ppm and above. Some sensitive individuals even experience irritation at 1-2 ppm and sometimes experience severe bronchial spasms on exposure to 5-10 ppm sulphur dioxide. Sulphur dioxide effects mucosa and cilia of the throat, bronchial tube and lungs, with the resultant damage turning into chronic bronchitis over the years. Its immediate effect on most people are negligible. However, a 1 hour exposure to 5 ppm causes choking and 1 hour to 10 ppm produce great distress. A few individuals experience severe bronchopasm at 1-2 ppm (ACC, 1965). Along with comes varying degrees of lung depression with the heart being affected, resulting in pulmonary emphysema. Of course, when the pathological destruction has reached this level, the only outcome is slow and lingering death (Ui, 1974).

Nitrous oxide (N_2O), Nitric oxide (NO) and Nitrogen dioxide (NO_2) are various oxides of nitrogen. Nitrous oxide is not known to be hazardous (Table 21). Nitric oxide too is not known to cause health hazards. Nitrogen dioxide is considerably less toxic than Sulphur dioxide and carbon monoxide. Like carbon monoxide, it forms bonds with haemoglobin and release oxygen transport efficiency. NO_2 has toxic effects on human beings especially causing respiratory diseases. NO_2 is known to irritate the alveoli, leading to symptoms resembling emphysema (inflammation) upon prolonged exposure to concentrations of 1 ppm. Inflammation of the lungs may follow edema (accumulation of fluid) and then to death. NO_2 has been shown to cause methamoglobinaemia. The biochemical mechanism of NO_2 toxicity are not clear. Probably cellular enzyme systems are susceptible to disruption by NO_2, including catalase and lactic dehydrogenase.

Table 21
Health hazards from NO_2 exposure (De, 1990)

Levels of NO_2 (ppm)	*Duration of exposure*	*Effect of human health*
50-100	upto 1 hour	Inflammation of lung tissue for 6-8 weeks

Contd.

Levels of NO_2 (ppm)	*Duration of exposure*	*Effect of human health*
150-200	—	Bronchiolitis fibrosa obliterans - fatal result within 3-5 weeks of exposure
500 or more	2 - 10 days	Death

Health effects of volatile organic compounds are as numerous as the compounds themselves. Formaldehyde in small amounts is harmless. At higher concentration, irritation of the eye, nose and upper respiratory tract (0.1–25 ppm), pulmonary damage (50-100 ppm), and death (100 ppm) can occur. Chronic exposure at levels below 5 ppm have been associated with respiratory diseases. Benzene exposure results in susceptibility to, lung cancer. Benzo (a) pyrene is extremely toxic and potent cancer inducing hydrocarbon pollutant.

The possible health effects of particulates are determined by their size and composition. The inhalable particles (less than 15 micron in diameter) are removed by the upper respiratory tract. The respirable particles (less than 3 microns in diameter) are deposited in the gas exchange areas of the lung, and then swallowed. Thus, they are not only absorbed in the blood stream from the lungs but can also be absorbed in the stomach or small intestine. These particles may cause respiratory diseases such as tuberculosis and cancer.

The trace elements absorbed on the surface of flyash particles are readily disorbed following inhalation and pose a potential threat inspite of their insignificant amounts.

The lead particles emitted during combustion of leaded petrol are inhaled. About 90 per cent of lead is deposited in the bones, while the blood retains slightly less than 1 per cent lead. Individuals with a low level of exposure have 10-30 microgram of lead per 100 gm of blood. Highly exposed individuals often have more than 100 microgram lead per 100 gm of blood. Lead inhalation in an average person varies from 0.03 mg/day in urban areas to 0.01 mg/day in rural areas. The toxic effects of lead include abdominal colic, obstinate constipation, loss of apitite, insomnia, neuritis, paralysis etc., because of its action on the central and the peripheral nervous system.

Chapter 4

Modern Fuels and their Environmental Impacts

A consensus is emerging in the world scientific community that the damage being done to the environment by current patterns of fuel consumption is the most urgent issue facing fuel decision makers today.

In the industrialized countries and increasingly in the developing world as well, the production of sulphur and nitrogen oxides from the combustion of fossil fuels is contributing to the acidification of forest, lakes and soils, destroying the delicate bio-chemical balance in many terrestrial and aquatic ecosystems. In many developing countries, the use of biomass fuels, particularly fuelwood, is damaging the health of women and children who live and work in smoke filled homes. These also bear the brunt of growing shortages of fuelwood due to rural deforestation: a problem which may be attributed, atleast in part, to certain sectors of fuelwood consumption. Neither acidification nor deforestation are automatic consequences of the use of fossil and biomass fuels, but the investment required to control them, in emission control and in sustainable land resource management, may be considerable. The necessary commitment from the world's fuel decision makers is only just beginning to emerge.

The most uncertain, but potentially the most serious impact of all, is the threat to global climate change due to the accumulation of carbon dioxide, methane and other relatively active ("greenhouse") gases in the atmosphere. Roughly 50 per cent of projected global warming caused by human activity over the coming 50 years may be primarily attributed to the combustion of fossil fuels. While increased use of oil and natural gas in place of coal would have a positive impact on carbon dioxide emission, atleast in the medium term, reducing overall fossil fuel consumption is an urgent priority, if the pace and consequences of global warming are to be kept within manageable bounds.

Debate continues on the question of whether animal oil has the potential to provide viable and acceptable alternative. Public concern over their pollution potential and increasing availability has increased considerably.

Responsibility of all these fuel-environment problems remains primarily with fully industrialised economies, although the importance of developing and newly industrialised countries is growing rapidly. The consequences, however, are increasingly felt by all, particularly through transboundary pollution and transport of waste products.

Although in the course of the 1980's, the short-term fuel price situation may have changed, the need for a transition to a more socially equitable and sustainable fuel future is more urgent than ever. The damage being done to the environment by current patterns of fuel use calls for a new approach to fuel planning and policy formulation: an approach which considers the full fuel system from fuel source to end user, and which takes into account the external environmental and social costs of fuel, not merely the internal costs of production and distribution.

The implications of new and renewable fuel sources (NRFS) such as biogas, hydrogen, ethanol, phyto-hydrocarbons etc are two fold: *First*, current evidence suggests that they are inherently more environmentally benign than conventional fuels—more specifically, they internalise a higher proportion of their costs. Thus, the inclusion of external costs into fuel decision making will, prima facie, substantially improve the case for a greater emphasis to be placed on modern fuels. This does not mean a uniform, across the board transition: modern fuels are, by nature, highly site-specific, particularly with regard to their external costs. Thus, the timetable and specific mix of actions required will both depend on an individual country's present fuel situation, and political priorities. While there is a global need for an increased use of modern fuels, action must be taken on a local or regional level. *Second,* the modern fuels are no longer to be advocated simply as alternative fuel sources, but as integral component of the sustainable fuel systems of the future. The modern fuels appear to be environmentally benign in a small scale application, it cannot be taken for granted that it will prove equally benign in large-scale application. Hence, it is essential that their full environmental impacts should be assessed and quantified, and that complex scale dependent impacts are identified as they emerge (Allen, 1989).

Methanogens and Biogas

Biological methane (CH_4) formation is a geologically important process that occurs in most anaerobic environments where organic matter undergoes decomposition: swamps, lake sediments, the intestinal tract of animals, and anaerobic sewage digestors. It results from the activities of a highly specialised group of bacteria that convert fermentation products by other anaerobes (notably carbon dioxide, hydrogen, format and acetate) to methane or methane and carbon dioxide. Since methane is a gas that is sparingly soluble in water, it escapes from the anaerobic environment. The methanogens are consequently terminal members of the anaerobic food chain, whose metabolic activity prevents the sequestering of large amounts of organic material in anaerobic ecosystems (Stainer et al, 1987).

Diversity of methanogens

Methanogens comprise atleast three major groups. Group I contains *Methanobacterium* and *Methanobrevibacter*, Group II contains *Methanococcus*, and Group III contains several genera including *Methanospirillum* and *Methanosarcina*. In morphology they are of different types such as cocci, bacilli, spirilli and sarcinea. Batch et al (1979) have classified the methanogens as follows:

Order I. Mechanobacteriales
  Family. Methanobacteriaceae
    Methanobacterium formicicum
    M. bryantii
    M. thermoautotrophicum
    Methanobrevibacter ruminatium
    M. arboriphilus
    M. smithii

Order II. Methanococcales
  Family. Methanococcaceae
    Methanococcus vanielii
    M. voltae

Order III. Methanomicrobiales
  Family 1. Mathanomicrobiaceae
    Methanomicrobium mobile
    Methanogenium cariaci
    M. marisnigri
    Methanospirillum lungatei
  Family 2. Methanosarcinaceae
    Methanosarcina barkeri

The characteristic lipids of Group I are both diphytanyl diethers and dibiphytanyl tetraethers. Group II contains exclusively diphytanyl diethers, while Group III contains some strains whose lipid profiles resemble that of Group I and some that resemble Group II.

The cell wall among the methanogens are of atleast three different types. Group I, which is rigid and composed principally of pseudomarein, and is Gram positive. Group II has a flexible cell wall composed of protein with traces of glucosamine. Group III has the most complex cell wall. Its wall is a flexible envelope composed of atleast two layers—an inner, electron dense one of unknown chemical composition, and an outer one entirely of protein (Methanospirillum). On the other hand Methanosarcina contains a thick, rigid, lamellar wall of moderate electron density composed of an acidic heteropolysaccharide (made up of galanctosamine, neutral sugars, and uronic acids (Table 22).

Table 22

Characteristics of methanogens (Stainer, et al, 1987)

Group	*Representative Genera*	*Percent G + C*	*Characteristic Lipids*	*Cell Wall Structure*	*Gram Reaction*	*Motility*	*Substrates Used*
I	*Methanobacterium*	32–50	C_{20} diethers and	Pseudomurein	+	–	H_2; some use
	Methanobrevibacter	27–32	C_{40} tetraethers				formate
II	*Methanococcus*	30–32	C_{20} diethers	Protein (trace of glucosamine)	–	+	H_2; formate
III	*Methanospirillum*	45–47	C_{20} diethers and C40 tetraethers	Proteinaceous sheath; unknown structural material	–	+	H_2; formate
	Methanosarcina	38–51	C_{20} diethers	Heteropolysaccharide	+	–	H_2; formate; methanol; methylamine; acetate

Unique cofactors

Methanogens contain several cofactors not found in other bacteria. Three of them (methanopterin, methanofuran, and CoM) are carriers of the C_1 unit during its reduction from CO_2 to CH_4 (Figure 8). Factor 430 probably functions as a hydrogen carrier in these reductions, and factor 430 an unusual metal tetrapyrrole, is the prosthetic group of methyl—CoM reductase, the last enzyme in the reduction pathway. Some methanogens require CoM as a growth factor.

They are unique group of bacteria. They are obligate anaerobes and have a slow growth rate. They play a major role in breakdown of substrate into gas form. They are the only organisms which can anaerobically catabolize acetate and hydrogen to gaseous products in the absence of exogenous electron acceptors other than carbon dioxide or light energy. In their absence, effective degradation could cease because of accumulation of non-gaseous, reduced fatty acids and alcohol products of fermentative and other hydrogen using bacteria that have atmost the same energy content as the original organic matter (McInerny and Bryant, 1981). In morphology they are of different types such as cocci, bacilli, spirilli and sarcinea.

All these bacteria require hydrogen and formate (except *Methanobacterium bryantii, M. thermoautotrophicum,* and *M. arboriphilus)* for growth and methane production, whereas *Methanosarcinia barkeri* requires besides hydrogen, methanol (CH_3OH), methyl amine (CH_3NH_2) and acetate for their growth.

Biomethanation

Sustained availability and use of energy services are the key to development in our modern society. India has limited energy potential through fossil fuel and other resources and hence biomass can take its place. Biomass resources and potential in our country are immense (with 1150 herbaceous species and 200 woody plant species biomass millions tons/year). However, to realize this potential it is necessary to develop efficient conversion technologies that are capable of sustained use and do not erode this resource base. Biogas (biomethanation) technology have the potential to meet a large segment of the energy needs

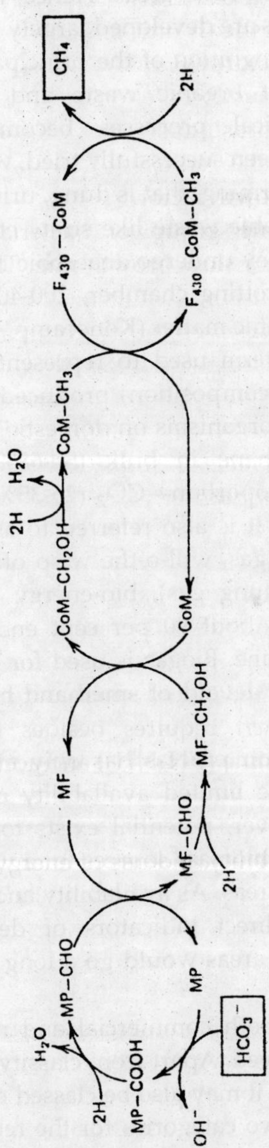

Figure 8

The pathway of methanogenesis: MP-methanopterin; MF-methanofuran; CoM-coenzyme M; F_{430} - Factor 430.

for development. It is important to note that 70 per cent of our population live in rural areas. Hence, it is necessary that these new technologies are developed largely for the rural application.

With the recognition of the principal of energy from waste, the recycling of organic waste and new energy resources through biological processes become appropriate. Biogas technology has been successfully tried, wherein the waste arising from livestock farming, that is dung, urine, farmyard manure etc as well as vegetable waste like straw etc are placed in air tight towers, where they undergo anaerobic rotting process. After 20-30 days in the rotting chamber, 200-400 litre gas is produced from 1 kg of organic matter (Konecamp, 1977).

Biogas is a term used to represent a mixture of different gases (in varied composition) produced as a result of action of anaerobic micro-organisms on domestic and agricultural wastes. It contains methane in bulk (50-68%) and other gases in relatively low proportion—CO_2 (25-35%), H_2 (1-5%), N_2 (2-7%) and O_2 (0-0.1%). It is also referred to as biofuel, sewerage gas, Klar gas, sludge gas, will-o-the wisp of marshlands, fool's fire, gobar gas (cowdung gas), bio-energy and fuel of the future (Dasilva, 1981). About 90 per cent energy of the substrate is retained in methane. Biogas is used for cooking and lighting in rural sector. It is devoid of smell and burns with a blue flame without smoke.

Biomethanation process has only a limited scope in India today, due to the limited availability of dung in appropriate quantities. However, potential exists today to produce biogas from many other biomass sources for a variety of applications in rural and urban areas. As availability and efficient use of energy is one of the direct indicators of development, making it available in rural areas would go a long way in achieving rural development.

In a village both commercial and non-commercial energy sources are deployed. Apart from classifying energy based on its commercial value it may also be classed on the basis of demand. Thus, there are two categories for the requirement of energy in rural areas. First peak demands such as that for seasonal agricultural operations involving ploughing, pump irrigation, agro processing, transport etc. On the other hand, there is a constant energy demand for domestic activities such as cooking, water supply, illumination, milling etc.

The major energy used in rural areas is for cooking. Burning of twigs, branches, agrowastes accounts for 77 per cent of the total energy used in a typical village. These inferior fuels are burnt at a low thermal efficiency (8-14%). Biogas is best suited to meet constant energy demands. This is because, unlike other energy options, biogas production is a steady and slow process. Its output cannot be significantly stepped up to meet peak demands at short notice. Therefore, it is best, to cater to constant demand (load) types of energy services.

Biomethanation technology is attracting world-wide attention as a decentralized energy system (Nair et al, 1982). It is an anaerobic digestion of organic matter which has opened the door to the exploration of alternative processes for energy production (Viraraghvan, 1980). Any organic material consisting of carbohydrates and fats could be fermented in complete absence of oxygen to yield biogas as a final product. Comparable gas yield indicate that cattle dung yield 253 litre/kg of dry matter, pig dung 385 litre/kg, cattle manure (dung + urine + straw) 258 litre/kg and wheat straw 289 litre/kg (Konecamp, 1977). The advantage of biogas plant is that it produces fuel in the form of methane gas simultaneously with high quality manure. Biogas burns with a smokeless, non-luminous blue flame even without prior mixing with air, although mixing with air is preferred to economise gas. Biogas burns with a calorific value of 5400 kCal/m^3. This energy is being used for heating and lighting.

Biogas is produced by anaerobic digestion of biomass i.e., plant and animal waste. Biomass from plant origin is aquatic and terrestrial ones, derived from various sources. Biomass from animals are cattle dung, manure from poultry, goats and sheep, and slaughter house and fishery waste.

Biogas is produced by anaerobic digestion of plant and animal waste. It involves tree steps (i) hydrolysis, which converts organic polymers into monomers (with the help of hydrolytic bacteria); (ii) Acidogenesis, which involves conversion of monomers into simple compounds such as CO_2, NH_3, and H_2, using a group of acetogenic bacteria, and (iii) Methanogenesis, which involves conversion of simple compounds into methane utilizing anaerobic methanogenic bacteria.

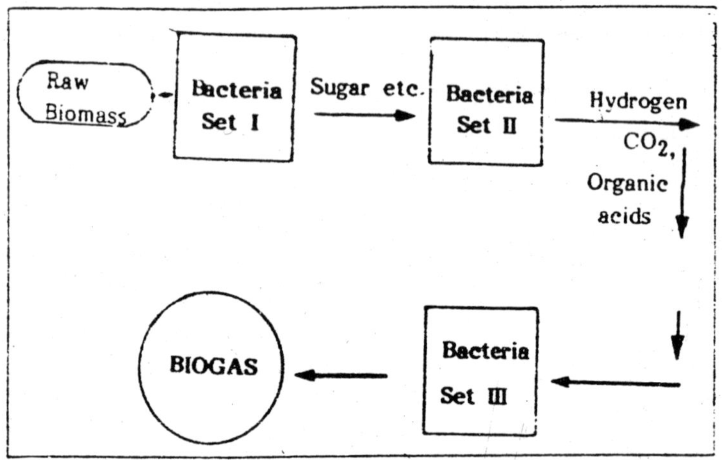

Figure 9
Schematic representation of the stages of methanogenesis

Hydrolysis

At this initial stage, the feed stock is solubilized by water and enzymes to make slurry. The complex polymers are hydrolysed into organic acids and alcohols by hydrolytic fermentative bacteria which are mostly anaerobic (Figure 10).

Digestion

Anaerobic digestion is carried out in an air tight cylindrical tank which is known as digester. A digester is made up of concrete bricks and cement or steel. It has a side opening (charge pit) into which organic materials for digestion are incorporated. There lies a cylindrical container above the digester to collect the gas (Figure 11). In India single stage digester is set up in gobar gas plant. However in other countries single stage, two stage and multistage digester(s) are set up to accomplish digestion at high rate. Fresh cattle dung is deposited into a charge pit, which leads into the digestion tank. Dung remains in tank, after 50 days, sufficient amount of gas gets accumulated in the gas tank, which is used for household purposes. Digested sludge is removed from the basin and is used as fertilizer. In tropical areas digesters are buried in soil in order to provide insulation. In temperate areas they are kept above ground so as to enable their heating.

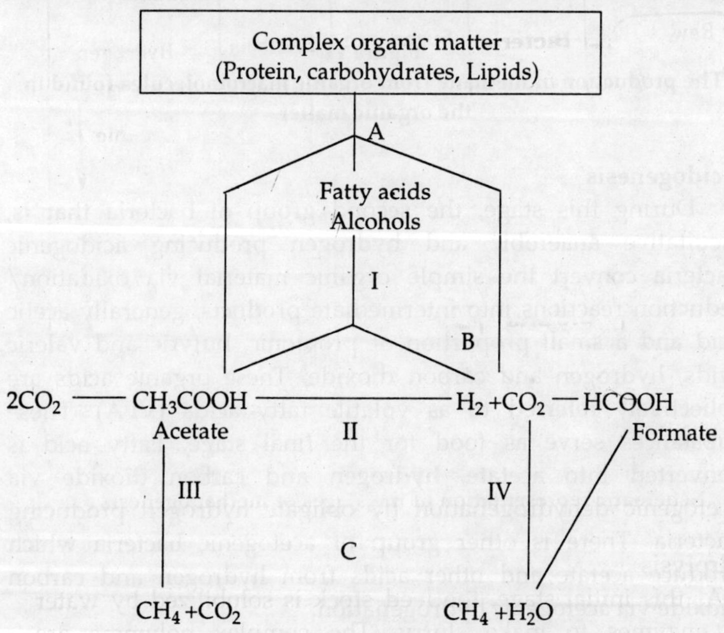

Figure 10

Anaerobic digestion of organic matter and production of methane. A-hydrolytic and fermentative bacteria; B-acetogenic bacteria (i) acetogenic dehydrogenation by proton reducing acetogenic bacteria; II-acetogenic dehydrogenation by acetogenic bacteria); C-methanogenesis by acetoelastic methanogens (i.e. acetate respiratory bacteria) (iii), and hydrogen oxidizing methanogens (IV).

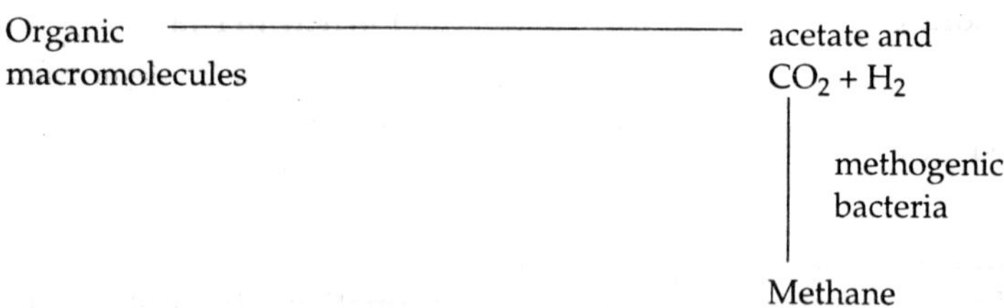

Figure 11
The production of methane from organic macromolecules found in the organic matter.

Acidogenesis

During this stage, the second group of bacteria that is, facultative anaerobic and hydrogen producing acidogenic bacteria convert the simple organic material via oxidation/ reduction reactions into intermediate products, generally acetic acid and a small proportion of propionic, butyric and valeric acids, hydrogen and carbon dioxide. These organic acids are collectively referred to as volatile fatty acids (VFA). These substances serve as food for the final stage. Fatty acid is converted into acetate, hydrogen and carbon dioxide via acetogenic dehydrogenation by obligate hydrogen producing bacteria. There is other group of acetogenic bacteria which produce acetate and other acids from hydrogen and carbon dioxide via acetogenic hydrogenation.

Methanogenesis

This is the final stage of anaerobic digestion where acetate and hydrogen plus carbon dioxide are converted by methane producing bacteria into methane, carbon dioxide, water and other products.

In case of dung fermentors, the rate of methanogenesis is faster than acidogenesis. There are very few simple saccharides left in the dung after digestion in the cattle rumen. However, when a biomass substrate containing a large fraction of pectin, starches, poorly lignified cellulosics is used as feedstock the acidogenesis rate surpasses the methogenesis rate by as much as 8-10 times. As a result, intermediate VFAs accumulate and bring down the pH of the digester. Methanogenesis fails when a threshold of 6 g/l of VFA accumulate or when pH falls below 5.5-6. To ensure trouble free fermentation it is necessary that biogas fermentors always remain within the allowable tolerance for pH and VFA levels.

Mechanism of methane formation

The mechanism of methane formation is not well understood. However, it is obvious from the work of Ralph Wolfe (1979) that several new coenzymes are involved, which are not present in any other group of bacteria. These coenzymes are methyl coenzyme M, hydroxy methyl coenzyme M, coenzyme F 420, coenzyme F 430, Component B, corrinoide, methylfuran or carbon dioxide reducing factor and methanopterin and formaldehyde activating factor.

Primary reaction in which carbon monoxide takes part is as below:

$CO + H_2O \rightarrow CO_2 + H_2$

Secondary reaction takes places in the presence of sufficient hydrogen:

$CO_2 + 4H_2 \rightarrow Ch_4 + 2\ H_2O$

Other reactions showing methane formation from various substrates are given below:

$4\ CH_3\ OH \rightarrow 3CH_4 + CO_2 + 2\ H_2O$

(Methanol)

$4\ HCOOH \rightarrow CH_4 + 3\ CO_2 + 2\ H_2O$

(Formate)

$CH_3\ COOH \rightarrow 12\ CH_4 + 12\ CO_2$

(Acetate)

Attempts have been made to improve biogas production in order to make anaerobic digestion process more attractive in rural areas. One method that has been used to upgrade the digestibility of cattle dung is treatment with sodium hydroxide (Smith, 1969). Addition of 200 ml of urine to 0.5 kg of dung was found to double the gas production (from about 17.21 to 31.51) at

7°C (Annonymous, 1977). Novak and Ramesh (1975) demonstrated that the degradation of acetic acid by methanogenic bacteria can be enhanced (upto 52%) in the presence of growth intermediates (lysed methane bacteria). Acetic and butyric acid fermentation were reported to be enhanced by the addition of dried supernatant solids from a digester (McCarty and Vath, 1963). Massey and Pohland (1978) observed that the process stability and control of anaerobic digestion may be enhanced by separating the acid and methane producing groups of micro-organisms. Provision for mixing the digester contents (Sen and Bhaskaran, 1962) and thermophilic conditions (Buhr and Andrews, 1977) have been reported to improve the biogas production to a considerable extent. Cobalt addition (5 to 25 mg/litre) improved the overall performance of anaerobic digester on biogas production from 41.97 to 43.06 ml using cattle dung as substrate (Wate et al, 1983). Attempts are now being made to identify other micro-nutrients besides cobalt, which are involved in the biochemical reactions leading to methanogenesis.

Under the national project for biogas development (NPBD), the Department of Non-conventional Energy Sources (DNES) has been promoting a variety of activities for the widespread use of biogas. Various designs of biogas plants have been tested. Small types to large ones and even institutional type plants have been used. Over 15 lakh family size biogas plants have been installed all over the country, 401 community type biogas plants, 359 institutional biogas plants and 59 nightsoil biogas plants have been completed (Table 23). These plants are capable of producing 1488 million cubic meters of gas, the fuel value of which is equivalent to 5 million tonnes of firewood valued at 200 crore of rupees per year. In addition, these plants produce high quality enriched manure which is estimated to be equivalent to manure costing rupees 174.5 crore.

Biogas as a Cooking Fuel

Biogas is a gaseous fuel which may be directly burnt in a gas burner for use as a cooking fuel or source of direct heat for thermal applications. Biogas has a calorific value of 5871 kCal/

m^3 and is comparable to coal in India. Heat value of biogas can be improved by about 30 per cent by reducing carbon dioxide contents. It takes nearly 350 kg/m^3 pressure to liquify it for easy storage.

Table 23

Biogas plants installed in India upto March 31, 1992

System	*Number installed*
Family size biogas plants	1575000
Community size biogas plants	401
Institutional biogas plants	359
Nightsoil biogas plants	59

India has a cattle population of about 300 million Singh and Bhullar (1990) estimated that our cattle are about 22 million and that of buffaloes about 8 million. Neelakanton (1975) estimated that the average dung excretion per animal per day at 11.3 kg and 11.6 kg for cattle and buffaloes respectively. The dung recovery has been assumed at 70 per cent, because of the Kachha floor of most of the cattle sheds. About 30 per cent of the dung gets puddled by the animals which is technically not usable for the biogas plants. The total availability of dung from cattle and buffaloes for biogas production (Table 24) works out to 2087000 metric tonnes (1519000 metric tonnes for cattle and 568000 tonnes for buffaloes.

Table 24

Daily availability of dung for biogas production in India (Singh and Bhullar, 1990)

Breed	*No. of animals (000')*	*Dung prod/ 000' animals (M.T.)*	*Total dung prod. (000' M.T.)*	*Recovery (per cent)*	*Useful dung for biogas (000' M.T.)*
Cattle	192068	11.30	2170	70	1519
Buffaloes	69866	11.60	811	70	568
Total	261934	11.38	2981	70	2087

Table 25

Daily potential gas production and number and percentage of potential beneficiaries in different seasons (Singh and Bhullar, 1990)

Season	*Dung prod.* (000'M.T.)	*Gas prod. kg.*	*Total daily gas prod. Mill. M^3*	*gas requirement/person M^3*	*No. of potential beneficiaries (Million)*	*Potential rural population coverage (per cent)*
Summer	2987	3.3	195	0.57	341.5	64.99
Monsoon	2087	2.0	119	0.57	208.7	39.71
Winter	2087	1.3	78	0.57	135.6	25.80

Biogas production potential per unit of dung varies widely in different seasons which results in variation in the number of potential beneficiaries (Table 25). It would be seen that total daily gas generation potential works out to 195 million cubic meters for summer, 119 million cubic meters for monsoon, and 78 million cubic meters in winter. In this way, the installed capacity is bound to remain under utilized in winter and monsoon seasons. So the technology needs to be refined to maintain the uniformity of gas production. Further, the per capita requirement of 0.57 cubic meters of gas for cooking can be reduced by introduction of modern cooking devices.

In addition to cattle dung, many organic materials such as leaf biomass, crop wastes, straws, leaf litter, aquatic and terrestrial weeds, urban garbage etc. can form good feedstocks for biogas production. In India, the annual productivity of non-woody biomass has been estimated to be about 1150 million dry tonnes (Jagdish, 1994). After meeting the cattle needs, a substantial amount could be made available for biogas production. Even if 0.5 Mt are available daily (180 Mt/year), the gas production from this would be around 150 million m^3/day, adequate to meet the cooking energy needs of 750 million people. Therefore, if these wastes are collected and converted to biogas efficiently, there will be surplus cooking fuel in our rural and urban areas. The limiting factor would then be the efficiency of collection, storage techniques and conversion technologies.

Biogas as Engine Fuel

Biogas is a reasonably high calorific value gas and can be burnt in engines instead of direct combustion for thermal applications. Both diesel (compression ignition engines) and petrol (spark ignition engines) can be modified to use biogas as the main fuel. Once engines are run on this fuel, the shaft power application are plenty. Owing to reduced calorific value of gas, the engines are to be derated by about 30-50 per cent by direct shaft power applications. On the other hand this shaft power may be converted to electricity for providing energy services such as domestic and street illumination, supply of drinking water, power for small scale commercial applications such as flour milling etc.

The current design of biogas installations though simple occupy a large space and subject to avoidable leakage of valuable gas. Again, each plant is fabricated *in-situ* and once installed becomes immovable. This could very well inconvenience the needs of future development around that site. However, the simplicity of current design suits the needs of the rural agriculturists who are unskilled in the operation of complicated apparatus. The raw material, namely cowdung and other organic refuge are plentiful in rural areas and space is not a problem. Under such circumstances the current design has proved a filling starting point (Krishnaswamy et al, 1981).

Classification of Biogas Plants

Biogas fermentation could be brought about either in a continuous/semi-continuous system, or in a batch fermentation system. These two systems are sometimes referred as wet process and semi-wet process respectively.

Continuous fermentation system

It generally consists of a single digester in which 1/50 of the amount is charged and discharged daily without completely emptying the digester. Thus, the digester always remain full and the fermentation continues. The common "gobar gas plant" is a good example of this system. However, fibrous plant products cannot be changed in the cattle dung biogas digester due to some bioengineering problems (floatation, movement and chock-up).

Goswamy and Choudhary (1966, 1967) designed a straw gas plant based on continuous fermentation system which operate with 95 to 328 lit/kg paddy straw. Wheat straw were evaluated for their biogas production potential in laboratory set up (Rajor and Goswamy, 1983). There is lot of scope for utilizing water hyacinth for biogas production along with cowdung. Eswara Rao and Nagendra (1982) found that yield of biogas from water hyacinth increased with increase of temperature. Krishnaswamy et al (1981) observed that water hyacinth alone yielded at the rate of 41 litre of gas per kg, whereas cowdung yielded only about 7 litres of gas/kg. Nair et al (1982) demonstrated that microalgae such as *Scenedesmus obliqua* available in substantial quantities from high rates sewage stabilization ponds, and macrophytes such as *Ceretophyllum, Chara* and *Eichhornia* can be used as the feed stock, to generate biogas. Efficiency of biogas utilization for cooking was also studied as a function of the gas supply line pressure (Goswamy and Rajor, 1983).

The acidogenic phase in a continuous fermentation system could be made to occur in the freshly added feedstock at the top of the fermentor. The volative fatty acid intermediates could be encouraged to gradually move through already decomposed biomass on which methogenic bacteria have been encouraged to grow. Volatile fatty acid intermediates would be converted to biogas in the lower 'packed bed' portion of the reactor. Depending on the decomposition rates and the presence of easily decomposable fractions in the biomass fed, the different zones in the solid biomass bed would stratify into acidogenic and methogenic zones. To ensure that the reactor does not suffer from volatile fatty acid buildup it would be necessary to frequently sprinkle the bed with a small amount of digester liquid recycled continuously. Such fermenter designs have been shown to be viable for biomass and urban garbage feedstock.

Batch fermentation system

A battery of 8-10 digesters are charged and discharged each at a time in synchronised manner with chopped plant wastes mixed with 12 per cent cattle dung, 2 per cent urea, 3 per cent bone meal, 3 per cent calcium carbonate and 70 per cent moisture. After 10-15 days, the generated gas is transferred in a common gas holder for daily use. The digester in then emptied after 40-45 days of gas collection. Such a plant of large size capacity (common in Europe) was installed at National Sugar Institute, Kanpur in 1962. However, a batch fermentation system is expensive both in installation and operation costs.

Family Size Biogas Plants

Most of the biogas plants (more than 90 per cent) disseminated in our country are of the family size versions. Two basic designs of biogas plants that have been employed—namely the Indian floating type and the Chinese fixed dome type. Both the designs are tailored to use cattle dung as feedstock. Cattle dung and water is mixed in the ratio of 1:1 and the resultant slurry containing 9 per cent total solids is fed to the biogas plant by gravity. This slurry is retained in the digester for a period of 35-50 days. During this retention period about 35-40 litres of biogas is recovered per kg of dung fed. As a result of this most biogas plants produce atleast 0.5 m^3/m^3 of digester volume.

Biogas generates from the disintegration of organic substrate, and being lighter than air rises upwards. This gas is collected in various types of drums. Therefore, according to the method of gas storage, biogas plants are of two types:

(1) Floating dome biogas plants
   (a) KVIC or iron drum plant
   (b) Ferro-cement plant
   (c) Ganesh plant
   (d) Pragati plant
(2) Fixed dome biogas plant
   (a) Janta plant
   (b) Deenbandhu plant

Indian Floating Drum Biogas Plants

The design (Figure 12) consists of a tank or well with a partition wall to prevent shorting of influent fresh dung slurry with the outgoing spent slurry. The gas produced is trapped under a metallic (usually mild steel) or plastic drum. With continuous production more gas is trapped under this bell and the drum rises. This acts as a gas storage unit and when the tap above is released, the gas is discharged at more or less constant pressure. The main advantage of gas issuing out at a uniform pressure enables simpler gas utilization devices such as biogas burners to be used in the kitchen, mantle lamps requiring little attention, simple inlets to duel fuel and spark ignition engines etc. This design also suffers from a disadvantage that the MS steel gas holder frequently corrodes and therefore requires replacement. Recent development attempts have resulted in replacing the MS steel gas holder with long lasting SI, ferrocement or fibre reinforced plastic (FR) as construction material.

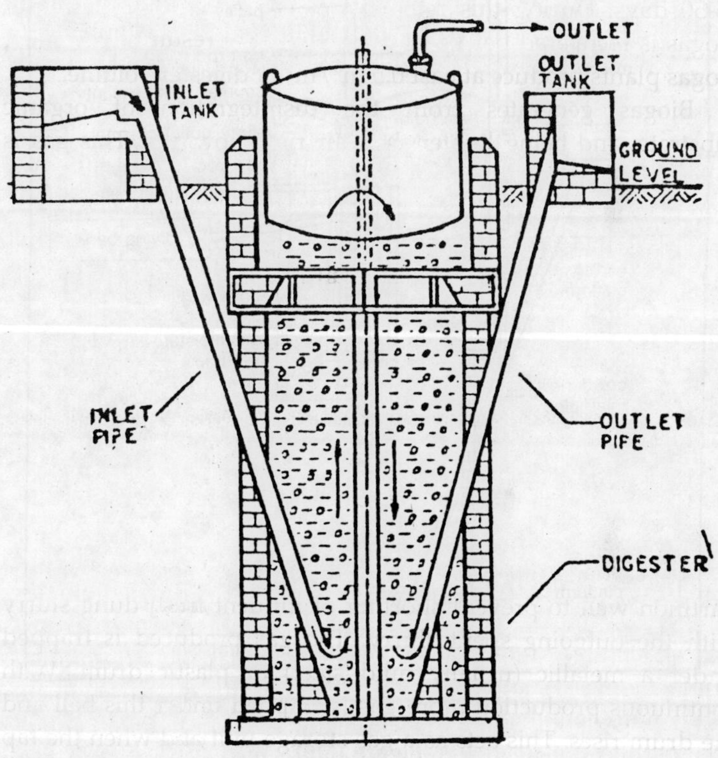

Figure 12(a)
Different parts of floating gas holder type (KVIC) biogas plant

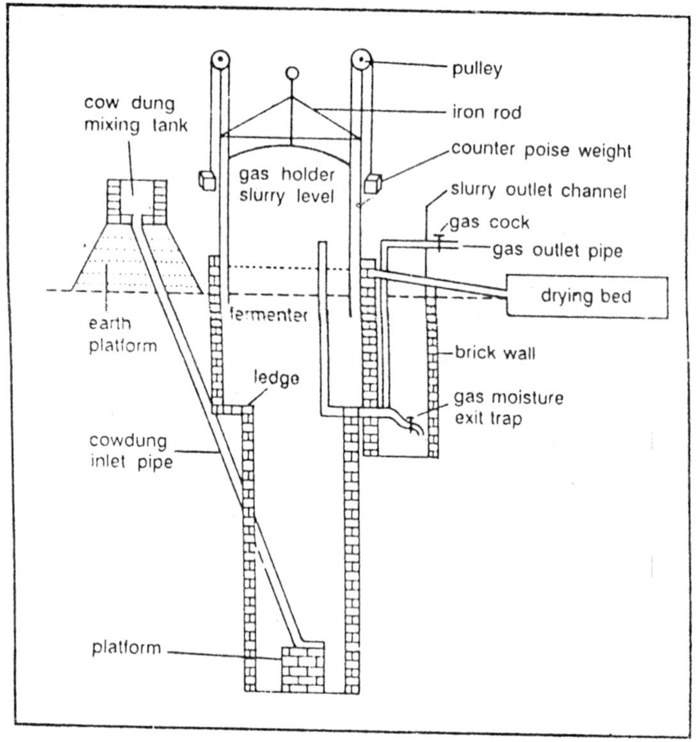

Figure 12(b)
Different parts of floating gas holder type (IARI) biogas plant

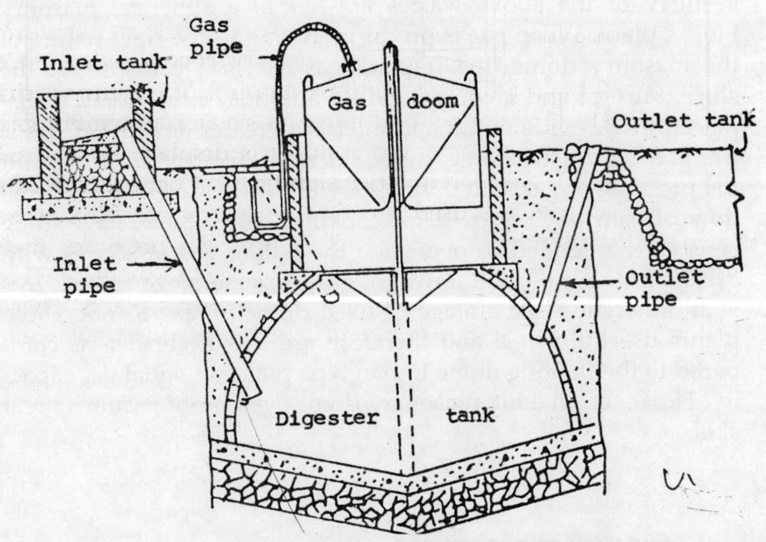

Figure 13
Different parts of floating gas holder type Pragati biogas plant

Today, nearly 2 million biogas plants have been constructed in India mostly of the floating drum design. The smallest biogas plant constructed are of 3 m^3 biogas per day size. Biogas plants in the range of 2-6 m^3/day are termed as family sized units. The small plants requires atleast 50 kg dung as daily input which corresponds to the output of about 5 stall fed adult cattle. When grazed animals or calves serve as the source of dung corresponding correction in the number of cattle heads required have to be made to overcome the shortfall from such animals. The major use envisaged for these plants is for the supply of cooking gas for homes and to some extent domestic illumination through the use of mantle lamps in rural areas.

Chinese Fixed Dome Biogas Plants

The Chinese evolved this design (Figure 14, 15) to reduce the steel required in the construction and therefore the overall costs of this biogas plant. In China, owing to the extensive handling of human and piggery wastes at farm level, biogas plants were introduced to contain the potential pathogens, provide energy for cooking and provide manure to fields. In the Chinese design, a slurry of the above wastes are fed to a spherical masonry biogas plant. When gas is produced owing to the rigid nature of the masonry dome, the trapped gas exerts a pressure on the slurry surface and a corresponding amount of the slurry is displaced into a wide outlet and inlet. As greater amount of gas is trapped under the dome, more slurry is displaced. As a result the pressure of gas stored varies significantly. Consequently, the downstream devices used for cooking, lighting etc. have to be constantly attended to overcome the falling gas pressures with greater amount of withdrawal of gas from the biogas plant. This is an inherent disadvantage of fixed dome biogas plants. These plants use little steel and therefore are less expensive as compared to the floating dome Indian type plants of equal gas capacity. However, making masonary domes gas proof requires great skill.

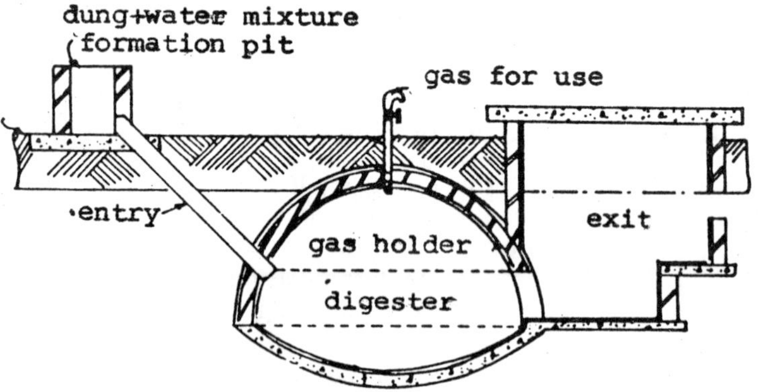

Figure 14
Different parts of Deenbandhu biogas plant

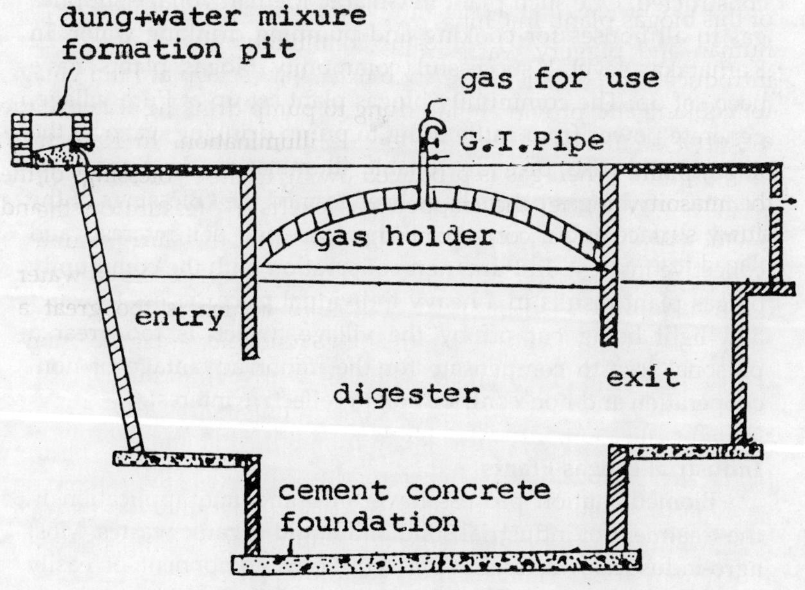

Figure 15
Different parts of Janta biogas plant

The biogas programme of China was oriented to improve sanitation and organic manure production in rural areas, while energy recovery was secondary in the programme. The gas produced was expected to be used for cooking and domestic illumination. However, experiments of running tractors, buses and other engines for shaft power is also documented.

Community Biogas Plants

Only a few households in villages own adequate number of cattles to possess enough dung to successfully run biogas plants. As a result it appears that only the rich who own cattle in sufficient numbers can benefit from this technology. On the other hand, when all the dung resources of the village is pooled, it is possible to provide several energy services to all the village households. Such biogas plants are often called community sized biogas plants. In India, about 450 such biogas plants have been constructed. One such plant in Gujarat, Methan vollage supplies gas to all houses for cooking and pumping drinking water. In Karnataka, several successful community biogas plants have been set up. The community biogas plant set up at Pura village, generate power from cattle dung to pump drinking water to the village, and also provide reliable illumination. In Karnataka, community biogas plants have been termed the "Blessings of the Commons", based on the coincidence of self-interest and collective interest. In Pura, non-cooperation with the community biogas plant results in a heavy individual price, access to water and light being cut off by the village, which is too great a personal loss to compensate for the minor advantage of non-cooperation and non-contribution to collective interest.

Industrial Biogas Plants

Biomethanation process have an immediate application in the treatment of industrial solid and liquid organic wastes. Most agro-industrial wastes have a significant component of easily degradable organics. Distilleries fruit and vegetable processing industries, paper and pulp industries, sugar, textile production etc discharge highly degradable organic wastes. Anaerobic digestion is the most favoured primary treatment. Biomethanation removes 70-95 per cent of the BOD without producing too much microbial biomass as in the case of activated sludge. Most of the carbon is converted to energy rich methane which in most cases is adequate to meet the treatment plants energy requirement for operation and does not require high inputs of energy for aeration.

Dairy industries wastes could be processed to generate 0.85 m^3/kg in 6 day retention time. Disposal of whey is the most serious problem for the cheese manufacturing units. One tonne of cheese gives rise to 10 tonnes of whey. Each cubic meter of whey produces approximately 38 m^3 of biogas. Methane contents of biogas produced are 62 per cent.

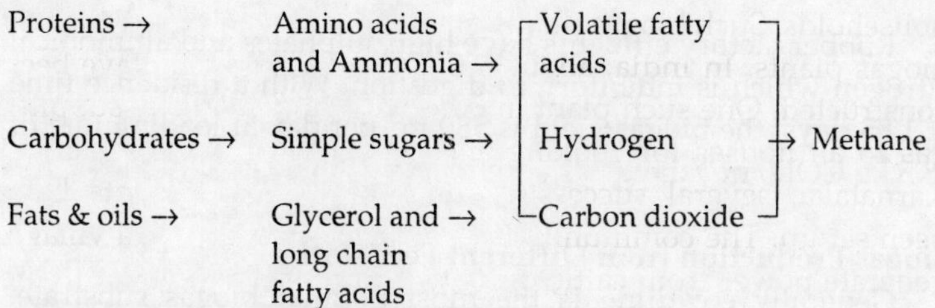

Figure 16
Formation of methane from whey components

Pea canning wastes can produce 0.87 m^3/kg biogas within 3-5 days retention time.

Waste water from manufacture of wheat gluten, starch from flour contain proteins, carbohydrates, mixture of amino acids, hemicellulose, pentose gums, suspended starch granules etc. Methane yield is 0.33 m^3/kg. Methane content in the biogas is 65 per cent.

In the citrus processing industry, peels can be used for anaerobic fermentation. Oil is first removed from them, since it is inhibitory to microbes. 0.5 m^3/kg biogas with 50-55 per cent methane is then produced.

Distillery wastes with a residence time of 8-10 days pH of the digester 7.2 and temperature around 48-52°C generate 40-50 m^3 biogas/m^3 effluent.

Jute caddies, the unspinnable short fibres deposited by jute mills looms can generate biogas when fermented. Jute caddies is lignocellulose waste. The alkali treated caddies are used instead of raw caddies. The remaining slurry after biogas production is rich in NPK nutrients and is comparable to farmyard manure.

Textile industry generates willow dust which is solid cellulosic waste material. Biogas plant produces 17 m^3 biogas from 100 kg willow dust in 30 days. A large scale trial was taken by the Apollo Mills, Bombay, where with the help of 6 digesters, 12 tonnes of willow dust were digested per month. 350 m^3 biogas was obtained in each digester handling 2 tonnes of willow dust with 90 days retention period.

Pulp and paper industry waste water yield biogas 1.1 m^3/day. The gas contain 81 per cent methane.

Rubber factory effluents have high sulphates and ammonical nitrogen which is inhibitory to digestion. With a residence time of 1.85 days, the biogas yield is 580 m^3 per day at loading rate of 18.5 kg COD/m^3/day.

Biogas Production from Different Feedstocks

Generally cowdung is the most popular biogas substrate. Other materials which are now being used include—aquatic weeds, municipal waste, industrial wastes. Although these materials differ in their chemical composition yet they yield biogas.

Biogas from Aquatic Weeds

Azolla pinnata is a common aquatic weed containing a very useful nitrogen-fixing alga *Anabaena azollae* within its leaves. This weed is already used as a biofertilizer in the waterlogged rice fields. Its prolific growth (333 t/ha/yr), ease of harvesting and possession of high amount of carbohydrates, proteins and fats, coupled with negligible legnin content, make it a good source for biomethanation process. Observations (Table 26) indicate that substrates containing either cowdung or *Azolla* or mixtures of cowdung and *Azolla* in different proportions produced biogas. Cowdung and *Azolla* in 1:0.4 proportion was found to give maximum rate of methane production between 8th and 11th day of fermentation. Biogas production from the above mixture was 1.4 times that of cowdung alone.

Table 26
Biogas generation from *Azolla pinnata* (Das et al, 1994)

Substrate	*Parameters*		
	Average methane content (% v/v)	*Gas yield (L)*	*Time required for maximum gas production (d)*
Cowdung	64	100.2	10
Azolla : Cowdung mixure			
0.6 : 1	63	101	11
0.4 : 1	62	140	08
0.2 : 1	68	111	09
Azolla pinnata	56	65	11

Salvinia, a member of Pteridophyta, is commonly known as water fern. It grows luxuriently in stagnant water. *S. molesta* is the world's worst weed. In India it dominates in Kerala, Kashmir and North-east states. This fern can be a feedstock material for biogas production. Fermentation of *Salvinia* starts within 7-9 days. Biogas yield is about 0.1 litre/kg fresh weight for 4 weeks. Air dried weed produces about 1 litre/kg for 90 days. Thereafter, gas yield gradually declines. Special advantage of using *Salvinia* is that unpleasant odour do not come out.

Eichhornia, popularly known as water hyacinth, is another world's worst weed. Its decomposition rate is higher than that of cowdung. It gets totally decomposed within 3 days in summer while cowdung takes 8 days. 1 kg of dried water hyacinth produces 374 litre of biogas. One hectare area can produce 600 kg of dry biomass per day, which can generate 229300 litre of biogas/ha/day. Increase in recreational value of water bodies from which *Eichhornia* is to be removed can be an additional advantage of exploring it for gas production. The *Eichhornia* biomass contains substantial amounts of nutrients, the removal of which may keep a water body clean and less eutrophic.

Ipomea aquatica is a tropical trailing herb, habitating muddy stream banks, but is also found in fresh water ponds and marshy areas. The components of the plant which are submerged show some resistance to adverse conditions ensuring perpetuation of the species under favourable conditions (Jahan and Kaur, 1992).

Biogas from Plant Waste

The "Kachara gas" plant has been designed to ferment any plant material which is non-lignified along with or without cattle dung, in any proportion. All plant waste (leaves, grasses, weeds, trees or bush leaves, aquatic weeds, fruit peels, sugarcane baggasse, vegetable dressings, waste paper or cardboards etc) must be chopped to less than 5 cm size and stuffed in dry through the feeder hole. Unlike the gobar gas plant, there is no slurry handling involved and no water need to be added except to replenish the loss due to evaporation. Thus, water requirement for its regular operation is negligibly small. The benefit of using aquatic plants in biogas plants is because of their high moisture content which is an important asset in biogas production.

Biogas from Kitchen Refuge

Kitchen refuge has a C/N ratio ranging from 26 to 32 and therefore it satisfies the requirement for anaerobic digestion. The kitchen refuge from the Government College of Technology, Coimbatore was on an average quantity of 235 gm/person/day. For a hostel strength of 130 students, the volume of biogas production would be 2.846 m^3/day. This volume of 2.846 m^3/day of biogas would serve for 12.5 person/day, at the rate of 0.227 m^3/person/day. Fully shredded kitchen refuge having 3 per cent total solids were added to the digesters. After 25 days the methane gas started yielding (Table 27). Methane production was maximum (75.38%) at 6.15 per cent total solid and minimum (61.18%) at 3.7 per cent total solid loading in the digesters.

Table 27
Biogas production from hostel kitchen refuge
(Sunderarajan et al, 1996)

Total solids in the influent (mg/lit)	3.7	4.5	4.7	6.05	6.15
Biogas (m_3/m_3 of dining days)	0.830	0.311	0.415	0.493	1.038
Methane (%)	61.18	74.24	67.25	63.97	75.38

The biogas production from army camps of Egypt using two stage biogas digester with a total capacity of 190 m^3 fed at average rate of 480.9 kg/day generated a biogas yield of 1.009 m^3/kg added per day (El-Halwagi, 1984). The payback period of less than 6 months encouraged the authorities of the army camps of Egypt to finance the programme and to popularise the biogas technology.

Biogas from Sewage

Biogas can be generated from sewage. There are 142 class I cities in India, which produce around 9000 million litres of sewage per day. If 9000 million litres of sewage is converted into biogas per day from major cities, 20 per cent of their energy demand could be met. The Okhla Sewage Works, New Delhi receives 25 cubic meter sewage per second and generates 180000 cubic meters of gas. From the plant, gas pipe lines are given to about 700 families of the surrounding villages (Sharma, 1984).

In big cities with facilities to generate biogas from sewage and distribute it, the gas is supplied as a fuel for domestic consumption or to hospitals, hotels etc. But even where distribution lines are lacking, biogas could be used by separating methane from it and supplying the methane in cylinders. It will, in fact, be possible to transport methane economically over a larger distance than through pipes. The gas, however, needs to be compressed.

Biogas contains 60 per cent methane, 40 per cent carbon dioxide and upto 20 ppm hydrogen sulphide. Under high pressure, carbon dioxide separates out; this enhances the calorific value of the remaining gas. Upto 90 per cent pure methane could be obtained at high pressures and low temperatures (CO_2 solidifies at a pressure around 46 atm (673 psi) and at –68°C. The gas composition in equilibrium with solid CO_2, or dry ice, at this temperature and pressure, is 90 per cent methane and 10 per cent carbon dioxide). The methane can be separated and stored in cylinders. Methane thus bottled can be used as a supplement to petroleum fuels. Mechanical compression of biogas is done in three stages followed by refrigeration. Intercoolers are to be provided in between the compression stages (Figure 17).

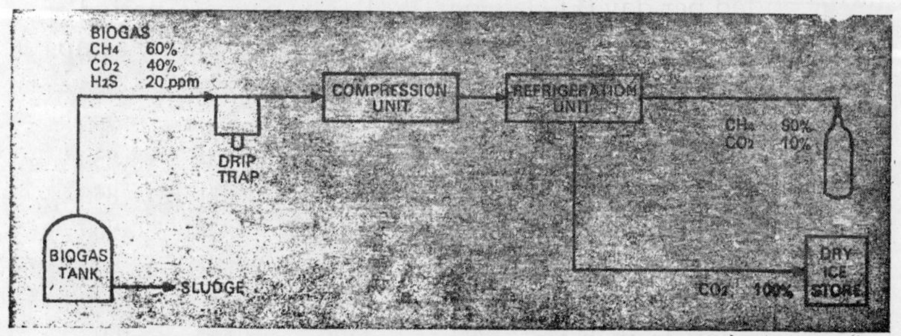

Figure 17: Methane and carbon dioxide separation from biogas

Bottled methane was successfully used in Germany for running tractors and other equipment. Bottled methane plants have also been reported from USA. A system of separating CO_2 from biogas was used at the Dadar Sewage Treatment Plant in Bombay. The gas was compressed to 5000 atm in three stages. After the second stage, CO_2 was scrubbed in water at 14 to 14.7 kg/cm^2 (this brought down the CO_2 content from 30 to 5-7%). Methane was then dried (by calcium chloride), compressed to 5000 psi, and stored in 0.59 m^3 cylinders and used as fuel for trucks. Each truck carried about 4 such cylinders—the equivalent of 160-204 litres of petrol (Pathak and Colah, 1976).

Biogas from Poultry Litter Waste

Poultry farming has become an important industry in India, and these farms generate large quantities of waste. Excessive land application of these wastes causes pollution problems, since they release obnoxious odours and become a breeding ground for insects and worms.

Table 28
Biogas production (ml) at different total solids content of poultry litter waste at 37 ± 1°C (Shivraj and Seenayya, 1994)

Days/ Total solids	*5%*	*6%*	*7%*	*8%*	*9%*	*10%*
1	300	350	400	500	400	400
2	400	500	500	550	500	500
3	600	400	600	550	500	500
4	650	500	700	800	700	700
5	700	550	700	800	700	700
6	700	600	800	940	700	700
7	700	600	800	940	800	800
8	600	700	900	950	800	850
9	600	750	900	950	900	850
10	600	750	950	1000	900	900
11	500	800	950	1100	950	900
12	500	850	1000	1100	1000	900
13	400	800	1000	1150	1000	1000
14	300	600	900	1150	1000	1000

Contd.

Days/ Total solids	*5%*	*6%*	*7%*	*8%*	*9%*	*10%*
15	200	400	800	1200	1100	1100
16	180	350	750	1100	1200	1200
17	—	—	—	1000	1300	1300
18	—	—	—	900	1200	1300
19	—	—	—	800	1000	1200
20	—	—	—	700	900	1000

Observations on biogas production from poultry litter waste (Table 28) indicate that biogas production increased with increasing time. The digester containing 5% total solids produced maximum biogas on the 7th day. In the digesters with 6 and 7% total solids maximum biogas generation was observed on the 15th day, and the digesters with 9 and 10% total solids produced maximum biogas on the 17th and 18th day respectively. Thus with increasing total solids in the digesters the retention time also increased. The biogas generation upto 15th day throughout the anaerobic digestion process at 8% total solids was higher than the biogas generated at higher as well as lower total solids. However, there was a marginal increase in the biogas generation in the digesters containing more than 8% total solids with increasing time.

A constant daily gas production was observed on daily feeding (6 gm of poultry litter waste in 56 ml of water) after removing equal volume of effluent, so as to maintain the total solids constant in the digester.

Environmental Aspects of Biogas Production

In the wake of the primary and secondary energy crises, attention has been paid to utilize all possible substrates for biogas production. Until now the huge biomass waste resource was considered as menace because it impedes water flow and drainage, hamper fish production, nevigation, mar recreational value of the water bodies. Both chemical and biological methods were used all over the world to eradicate weeds, but economic utilization of these so-called weeds seems to be the only feasible alternative in the years to come, because if they are sued for biogas production, they help not only to check the pollution of water bodies but also provide energy and fertilizers.

When a feedstock such as dung or biomass is used, the biogas producing bacteria convert mainly the carbon portions to biogas. The nitrogen utilized by growing bacteria is small as compared to other aerobic fermentations. Therefore, a large portion (85%) of the nitrogen in the feedstock remains intact by this process and all this can be recovered in the spent feedstock of a biogas plant. As a result of this, biogas production process conserves over 85% of the nitrogenous compounds in the feedstock compared to only 40-50% in the conventional compost process. Very little of the nitrogen is lost from the fermenter because the redox potential in the fermenter is very low (–200 to –600 mV) and is not conducive to oxidative losses.

The amount of wood saved by using biogas for domestic purposes will depend upon temperature variations and number of rainy days in a year. Assuming that summer has 122 days, winter 120 days and monsoon 123 days, the saving of wood has been estimated at 66 million tonnes per year. Apart from saving of the fuel wood, biogas generation can reduce the pollution caused by smoke produced during burning dung and wood as fuel.

From the farmers point of view, biogas production is a process for energy recovery without concomitant loss of the 'soil conditioner' or 'fertilizer value' of biomass feedstocks. From a national perspective, the extensive use of biogas substitutes a need to import valuable fossil fuel—petroleum products. Secondly, by virtue of containing wastes (with pathogens) in a reactor, the incidence of insect vector menace and concomitant disease spread is controlled and the overall hygiene in villages is therefore vastly improved. Spent material from a biogas reactor does not attract insect vectors. They would reflect as greater number of productive working hours in rural areas as well as reduced expenditure on health care.

Biogas production is a convenient way of agricultural waste disposal for more than one reason. Substrate detoxification, deodorization, inactivation of pathogens, dehelminthization occur along with biogas production and fertilizer or humus forming substances as a byproduct.

The spent material from biogas plants, atleast partly, can be converted to value added products. In case of biomass based biogas plants, the spent feedstock is stripped of simple saccharides and celluloses. The remaining material is rich in lignin. Such a material is rich substitute for the cultivation of edible mushrooms, as compared to conventional straw and yields of mushrooms are more than double.

The spent biomass feedstock also form ideal material for raising earthworms and the generation of vermi-compost. Earthworms are known to prefer partly decomposed biomass substrates. The spent feedstock from biomass plants are dung spent slurry mixed appropriately can be used to produce vermi-compost which is in great demand in several cash crops.

Beneficial Role of Biogas

Among the different renewable energy technologies relevant for rural applications, the biogas route has several benefits, some of which are: (1) among renewable technologies, biogas technology requires the lowest financial input per unit output of energy and the technology is among the most mature; (2) the anaerobic digestion is a natural process and crucial nutrients such as nitrogen, phosphorus and potassium are conserved and recycled; (3) anaerobic digestion is a versatile process and can utilize a wide variety of organic feedstock; (4) the digestion process, when utilizing animal and human waste, reduces plant, animal and human pathogens and thereby results in the associated health benefits; and (5) biogas (methane) is a clean burning fuel and hence reduces exposure to emissions associated with the use of unprocessed fuels (Stuckey, 1986).

Microbial Hydrogen Production

Hydrogen is the simplest molecule present in the universe. It is a renewable fuel, because the raw material in the form of water is abundant. It can be easily collected, stored (as gas, liquid or hydrides of metals) and transported (by trucks, ships or trains). Furthermore, when used as a fuel, hydrogen gas causes no pollution and forms water thus renewing the raw material. In view of this, hydrogen production is emerging as an important area of biotechnological research with respect to environment friendly non-conventional and renewable source of energy. In achieving this objective, the principle used in photosynthetic machinery by green plants is used. The chlorophyll helps in trapping solar energy in the form of ATP, which is utilized in photosynthesis. In this process, water is subjected to photolysis, which leads to splitting of water molecules into (1) oxygen, (2) electrons, and (3) hydrogen ions. The hydrogen ions do not get a chance to form hydrogen gas, but are used to form energy rich compounds like glucose. However, if these hydrogen ions can be converted into hydrogen gas, the latter can be collected and used as a fuel. Enzymes hydrogenase and nitrogenase have been used to convert hydrogen ions into hydrogen gas.

Hydrogen gas production using hydrogenase

Some microbes posses the enzyme hydrogenase (Table 29), which helps the two electrons join two hydrogen ions to produce one molecule of hydrogen gas. In the visible light, hydrogenase separates high energy electrons from ferredoxin and facilitate their transfer to hydrogen ion, and ultimately hydrogen gas is evolved. Those plants which produce carbohydrates lack hydrogenase. Sensitivity of hydrogenase to oxygen varies with species. It is however, unlikely to develop an efficient system for hydrogen production if considerable amount of oxygen is produced.

Table 29
Algal species containing hydrogenase (Dubey, 1993)

Groups		*Species*
Blue-green algae (Cyanobacteria)	:	*Anabena azollae, A. cyllindrica, Anacystis elongata, Nostoc muscorum, Spirulina platensis, Synechococcus elongatus*
Green algae	:	*Chlamydomonas moewusii, Chlorella fusca, C. homosphaera, C. kessleri, C. sorokiniana, Ulva lactuca*
Brown algae	:	*Ascophyllum nodosum*
Red algae	:	*Ceramium rubrum, Chondrus crispus, Corallina officinalis, Porphyra sp. Porphyridium cruentum*

The above microbes may be used for the isolation of hydrogenase enzyme, which can be used with isolated chloroplasts, so that the hydrogen ions and electrons liberated due to the use of solar energy by the chloroplasts are made to combine into hydrogen molecule. The hydrogen gas thus formed, will bubble out of the solution and can be easily collected.

An alternative approach is the three-stage procedure for making hydrogen from water:

The enzyme responsible for making hydrogen gas (joining two hydrogen ions and two electrons to form a molecule of gas) is hydrogenase. This enzyme has so far been extracted from 15 species of bacteria and algae. *Clostridium butyricum* supplied with sugars produces hydrogen, but the system are unstable. Immobilised C. *butyricum* cells produce hydrogen gas for a month instead of few hours when fed with waste water containing sugars from an alcohol factory. Water when split by algae during photosynthesis produces hydrogen ions, but not hydrogen gas. Efforts are on to see if hydrogen gas can be formed. In normal case, in plant cell, the hydrogen ions do not get a chance to combine to form hydrogen gas, but are used to manufacture energy-rich compounds. To produce hydrogen gas, this process must be subverted. By using hydrogenase enzyme obtained from isolated chloroplasts, hydrogen molecules are produced (Figure 18).

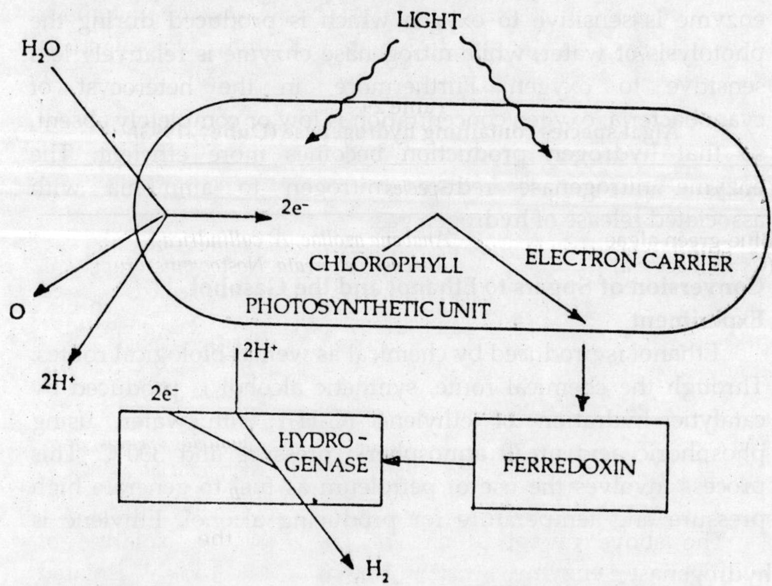

Figure 18: Coupling solar energy to produce hydrogen gas

Glycolate is collected by breaking apart algae cells. Glycolate is fed into immobilised enzyme glycolic oxidase from plants. Immobilised bacteria supply formic hydrogenlyase.

(i) $$2CO_2 + 2H_2O \xrightarrow[\textit{Chlorella pyrenoidosa}]{\text{Light}} H-\underset{\underset{\text{OH}}{|}}{\overset{\overset{\text{H}}{|}}{C}}-COOH + 1.5O_2$$

glycolate

(ii) $$H-\underset{\underset{\text{OH}}{|}}{\overset{\overset{\text{H}}{|}}{C}}-COOH + O_2 \xrightarrow{\text{glycolic oxidase}} O=\overset{\overset{\text{H}}{|}}{C}-COOH + H_2O_2$$

(iii) $$O=C-COOH \xrightarrow{\text{nonenzymatic}} HCOOH + CO_2 + H_2O$$

(iv) $$HCOOH \xrightarrow{\text{formic hydrogenlyase}} H_2 + CO_2$$

In total, $2H_2O \rightarrow 2H_2O + O_2$

Hydrogen gas production using nitrogenase

In cyanobacteria nitrogenase enzyme is the chief hydrogen producer. This is considered better, because hydrogenase enzyme is sensitive to oxygen which is produced during the photolysis of water, while nitrogenase enzyme is relatively less sensitive to oxygen. Furthermore, in the heterocyst of cyanobacteria, oxygen concentration is low or completely absent, so that hydrogen production becomes more efficient. The enzyme nitrogenase reduces nitrogen to ammonia with associated release of hydrogen gas.

Conversion of Sugars to Ethanol and the Gasohol Experiment

Ethanol is produced by chemical as well as biological routes. Through the chemical route, synthetic alcohol is produced by catalytic hydration of ethylene (C_2H_2) with water, using phosphoric acid at 70 atmospheric pressure and 300°C. This process involves the use of petroleum as fuel to generate high pressure and temperature for producing alcohol. Ethylene is derived from both natural and coke oven gases, and the waste gases released in refining petroleum to produce gasoline.

Sugary materials

A variety of crops can be used as feedstock for production of ethanol from fermentable sugars using yeast, such as sugarcane, sweet sorghum, cassava and various cereal crops (Table 30).

Table 30
Ethanol yield from selected sugary materials (Sinha and Kishore, 1991)

Raw material	*Possible production (t/ha)*	*Carbohydrate content (%)*	*Ethanol yiels litre/t)*
Beet	40-50	16	90-100
Sugarcane	50-100	13	60-80
Maize	4-8	60	360-400
Wheat	2-5	62	370-420
Barley	2-4	52	310-350
Grain sorghum	2-5	70	330-370
Potatoes	20-30	18	100-120
Sweet potatoes	10-20	26	140-170

The most widely used feedstocks, however, are sugars. When sugar crops such as sweet sorghum and sugarcane are used for ethanol production, sugary juices can be tapped from the plant and fermented directly.

Sugar crops for ethanol production have potential in regions where large areas of reasonably fertile land are under-utilized.

The main disadvantage of sugarcane is that it requires fertile land and adequate irrigation for high yield (Hall et al, 1982; Vimal and Tyagi, 1988).

Another potential feedstock for ethanol production is surplus molasses from existing sugar production facilities. Every tonne of sugarcane produced, results in approximately 190 litres of molasses as a by-product. This contains 50-55 per cent fermentable sugars and yields about 280 litres of ethanol per tonne of molasses when fermented (Hall et al, 1982).

Ethanol is generally produced by fermentation of some sugar rich product such as sugar cane, sugar beet, tapioca, sweet potatoes, fruit juice, sweet sorghum etc. with the help of yeast, *Saccharomyces carevisiae* or sometimes with *Kluyveromyces fragilis.* Several other organisms (fungi and bacteria) are also known to produce small quantities of ethanol. However, in these later cases in addition to ethanol, other undesirable products are also produced so that often they are not used for alcohol production. Fermentation is often most active under anaerobic conditions, when carbohydrates are converted into ethanol via pyruvate and acetaldehyde. Under aerobic conditions on the other hand very little alcohol is produced, most of the carbohydrate getting oxidised to CO_2 + H_2O. Furthermore, ethanol is inhibitory for yeast growth at high concentration, although yeast strains differ in their tolerance to ethanol's inhibitory action. The steps involved in ethanol formation by this method are sown in Figure 19.

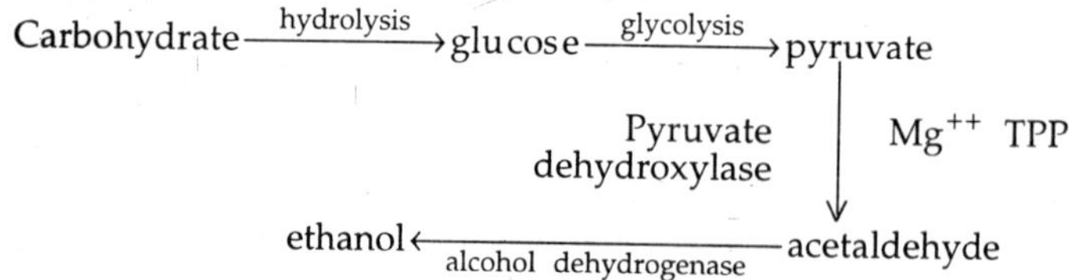

Figure 19
Different steps involved in the manufacture of ethanol from a carbohydrate via pyruvate and acetaldehyde

Petite yeasts yields upto twice as much alcohol as normal relatives. If a normal yeast strain (IZ-1904) produces 41 per cent alcohol then Petite version of this yeast will produce 83%. *Zymomonas mobilis,* a bacterium, is the alternative of yeast. *Z. mobilis* is used as agave fermenter in Central America. It ferments sugars more efficiently to alcohol (Ignacimuthu, 1996).

Starchy materials

Starchy materials used in ethanol production are tapioca, maize, wheat, barley, oat, sorghum, rice and potatoes, but tapioca and corns are the two major substrates. It has been estimated that 11.7 kg of corn starch can be converted into 7 litres of ethanol (Chahal and Overend, 1982).

Starch → hexoses → glucose → ethanol

In United States, producing ethanol from crops like corn, poses no direct competition to its food supplies, because about 60 per cent of its corn is fed to animals and there is surplus of grains; but some argue that it will mop up grain that could otherwise be sold to other countries, or that it will affect international trade.

Cellulose materials

The aquatic weed *Eichhornia crassipes* is a promising source of x-cellulose (42%) and very little lignin (16%). This parameter is advantageous in its conversion to ethanol via saccharification by either enzymatic or acidic processes and subsequent fermentation of sugars using yeast. However, the enzymatic method is advantageous over acidic process as it leaves no unwanted byproducts as obtained by the use of acids. The enzymatic process involves microbial technology using cellulolytic microbes including fungi and bacteria for its initial breakdown into simple sugar like monosaccharides and disaccharides followed by the fermentation of the resulting hydrolyzates into ethanol using yeasts like *Sachharomyces cerevisiae* and *Candida tropicalis.* It has been estimated that one tonne of dried water hyacinth yields about 13 gallons of ethanol by this process (Jahan and Kaur, 1992).

Direct fermentation of cellulose to ethanol is of current interest. A thermophilic bacterium—*Clostridium thermocellum* and filamentous cellulolytic fungus—*Monillia* sp can produce ethanol directly from cellulose. However, in both cases, the fermentation rate is slow and the final ethanol concentration remains low.

Lignocellulose

Primary studies have concerned with chemical or thermal modification of wheat straw for subsequent enzymatic hydrolysis to glucose for ethanol fermentation by *Saccharomyces uvarum*.

A column reactor vessel was used for experiments with 300-500 gm of straw. Various chemical reagents were pumped continuously through the temperature-controlled column with aeration to optimize delignification (Figure 20). Saccharification of delignified wheat straw was achieved by treatment with *Trichoderma viride* cellulase (10 IU/g dry weight residue).

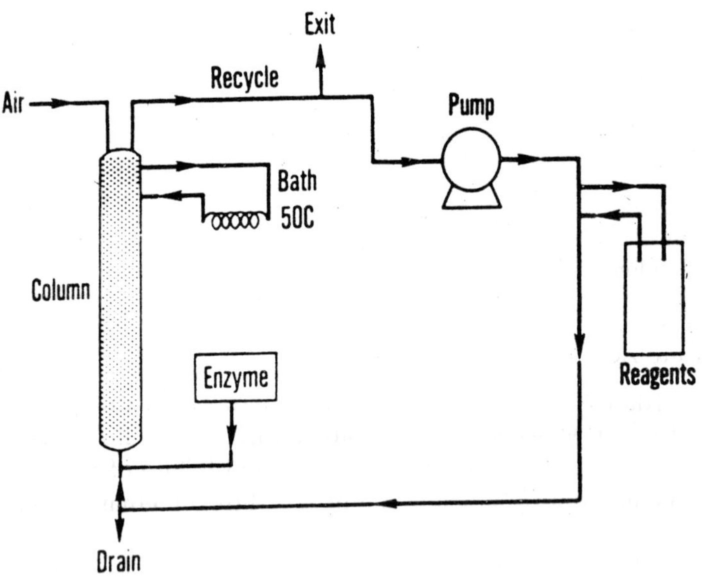

Figure 20
Reactor column for pretreatment, saccharification and fermentation

When 300 gram of straw was treated with 4.0% NaOH for 6 hours, 15-18% of the biomass was lost. Conversion of the cellulose of wheat straw to fermentable glucose in the various trials ranged between 30-50%. The crude fungal cellulose preparation used to treat the modified straw yielded also a substantial quantity of the fermentable 5-C sugar, xylose.

Ethanol production from the sugars in the column inoculated with *S. uvarum* at 10^8 cells/ml was 30-42% theoretical yield. The lower fermentation values may be due partially to the presence of endogenous substances generated in the alkali modification process, coupled to dilute sugar concentrations of 1-3% in the column reactor aqueous phase.

Table 31
Ethanol production from ligno-cellulosic raw material

Raw material (hydrolytic catalyst)	*Dry matter t/ha*	*Ethanol yields litre/t*
Soft wood		
dilute acids	9-15	190-220
concentrated acids	9-15	230-270
Hard wood		
dilute cids	9-15	160-180
concentrated acids	9-15	190-220
Straw		
dilute acids	1.5-3.5	140-160
concentrated acids	1.5-3.5	160-180

Source: Adapted from OECD, 1984.

The other sources of lignocellulose are agricultural wastes and wood. However, yield of ethanol from ligocellulose is low because of lack of suitable technology and failure of conversion of pentoses into ethanol. About 400 litre of ethanol can be produced from one tonne of lignocellulose (Chahal and Overend, 1982).

Legnin is a complex and high molecular weight polymer. It is composed of higher complex aromatic units. It is formed by dehydrogenation of p-hydroxycinnamyl alcohols, such as p-coumaryl, coniferyl and sinapyl alcohols. The presence of these alcohols differs in different plant groups. For example, gymnosperms legnin is formed from coniferyl alcohols, angiosperm legnin is formed from a mixture of coniferyl and sinapyl alcohol, and grass lignin from mixtures of coniferyl, sinapyl and coumaryl alcohols. Legnin is phenolic in nature; it is very stable and difficult to isolate. It occurs between the cells and cell walls. It is deposited during lignification of the plant tissue and gets intimately associated within the cell walls with cellulose and hemicellulose and imparts the plant an excellent strength and rigidity.

On the basis of technology available today about 409 litres of ethanol can be produced from one tonne of lignocellulose. Production of ethanol from lignocellulose follows the following: (a) hydrolysis, and (b) enzymatic pretreatments are used. Lignin hinders substrate hydrolysis. When the amount of lignin is high in the substrate, the yield of ethanol is low. Treatment of substrate with alkali (2% NaOH at 70°C for 90 minutes, washed and sterilised) would be more effective than without treatment. Moreover, stream explosion (i.e., steam treatment at 190°C) of lignocellulosic materials is another pretreatment which has proved to be the best for hydrolysis (Figure 21). Enzymatic hydrolysis of lignocellulosic materials is performed by using cellulases, hemicellulases, pectinases, lignases etc. These enzymes are produced on a large scale from micro-organisms (Figure 22).

Hemicellulose

Pachysolen tannophilus strain NRRL 2460 is capable of an ethanol fermentation of xylose from crude wheat straw under initial aerobic conditions for generation of high cell population. Detroy et al (1982) reported batch fermentation with 70 gm/I.D.-xylose, resulting in 0.3 gm ethanol/gm of pentose metabolized in 6 days. Cell populations double every 24 hours approximately 3 days. Ethanol concentrations of 2.0% were achieved by 6 days with complete utilization of the xylose available.

Figure 23 is of an overall scheme with an initial chemical pretreatment to yield a xylose/pentosan component plus the cellulosic residue for ethanol production. From 500 gm of wheat straw, one obtains 400 gm cellulosic pulp after chemical pretreatment with 4% NaOH for 6 hours. The liquor contains some 40 gm of fermentable D-xylose, which supports the *P. tannophilus* 5-C fermentation. Treatment of the cellulosic pulp with cellulose (10 IU/gm) for 6 hours yields 105 gm of fermentable sugars, which is only 60% of the available glucose in the pulp. Addition of *S. uvarum* cells to the saccharified material yields 42 gm ethanol in 48 hours.

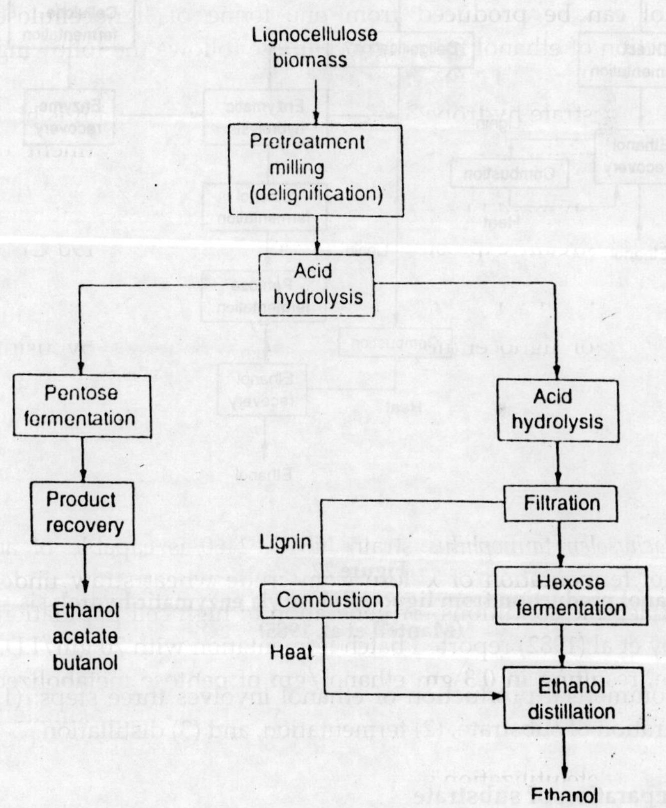

Figure 21
Ethanol production from lignocellulose via acid hydrolysis (Mantell et al, 1985)

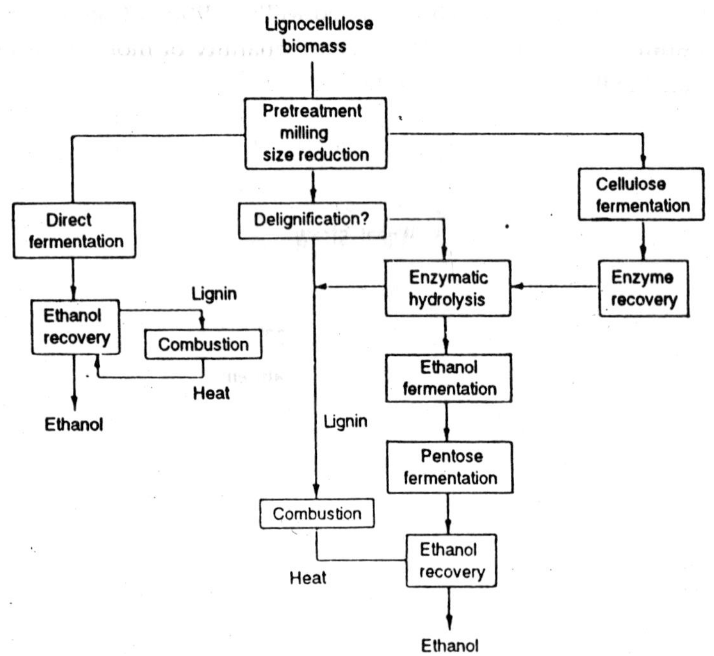

Figure 22
Ethanol production from lignocellulose via enzymatic hydrolysis (Mantell et al, 1985)

Commercial production of ethanol involves three steps: (1) preparation of substrate, (2) fermentation, and (3) distillation.

(1) Preparation of substrate

For commercial ethanol production, following three substrates are used as renewable raw material: (a) roots, tubers or grains that rich in starch, (b) molasses or juice derived from sugarcane, palm or sugarbeet, and (c) wood or waste products rich in cellulose (Figure 24, 25). In India, for the current licensed production capacity of about 1500 million litres of alcohol per annum (by 193 distilleries) adequate quantity of molasses is not available. Therefore, other substrates have often been used.

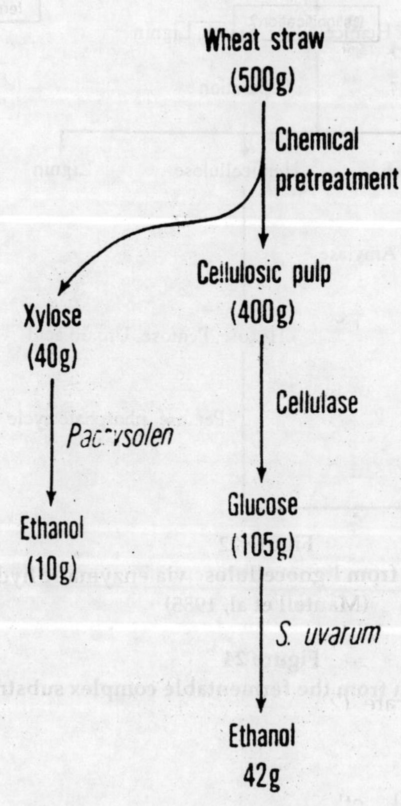

Figure 23
Fermentation scheme for ethanol from wheat straw

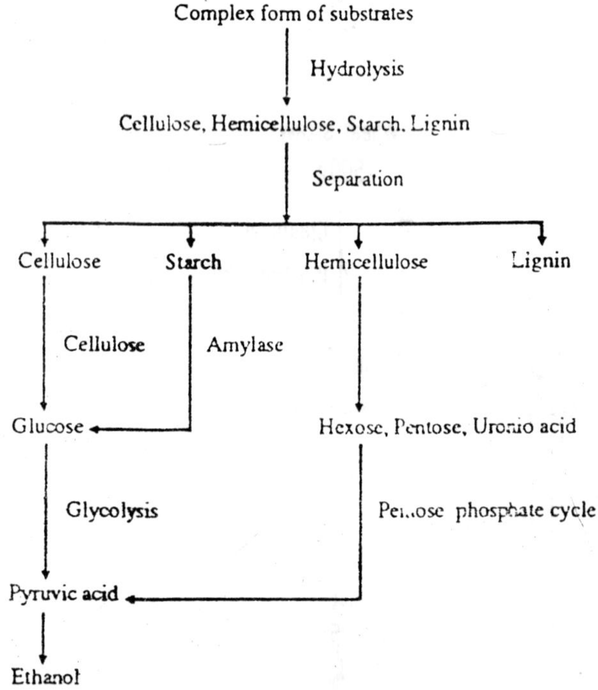

Figure 24
Ethanol formation from the fermentable complex substrates

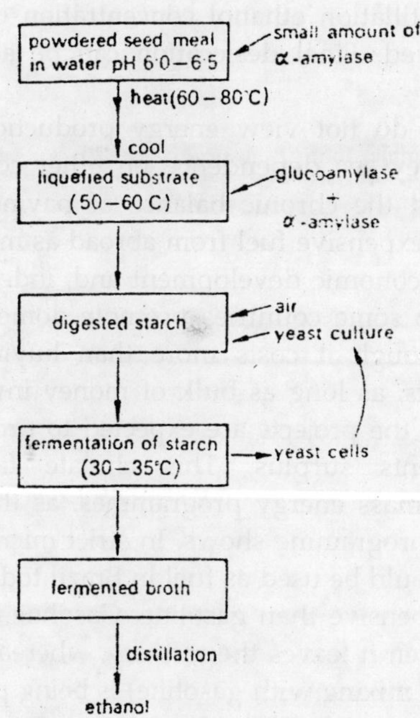

Figure 25

A flow diagram showing various stages in the manufacture of ethanol from starchy seed

(2) Fermentation

The soluble hexoses and pentoses obtained in solution after hydrolysis of sugar starch-cellulose are subjected to anaerobic fermentation by using bacteria, yeast and filamentous fungi. Various types of continuous fermentation plants are available for producing ethanol. In a typical fermenter using molasses as a substrate, diamonium phosphate (as a source of nitrogen for the yeast) is mixed up with the substrate. The pH is maintained at 5.0 and fermentation is carried out at 30-35°C. After fermentation yeast cells are separated by centrifugation or sedimentation and used again.

(3) Distillation

Ethanol from the fermented mixture is recovered by distillation. The first distillation yields about 80% pure ethanol, but by repeated distillation ethanol concentration of 99.4% or more can be achieved. Final desiccation can be achieved by chemical desiccants.

Many countries do not view energy production in strict economic terms. They see dependence on other countries for energy supplies and the chronic balance of payment deficits incurred by buying expensive fuel from abroad as major threats to their social and economic development and, indeed, to their political stability. So some countries promote domestic energy production even though it costs more than buying fuel on international markets, as long as bulk of money invested is in their currencies and the projects are expected to produce a net balance of payments surplus. This climate favours the development of biomass energy programmes, as the Brazilian alcohol production programme shows. In strict micro-economic terms no ethanol should be used as fuel in Brazil today, because it is much more expensive than gasoline. Gasoline costs about US $ 45 a barrel when it leaves the refinery, whereas waterfree ethanol (suitable for mixing with gasoline) is being produced at about US $ 68 (Vergara et al, 1981). Nevertheless, Brazil supports the use of locally produced ethanol fuel for wider national motives, particularly reducing energy dependence. The expanding industry has also provided jobs and helps regional development in many parts of the country. So the government subsidizes ethanol by taxing gasoline so that at the pump the home-brewed fuel costs about 65 per cent of the imported one, even though it is so much more expensive to produce.

Not every country in the world can resort to producing ethanol by fermentation for fuel. Only those that produce enough food, but too little energy, for their needs, may embark on a programme for making alcohol fuel from biomass, and then only if their climate, soil, water supplies and other environmental conditions are favourable for the programmes.

Gasohol Experiment

The use of alcohol as liquid fuel is fairly in operation in several countries. This renewable source of energy could partially substitute petroleum products, making a case for possible replacement of petroleum. The development of simple and cheaper technology for production of alcohol from various biomass, and its transfer to rural areas in small scale industry will help solve fuel crises. This will provide opportunities for employment and rural upliftment.

A proper balance between the production of energy and food crop, and selection of areas and the use of disposal farm waste will improve the production of food, fibre and oil besides, improving the environment. The development and use of alcohol has been appreciated by and large, but there are claims and counterclaims for its profitable adoption. During World War II, mixing of methanol and ethanol in different amount with petrol was carried out in Europe. Use of 5 per cent alcohol in petrol was made compulsory in Brazil in 1931. Thereafter, blending of alcohol with petrol to be used as transport fuel started after the oil crises in 1973. In 1985, Brazil has launched a programme of using gasohol mixture of 75 to 80 per cent inferior grade gasoline and 25 or 20 per cent ethanol, and thus saved 40 per cent of its petrol consumption. The comparative physical/ chemical properties of hydrocarbon fuels affirm the practical possibility of reducing the percentage of hydrocarbon in the mixture (Table 32).

The gasohol is a mixture of petrol and alcohol. It uses 99.5 per cent anhydrous ethanol with upto 20 per cent gasoline. It is used as a motor fuel. The Brazilians are using gasohol for automobiles, railways and other heavy energy needed industries. The use of gasohol improves the octane number and combustion efficiency. The gasoshol use does not require alteration of the engine. However, an additional heat exchanger is to be installed, to start automobiles, as the ignition point of alcohol is higher than that of gasoline (Lones, 1980, Mathur, 1987).

Table 32
Properties of liquid fuels (Mathur, 1987)

Property	*Ethanol*	*Methanol*	*Gasoline*	*Diesel*	*Fuel Oil etc.*
Formula	CH_3CH_2BH	CH_3OH	C_4 to C_{12} Hydrocarbon	C_{14} to C_{19} Hydrocarbon	C_{20} Hydrocarbon
Mol. Wt.	46.1	32.0	100-105	204	—
C	52.2	12.5	85.88	85-88	85-87
H	13.1	12.5	12-17	12-15	10-11
O_2	34.7	50	Negative	Negative	Negative
Cal. Value Kcal	6390	4760	10500	10280	9560
Sp. Gr.	.79	.79	.72-.78	.83-.88	88-.98
Boiling Temp.	78	65	27-225	240-360	360+
Flast Point° C	13	—	43	38	66
Octane No.					
Research	106-111	106-115	79-98	—	—
Motor	89-100	82-92	71-90	—	—
Cetane No.	0-5	NA	5-10	45-55	NA
Solubility in water	Infinite	Infinite	0	0	0

In 1980, United States commercialized the gasohol (20% alcohol added to petrol). Now at about 900 service stations, gasohol is commercially available. It has been estimated that replacing all the petrol consumed in United Stated by gasohol would require a production of atleast 50.6 billion litres of ethanol per year.

Computer-Controlled Fermentation

Fermentation classically consists of four components—the fermentor hardware or vessels, the sensors that monitor the system, the medium, and the organisms. A fifth component of modern fermentation is the Computer-control system.

Most classical fermentation engineers approach fermentation from the standpoint of manipulating the major bulk phases so that the "mass transfer" of nutrients and oxygen into the cells is maximized (Figure 26). The second concept of fermentation uses the principle of material balances to indirectly measure the productivity, on the other hand, approach the model with the paradigm of molecular level controls—single cell kinetics—microbial population dynamics-reactor productivity. This model is extremely important for fermentation of cells containing recombinant plasmids because the stability of the extrachromosomal recombinant DNA is required for product formation, and therefore productivity (Bailey et al, 1983).

An important aspect of developing a general model such as shown above is that it demonstrates the complexity of the fermentation process and the degree to which that process is made up of several subprocesses. If any of the subprocesses is left out of the control picture, the productivity suffers in the end. Just as control of fermentation must consider the subprocesses, the control system is only as good as its components.

A modern fermenter is designed to provide optimum environment in which micro-organisms or enzymes can interact with a substrate to form ethanol (Figure 27). The microbes are grown on nutrients placed in the fermentation vessel. The vessel is cooled by a water jacket. Air is pumped into the bottom of the liquid, and acid or alkali added as necessary. A stirrer keeps the contents well mixed. Steam lines are provided so that the vessel

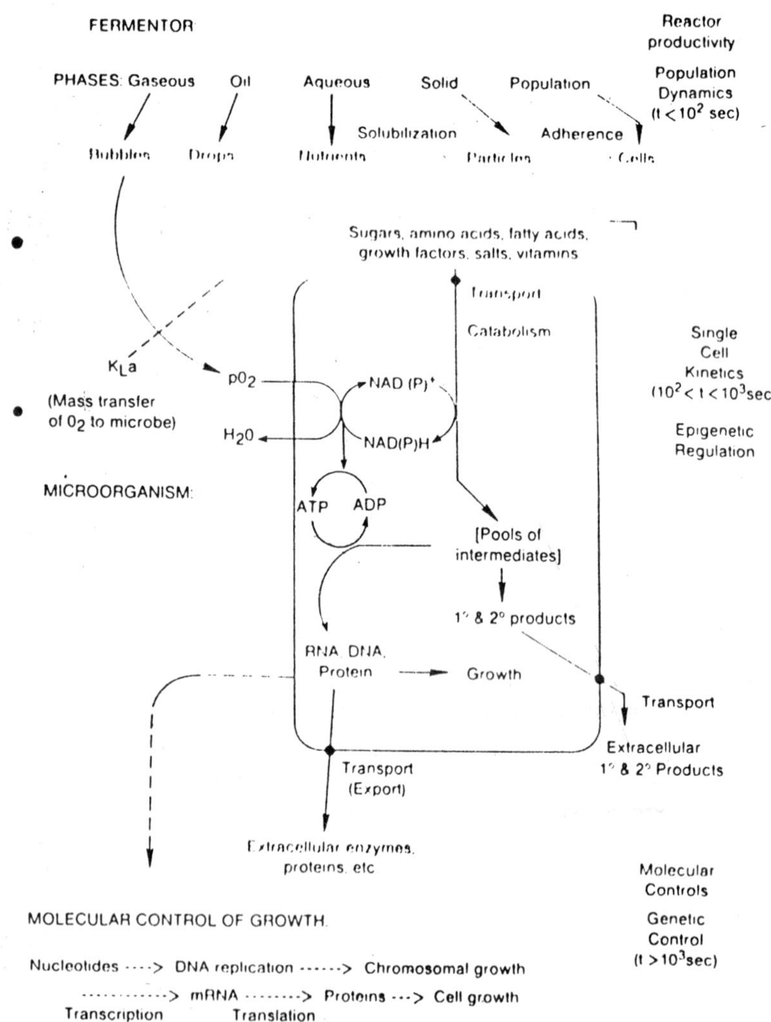

Figure 26
Fermentation process involving mass-transfer and molecular control of growth

can be sterilized after each fermentation batch. During fermentation it is necessary to regulate many factors such as oxygen, carbon dioxide, pH, temperature and media concentration. This is done with the help of sensors for monitoring and regulating these factors.

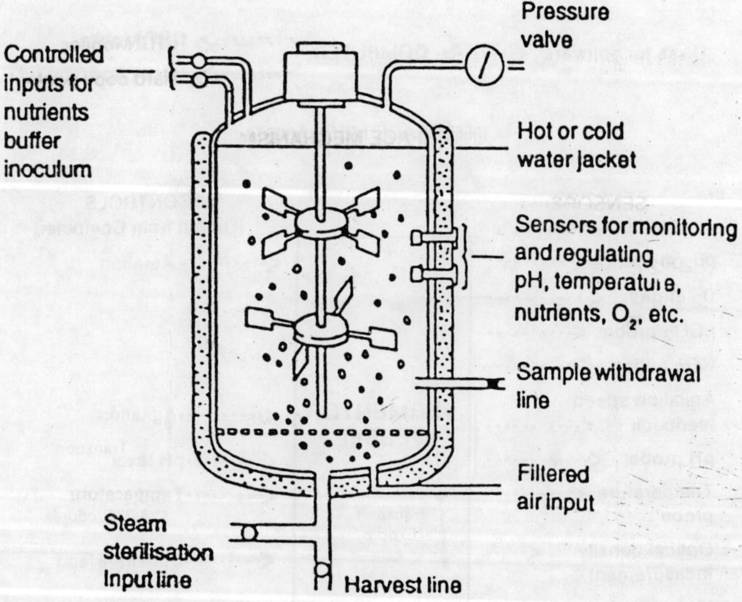

Figure 27
A continuously stirred reaction vessel (bioreactor)

The ability to control fermentation process is completely dependent on the degree to which various parameters of the fermentation can be measured. A proper fermentor should contain the input/output sensors. These sensors provide feedback to the computer for decision on control. The fermentor should be controllable using nutrient flow rates, agitation, aeration, pH, temperature and other additions (Figure 28).

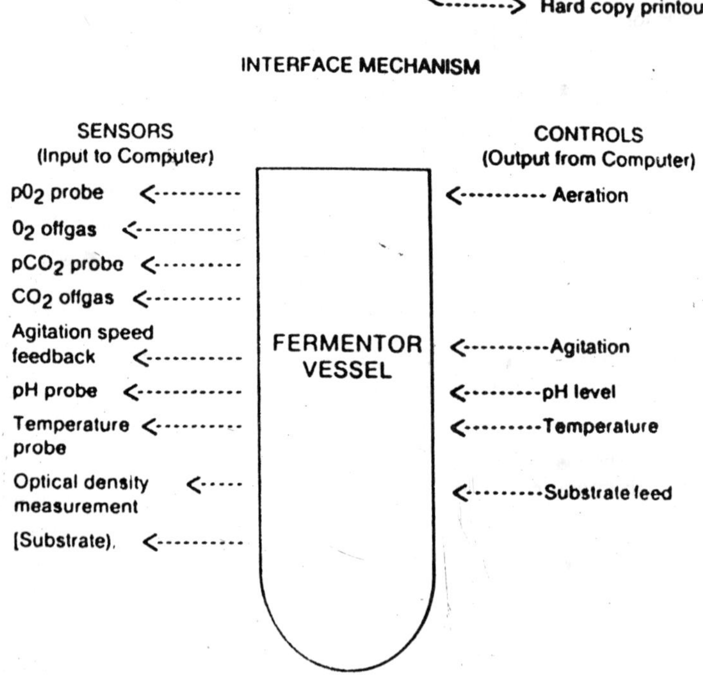

Figure 28
Computer-controlled fermentation

The computer language must have the capacity to do all functions within short-time and without completely dominating computer memory space. Most of the programming for computer control of fermentation has been done using FORTRAN or BASIC languages. In the future FORTH and PASCAL are finding wider usage in computer control, as they are more efficient computer languages for control purposes than either BASIC or FORTRAN.

Minicomputers have been used for standard computer control of fermentations in most industrial and biotechnology research institutions. These systems have the power, the speed, the versatility, and the disk-operating systems that are optimal for controlling various processes. Minicomputers are relatively inexpensive for industry, however, their return is potentially enormous in terms of productivity, energy savings, efficiency of operation and optimal control.

Microcomputer monitoring and control have been applied to several types of continuous fermentation processes. Mori et al (1983b) employed the ethanol-utilizing yeasts *(Candida brassicae)* and sensors that measured pO_2 and ethanol concentration in the vessel. Titus et al (1984) utilized a microcomputer with A/D-D/A system to control turbidostat—and chemostat-type continuous cultures of *Vibrio natriagens.* Forberg and Collaegues (1983) showed that immobilized non-growing cells of *Clostridium acetobutylicum* fed with pulses of limiting concentration of glucose and other nutrients, produced acetone and butanol with very high specific productivity.

Computer control may have great effect on the production of recombinent DNA products by bacteria containing controllable plasmid vectors. Cloning vectors have been derived at restrictive temperatures have low copy numbers. The growth of biomass can take place under those plasmid restrictive conditions. When the temperature of the culture system is raised to a permissive temperature, the plasmids start "run-away" replication, with concomitant production of enzymes or proteins that are encoded by the cloned DNA in the vectors. For computer controlled fermentations, these run-away plasmid cultures could be controlled at restrictive temperatures until certain culture densities were reached, at which point the computer would automatically raise the temperature to permissive run-away plasmid replication conditions for the production phase.

Environmental Impacts of Ethanol Production

Ethanol production can be very polluting. Every cubic meter of the fuel produced from sugarcane generates 12-13 cubic meters of effluents. If this is discharged without treatment into inland coastal waters, it can cause as much pollution, in terms of biological oxygen demand, as the sewage produced by 6000-6500 people in one day (Trindade, 1980). The potential impact of this pollution is so great that strict measures to control the discharge of this effluent, called sillage, must be used from the start. Delay in introducing them causes serious environmental damage. Discharging sillage is, any way, causing waste in more senses than one, for it can be a valuable raw material. It is normally not contaminated by disease carrying organisms or toxic compounds, so recovery of mineral and organic substances from it is a potentially attractive undertaking. It is technically feasible to turn it into methane gas to provide energy for the ethanol-producing process or other means, or to convert it into marketable products like fertilizers or feed additives (Costa Rebeiro and Branco, 1980; Branco and Costa Rebeiro, 1979). In Brazil, raw sillage is sprayed back into the sugarcane fields, reducing or eliminating the need for mineral fertilizers (Young and Trindade, 1978). But in some areas the soil is not suited to it.

Plant Based Petroleum Industry

There are some species of certain families which accumulate the photosynthetic products (hydrocarbons) of high molecular weight (10,000). They are commonly known as petroplants or petroleum plants. Calvin (1979) for the first time, collected and used photosynthetically produced hydrocarbons from plants. Furthermore, he suggested it as a substitute for conventional petroleum source.

Efforts are afoot to identify potential species (Table 33) as sources of liquid hydrocarbons which may offer substitutes for liquid fuels. These plants contain hydrocarbons which can potentially be converted into petroleum hydrocarbons.

Table 33
Plants producing hydrocarbons

Plant group/families	*Common names*	*Botanical names*
Algae (Chlorophyta)	—	*Botryococcus sp.*
		Chlorella pyrenoidosa
Euphorbiaceae	Hevea rubber	*Hevea brasiliensis*
	Rubber plant	*Euphorbbia abyssinica*
	" "	*E. resinifera, E. lathyris,*
	Sehund	*E. tirucalli*
Compositae	Guayule	*Parthenium argentatum*
	Russian dandelion	*Taraxacum koksaghyz*
Asclepiadaceae	Aak	*Calotropis procera*
Leguminoseae	-	*Copaifera langsdorfii,*
		C. mutijuga
	Samprani	*Hardwickia pinnata*
Dipterocarpaceae	Gurjun	*Dipterocarpus turbinatus*
Myristicaceae	-	*Dialynthera otoba*
Pittosporaceae	-	*Pittosporum resiniferum*

Interest in petro-crops was intensified during the seventies because of the sudden rise in the cost of petroleum products after the 1973 oil embargo, although the use of *Euphorbia* as a source of gasoline was known even in 1930's. It has been mentioned that at that time two efforts to cultivate hydrocarbon producing plants were initiated in Africa—one in Ethiopia and the other in Morocco. The Italions attempted to grow *Euphorbia abyssinica* in 1935-36 to use its latex for fuel production.

Important petro-crops could be grouped in the following four categories: (1) Seed oil plants, (2) Essential oil plants, (3) Laticiferous plants, and (4) Resinous plants. Plants in the last two categories viz., laticiferous and resinous can be further subdivided into: (a) Laticiferous plants suitable for latex tapping, (b) laticiferous plants suitable for solvent extraction, (c) Resinous plants suitable for resin tapping, and (d) Resinous plants suitable for solvent extraction.

Quite a large number of plants belonging to different categories as described above have been examined for their potential as petro-crops in different countries of the world.

The seed oil bearing plants that have received attention are *Pithosporum resiniferum, Cocos nucifera, Ricinus communis* and *Jatropha curcas.* There are a number of non-edible seed oil bearing plants such as *Azadirachta indica, Caesalpinia* spp, *Calotropis inophyllum, Jatropha curcas, Pongamia pinnata, Sapium sebiferum, Schleichera* spp which need examination as petro-crops.

Among essential oil bearing plants, species of *Eucalyptus* have received greater attention.

Laticiferous plants yielding latex by tapping are mostly the dendroid *Euphorbia* spp. Species available in abundance for latex tapping are *Euphorbia antiquorum, E. caducifolia, E. nerifolia, E. nivulia, E. royleana* and *E. trigona.*

There are a number of laticiferous plants that yield hydrocarbons by solvent extraction. Some important plants in this group are *Asclepias* spp, *Calotropis gigantea, C. procera, Cryptostegia grandiflora, Euphorbia lathyris, E. tirucalli* and *Pedileanthus* spp which are found abundantly in India.

Among the resinous plants, the ones yielding resins by tapping and considered important ones are *Cappaifera, Pinus* and *Sindrora* spp, while in India *Canarium* spp, *Dipterocarpus* spp, *Pinus* spp, *Shorea* spp and *Vateria indica* are important ones emenable to tapping.

Resin collection by solvent extraction method have been found suitable for *Chryosthamnus* and *Grindelia* spp. A large number of species of Compositeae may also yield resin by solvent extraction method.

India has a rich flora comprising of both wild and cultivated plants. Large number of plants belonging to different categories of petro-crops are available in the country. In order to search alternative liquid hydrocarbon material of about the same aggregate chemistry as the current petroleum products, a collaborative project entitled "Introduction, Screening and Cultivation of potential petro-crops and their conversion to petroleum hydrocarbons" was initiated jointly at National Botanical Research Institute, Lucknow and Indian Institute of Petroleum, Dehradun. During the first phase, a list of 386 species belonging to six laticiferous families viz., Euphorbiaceae,

Asclepiadaceae, Urticaeae, Apocynaceae, Convolvulaceae and Sapotaceae was prepared. Wide distribution, sufficient latex content and possibility of easy cultivation were the various criteria for selecting the species. The second phase of this project related to techno-economics of cultivation of promising species, building up of germ-plasm bank of latex bearing plants, development of agro-techniques for maximum biomass production of perspective plant species; standardization of techniques for hydrogenation/hydro-cracking of the hydrocarbon type of material together with determination of composition of the products as well as their evaluation (Khoshoo, 1982; Vyas, 1991).

Research and development has shown that Guayule *(Parthenium),* which is native to deserts in Texas and North Mexico can be grown in India but the rubber content is less. Jajoba plant *(Simmondsia),* a liquid was containing plant, is an alternative for petroleum hydrocarbon. This plant has just been introduced in India. Work on its introduction, evaluation, cultivation and propagation is in progress (Das, 1988).

In Italy, Euphorbia Gasoline Refinery was set-up to tap vegetative gasoline. *Euphorbia lathyris* is an annual herb and *E. tirucalli* is a perennial one. *E. lathyris* can produce 20 barrels of oil/acre/year. Similarly, *E. tirucalli* can give upto 5-10 barrels of oil/acre/year. Chemical analysis of this plant in organic solvents revealed that heptan extract and either soluble fraction constituted about 8 per cent terpenoid extract. By using catalyst, it could be converted into high grade transportation fuel. O the 85 per cent converted materials, about 10 per cent is in the form of natural gas and 75 per cent in gasoline—like fraction (Nemathy et al, 1980). Calvin and coworkers estimated that 10 tonnes of biomass could yield 5.3 barrels of crude extract convertable to gasoline.

In hot and dry parts of India *Calotropis procera* grows as a common shrub. It secretes latex which contains high concentration of extractable hydrocarbons. The ratio of C, H, O in the hexane extract has been found as 78.22%, 11.22% and 10.71% respectively. The ratio of C and H is similar to crude oil, fuel oil and gasoline. Hydrocarbon yield and energy value of *C. procera* are comparable to those of *E. lathyris*. Therefore, this plant can be used as a substitute of petroleum.

The dead algal scum of *Botryococcus braunii* (a unicellular green algae of Chlorococcales), contains about 70 per cent hydrocarbons. The algal hydrocarbons closely resemble the crude oil, and therefore, can be used as a good source of direct production of hydrocarbons.

B. braunii grows in fresh or brackish water as well as in tropical and temperate zones. The algae appears in two forms. The first form is of green colour and contains linear hydrocarbons with an odd number of carbon atom (25-31) low in double bonds. The second form is red in colour which contains hydrocarbons with 34-38 carbon atoms and several double bonds, the Botryococcenes. The significance of these two forms are not known (Sasson, 1984).

Hydrocarbon is accumulated as globules on outer walls and cytoplasm of the cells. On cell war, a major portion of hydrocarbon (95%) is located, where as, a small amount (0.7%) of it is within the cells. Hydrocarbons are recovered from the cell by centrifugation. The cells are again added in the fresh culture medium as inoculent. For the production of hydrocarbons in high amount, it is necessary to increase the algal biomass.

Chlorella pyrenoidosa, a fresh water unicellular algae, also is a source of hydrocarbons. Hydrogenation is done in a steel reactor at high temperature (400°C) and pressure (12000 p.s.i.—pound per square inch) in the presence of a catalyst—cobalt molybdate. The algae is suspended in a mineral oil in the reactor. Hydrogenation is carried out for about one hour. Consequently, 50 per cent of the algal biomass is converted into oil. Oil is a clear golden liquid which is separated from the reactor, blended with light gas oil in refineries and processed before its use.

Chapter 5

Transgenic Plants
Progress and Potence

The traditional plant breeding moved to noval breeding, the molecular breeding which uses recombinant DNA technology, gene cloning, gene transformation, protoplast fusion and in-vitro regeration. Use of the genetic vectors allows molecular breeders to remove pieces of DNA from an organism, study its function and insert the gene into the currently grown elite cultivars and thereby genetically rectify the defects as well as improve the genotype. Another major advantage of the molecular breeding is that when a particular gene has been isolated and reconstructed, it can be first tested in model plants and later it can be used in a variety of cultivars of different crops. The molecular breeding which uses the methods and concepts of biotechnology has been able to improve agronomic traits and has produced plants with increased vigour and yield, high degree of tolerance or resistance to pests, diseases or climatic stress. Such production of genetically manipulated plants using one or more foreign genes—the transgenics, is being profusely used and holds promise for the future. The transgenic plants are also used as an analytical tool to explore unique aspects of gene regulation and serve important focus for unifying the basic plant science research in plant breeding, pathology, biochemistry, physiology and molecular biology, as the production of transgenics needs the expertise of all these areas of life sciences. Due to its multi area based foundation, a large number of transgenics have been produced and many have been released after stringent environmental and field tests. The foregoing description reveals that the transgenics have greatly complemented plant breeding programs to meet the increasing demands of food production needed for the ever growing human population, especially in the developing countries.

Agricultural biotechnology promises to improve crop productivity, complimenting traditional breeding by decreasing our dependence on harmful chemicals, such as pesticides, fertilizers and antibiotics. Contemporary social and environmental trends emphasize improved safety and quality of agricultural products. Genetic engineering is a tool breeders are using to supplement conventional practices in order to improve crop performance. It is important for the breeder to identify goals and objectives for the genetic modification of the crops, which are both scientifically feasible and economically viable. In the past, agriculture has been an energy and labour intensive industry. Biotechnology offers the opportunity to reduce both these costs. In order to achieve this objective it is important to identify the genetic needs from crop improvement as defined by plant breeders. Factors including basic physiology and genetics of pest resistance, the large number of years and locations needed to evaluate and identify stress tolerance, and the long time (in generations) needed to breakup undesirable genetic linkages or to assemble desirable traits needed to be examined very carefully (Cullis, 1987). Even in cases involving successful separation of the introduced gene from linked deleterious genes, the gene's inheritance and expression may be unpredictably altered in the new genetic background. In order to apply genetic engineering successfully to crop improvement one should be able to identify and isolate agronomically useful genes, modify them according to strict specifications and transfer them between species, and ultimately recover them in mature plants that can be used in a breeding program. Success in plant genetic engineering will rely, to a great extent, on a thorough understanding of the molecular, genetic and metabolic characteristics and properties of traits to be transferred (Barton and Brill, 1983).

With traditional breeding methods, the available gene pool is restricted by the sexual incompatibility of many interspecific and intergeneric crosses (Nisbet and Web, 1990). Genetic manipulation and *in-vitro* culture provide a means for the genetic pool to be substantially broadened by allowing transfer of specific genes controlling well-defined traits from one organism to another, thus improving crops in a less haphazard way. By avoiding backcrossing programs, which may take years in some cases, considerable time and financial resources can be conserved (Christou, 1994).

Foreign genes have been successfully transferred in plant cells, tissues or organs. There are more than 50 plant species, where transgenic plants have been successfully produced (Table 34).

Table 34
Transgenic higher plants

I. Herbaceous dicotyledons

1. *Nicotiana tobacum* (tobacco)
2. *N. plumbaginifolia* (wild tobacco)
3. *Petunia hybrida* (petunia)
4. *Lycopersicon esculentum* (tomato)
5. *Solanum tubersum* (potato)
6. *Solanum melongena* (eggplant)
7. *Arabidopsis* thaliana
8. *Lactuca sativa* (lettuce)
9. *Apium graveolens* (celery)
10. *Helianthus annuus* (sunflower)
11. *Linum usitatissimum* (flax)
12. *Brassica napus* (oilseed rape; canola)
13. *Brassica oleracea* (cauliflower)
14. *Brassica oleracea var. capitata* (cabbage)
15. *Brassica rapa* (syn. B. campestris)
16. *Gossypium hirsutum* (cotton)
17. *Beta vulgaris* (sugarbeet)
18. *Glycine max* (soybean)
19. *Pisum sativum* (pea)
20. *Medicago sativa* (alfalfa)
21. *M. varia*
22. *Lotus corniculatum* (lotus)

23. *Vigna aconitifolia*
24. *Cucumis sativus* (cucumber)
25. *Cucumis melo* (muskmelon)
26. *Cichorium intybus* (chicory)
27. *Daucus carota* (carrot)
28. *Annoracia sp.* (horse radish)
29. *Glycorrhiza glabra* (licorice)
30. *Digitalis purpurea* (foxglove)
31. *Ipomoea batatas* (sweet potato)
32. *Ipomoea purpurea* (morning glory)
33. *Fragaria sp.* (strawberry)
34. *Actinidia sp.* (Kiwi)
35. *Carica papaya* (papaya)
36. *Vitis vinifera* (grape)
37. *Vaccinium macrocarpon* (cranberry)
38. *Dianthus caryophyllus* (carnation)
39. *Chrysanthemum sp.* (chrysanthemum)
40. *Rosa sp.* (rose)

Woody dicotyledons

41. *Populus sp.* (poplar)
42. *Malus sylvestris* (apple)
43. *Pyrus communis* (pear)
44. *Azadirachta indica* (neem)
45. *Juglan regia* (walnut)

III. Monocotyledons

46. *Asparagus sp.* (asparagus)
47. *Dactylis glomerata* (orchard grass)
48. *Secale cereale* (rye)
49. *Oryza sativa* (rice)
50. *Triticum aestivum* (wheat)
51. *Zea mays* (corn)
52. *Avena sativa* (oats)
53. *Festuca arundinacea* (tall fescue)

IV. Gymmosperms (a conifer)

54. *Picia glauca* (white spruce)

Initially, the production of transgenic plants was restricted to dicotyledons, but it has now been extended to several monocotyledons like wheat, maize, rice and oats. Progress in this exciting area has been so spectacular that by the turn of the century, we hope to grow crops which have been tailored to market specifications by the addition, substraction or modification of genes. Transgenes will also be important in increasing the efficiency of crop production. Transgenic plants resistant to herbicides insect pests, viruses and other stresses have already been produced. Transgenic tomato plant with bruise resistance and delayed ripening have been produced. Transgenic plants can also be used as *bioreactors* for the production of chemicals and pharmaceuticals. Transgenic plants have also been produced for identification of regulatory sequences for many genes, using gene constructs with overlapping deletions.

Methods for the Production of Transgenic Plants

Plant cells or protoplasts can be cultured and used for the regeneration of whole plant. Hence, plant cells or protoplasts can be used for the production of transgenic plants. Besides cells and protoplasts, other meristematic cells (immature embryos, or organs), pollen or zygote can also be used for the production of transgenic plants. The enormous diversity of plant species and the availability of diverse genotypes in a species, made it necessary to develop a variety of techniques, suiting different situations.

The first step in the production of transgenic plant is to select cells that are capable of giving rise to whole transgenic plant. Some of these cell type and the methods used for regeneration of whole plants from them have been listed in Table 35.

Table 35

Theoretical routes for the production of transgenic pants (Gupta, 1995)

Target cell type	*Method used for regeneration*
1. Cultured cells or protoplasts	Organogenesis or embryogenesis via callus phase
2. Meristem cells from immature Embryo or organ	In vitro plant regeneration from transformed cells

Contd.

Target cell type	Method used for regeneration
3. Cells in immature embryos, shoot and flower meristems	Normal development *(in vivo)* of embryo, shoot or flower followed by use of pollen (transformed) from chimaeric plant to produce transformed seed
4. Pollen	Pollen (treated with DNA) used for pollination leading to the production of transgenic plants
5. Zygote	Direct development *(in vivo)* of transgenic plants

Cultured cells or protoplasts are often used for gene transfer. This in unlike the situation in animals, because the plant cells are totipotent and can be stimulated to regenerate into whole plants *in-vitro* via organogenesis or embryogenesis. However, *in-vitro* plant regeneration imposes a degree of 'genome stress', especially if plants are regenerated via a callus phase. This may lead to chromosomal or genetic abnormalities in regenerated plants—a phenomenon referred to as somaclonal variation. In contrast to this, gene transfer into pollen may give rise to genetically transformed gametes, which if used for fertilization *(in-vivo)* may give rise to transformed whole plant. Similarly, insertion of DNA into zygote *(in-vivo* or *in-vitro)* followed by embryo rescue, may also be used to produce transgenic plants. Another alternative approach is the use of individual cells in embryos or meristems, which may be grown *in-vitro* or may be allowed to develop normally for the production of transgenic plants.

Agrobacterium Mediated Gene Transfer

The common features of vectors used for transformation of eukaryotic cells have been shown in Figure 29. Most vectors carry *marker genes,* which allows recognition of transformed cells (other cells die due to the action of an antibiotic or herbicide) and are described as *selectable markers.* A number of selectable markers are now known (Table 36). Among these marker genes, the most common selectable marker in *npt II,* providing kanamycin resistance.

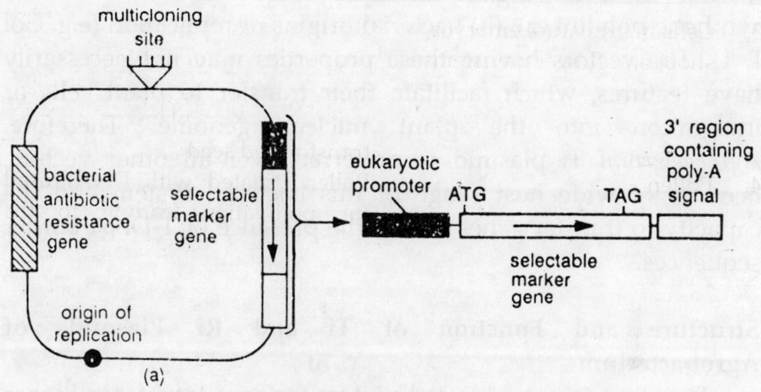

Figure 29

(a) Generalized vector for the transformation of eukarytic cells; (b) details of the selectable marker genes (redrawn from Grierson, 1991- *(Plant genetic engineering)*

Table 36

Selectable marker genes in vectors used for gene transfer

Selectable marker gene	*Substrate used for selection*
	I. Antibiotics
1. Neomycin phosphotransferase *(npt II)*	G418, kanamycin, neomycin
2. Hygromycin phosphotransferase *(hpt)*	Hygromycin B
3. Dihydrofoate reductase *(dhfr)*	Methotrexate trimethoprim
4. Gene *ble* (enzyme not known)	Bleomycin
5. Gentamycin acetyltransferase	Gentamycin
6. Streptomycin phosphotransferase	Streptomycin
	II. Herbicides
7. Mutant form of acetolactate synthase *(als)*	Chlorosulfuron immidazolinones
8. Bromoxynil nitrilase	Bromoxynil
9. Phosphinothricin acetyltransferase *(bar)*	L-phosphinothricin (PPT; known as bialaphos)
10. 5-enolpyruvylshikimate-3 -phosphate (EPSP) synthase *(aro A)*	Glyphosate (Roundup)

Other common features of suitable transformation vector include the following: (i) multiple unique restriction sites (a synthetic polylinker); (ii) bacterial origins of replication (e.g. Col E 1). The vectors having these properties may not necessarily have features, which facilitate their transfer to plant cells or integration into the plant nuclear genome. Therefore, *Agrobacterium* Ti plasmid is preferred over all other vectors, because of wide host range of this bacterial system and the capacity to transfer genes due to the presence of T-DNA border sequences.

Structure and Function of Ti and Ri Plasmids of Agrobacterium

The most commonly used vectors for gene transfer in higher plants are based on tumour inducing mechanism of the soil bacterium *Agrobacterium tumefaciens,* which is the causal organism for *crown gall disease.* A closely related species *A. rhizogenes* causes *hairy root disease.* An understanding of the molecular basis of these diseases led to the utilization of these bacteria for developing gene transfer systems. It has been shown that the disease is caused due to the transfer of a DNA segment from the bacterium to the plant nuclear genome (Figure 30). The DNA segment, which is transferred is called T-DNA and is part of a large Ti (tumour inducing) plasmid found in virulent strains of *A. tumefaciens.* Similarly, Ri (root inducing) megaplasmids are found in the virulent strains of *A. rhizogenes.*

Ti Plasmids

A. tumefaciens lives in soil and attacks many dicotyledonous plants, most probably at the level of soil surface. It enters the fresh wound, attaches with the intact cells and ultimately transfers its small part of plasmid (Ti) into the plant cells. Due to possession of tumour inducing plasmid, the bacterial plasmid is known as Ti plasmid. After infection the T-DNA (200 Kb) integrates with the plant DNA. The transferred fragment carries several genes which get expressed within the cell and affects the normal metabolism of plants. T-DNA encodes two enzymes involved in biosynthesis of auxin and one enzyme involved in biosynthesis of cytokinin (Ream, 1989).

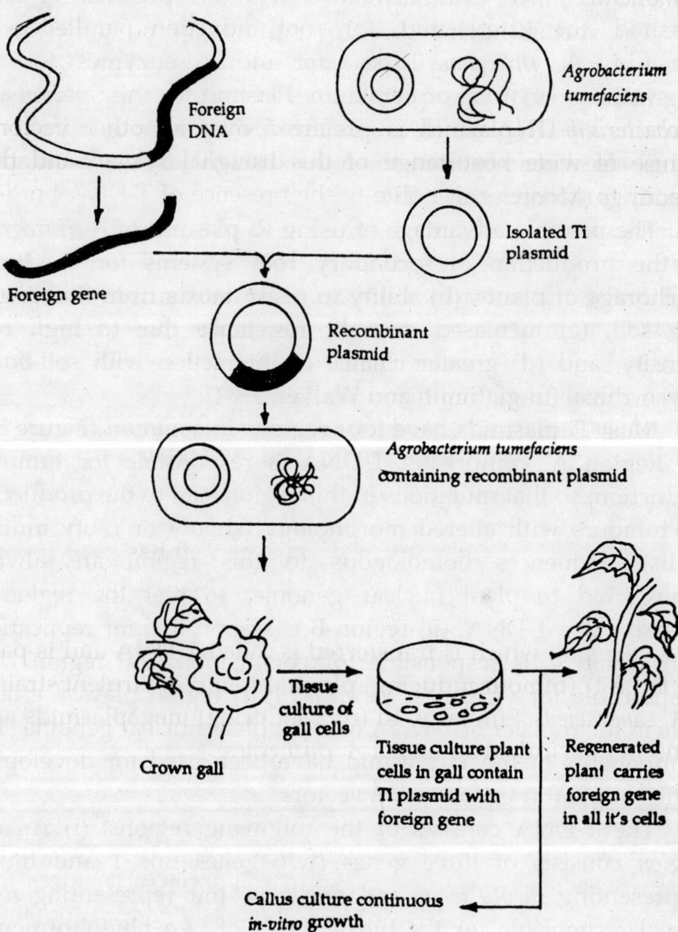

Figure 30
Plant genetic engineering with Ti plasmids of *Agrobacterium tumefaciens*

Ri Plasmid

A. rhizogenes causes formation of adventitious roots. This rhizogenicity has been correlated with the presence of large plasmid, the Ri plasmid, for root induction parallel to Ti plasmid. *A. rhizogenes* does not induce enzymes for the biosynthesis of auxin or cytokinin. Plasmid for this species also contains a T-DNA which causes the development of hairy roots. It has also been reported to confer drought tolerance on apple seedlings (Moore et al, 1979).

The possible advantage of using Ri plasmid of *A. rhizogenes* is the production of secondary root systems for (a) better anchorage of plants, (b) ability to resist anoxia from flooding of the soil, (c) increased drought resistance due to high root density, and (d) greater chance of interaction with soil-borne mycorrhizal fungi (Smith and Walker, 1981).

Most Ti plasmids have four regions in common (Figure 31). (i) Region A, comprising T-DNA is responsible for tumours induction, so that mutations in this region lead to the production of tumours with altered morphology (shooty or rooty mutant galls). Sequences homologous to this region are always transferred to plant nuclear genome, so that the region is described as T-DNA, (ii) region B is responsible for replication, (iii) region C is responsible for conjugation, (iv) region D is responsible for virulence. *Virulence* (vir) region plays a crucial role in the transfer of T-DNA into the plant nuclear genome. The components of the Ti plasmid have been used for developing efficient plant transformation vectors.

The T-DNA consists of the following regions: (i) An *onc region* consists of three genes (two genes tms 1 and tms 2 representing *shooty locus* and one gene tmr representing *rooty locus)* responsible for the biosynthesis of two phytohormones, namely *indole-acetic acid* or IAA (an auxin) and *isopentyladenosine 5-monophosphate* (a cytokinin). These genes encode the enzymes responsible for the synthesis of these phytohormones, so that the incorporation of these genes in plant nuclear genome leads to the synthesis of these phytohormones in the host plant. The phytohormones in their turn alter the developmental programme, leading to the formation of crown gall. (ii) A *nos*

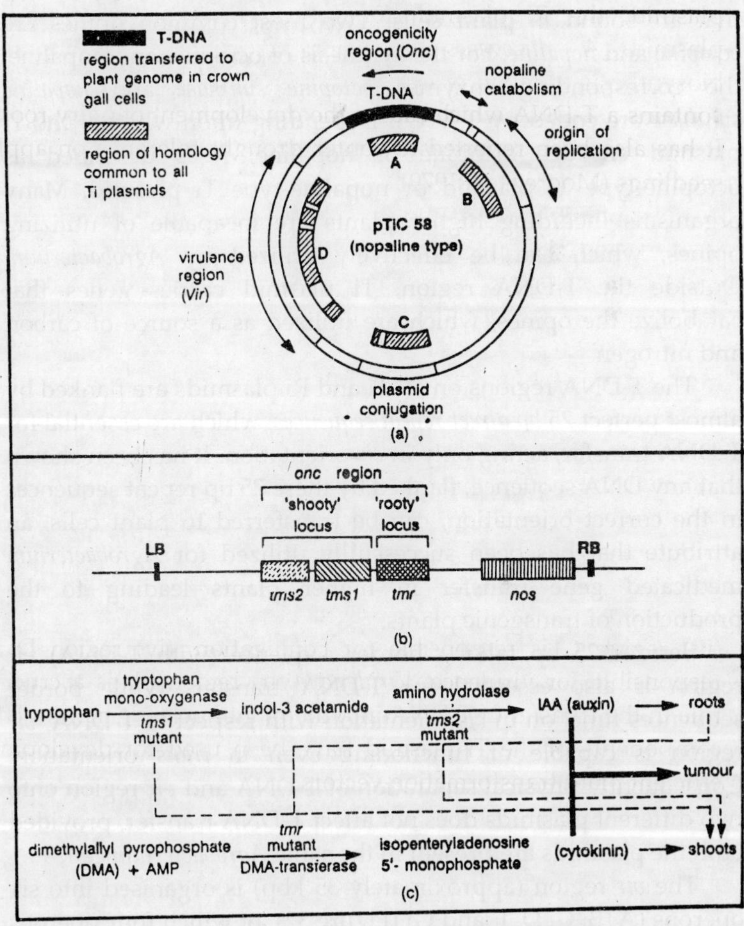

Figure 31

(a) General features of Ti plasmids; (b) general structure of the T-DNA; (c) phytohormone biosynthesis in crown gall tissues (Redrawn from Grierson, 1991)

region responsible for the synthesis of unusual amino acid or sugar derivatives, which are collectively called *opines.* Opines are derived from a variety of compounds (e.g. arginine + pyruvate), that are found in plant cells. Two most common opines are *octopine* and *nopaline.* For the synthesis of octopine and nopaline, the corresponding enzymes *octopine synthase* and *nopaline synthase* are coded by T-DNA. Depending upon whether the Ti plasmid encodes octopine or nopaline, it is described as octopine-type Ti plasmid or nopaline-type Ti plasmid. Many organisms including higher plants are incapable of utilizing opines, which can be effectively utilized by *Agrobacterium.* Outside the T-DNA region, Ti plasmid carries genes that catabolize the opines, which are utilized as a source of carbon and nitrogen.

The T-DNA regions on all Ti and Ri plasmids are flanked by almost perfect *25 bp direct repeat sequences,* which are essential for T-DNA transfer, acting only in *cis* orientation. It has been shown that any DNA sequence, flanked by these 25 bp repeat sequences in the correct orientation, can be transferred to plant cells, an attribute that has been successfully utilized for *Agrobacterium* medicated gene transfer in higher plants leading to the production of transgenic plants.

Besides 25 bp falking border sequence (with T-DNA) *vir* region is also essential for T-DNA transfer. While border sequences function in *cis* orientation with respect to T-DNA, *vir* region is capable of functioning even in *trans* orientation. Consequently physical separation of T-DNA and *vir* region onto two different plasmids does not affect T-DNA transfer, provided both the plasmids are present in the same *Agrobacterium* cell.

The *vir* region (approximately 35 kbp) is organised into six operons (A, B, C, D, E and G) (Figure 32), of which four operons (except A and G) are polycistronic. Genes *vir* A, B, D, and G are absolutely required for virulence; the remaining two genes *vir* C and E are required for tumour formation. The *vir* A locus is expressed constitutively under all conditions. The *vir* G locus is expressed at low levels in vegetative cells, but is rapidly induced to higher expression levels by exudates from wounded plant tissue. The *vir* A and G gene products regulate the expression of other *vir* loci. The *vir* A product is located on the inner membrane of *Agrobacterium* cells and is probably a chemoreceptor, which senses the presence of phenolic compounds (found in exudates of wounded plant tissue), such as *acetosyringone* and *p-hydroxyaceto-syringone.* Signal transduction proceeds via activation (possibly phosphorylation) of *vir* G (product of gene *vir* G), which in its turn induces expression of other *vir* genes (Figure 33).

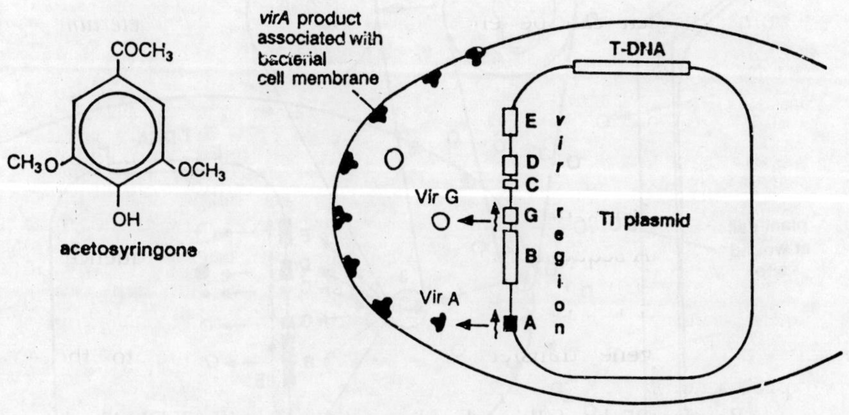

Figure 32
Ti plasmid gene expression in vegetative *Agrobacterium* prior to *vir* gene induction by the plant phenolic compound acetosyringone (Redrawn from Grierson, 1991)

T-DNA Transfer Process

An early event in the T-DNA transfer process is the nicking of Ti plasmid at two specific sites, each between the third and fourth base of the bottom strand of each 25 bp repeat. This initiates DNA synthesis from the nick in the righthand 25 bp repeat sequence in 5′-3′ direction, thus displacing a single T-DNA strand. This T-DNA single strand forms a complex with protein *vir* E and gets transported to the plant nucleus (Figure 34). The *vir* D operon encodes an endonuclease that produces the nicks in the border sequence. Several gene products of the *vir* B operon have been identified in the bacterial envelope, a location, which suggests that they may play a role in directing T-DNA transfer extracellularly. The functions of several other *vir* gene products are largely unknown.

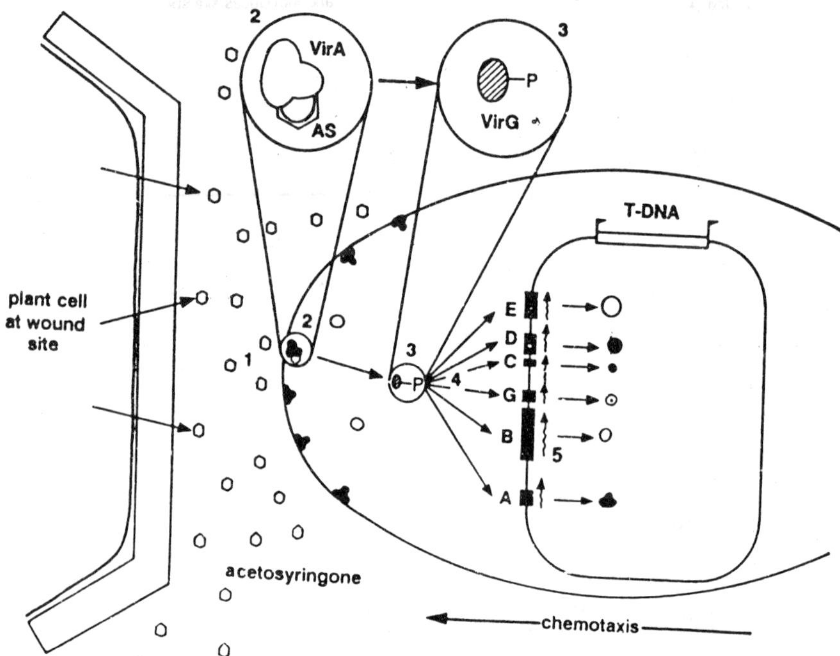

Figure 33
Key events in the induction of the Ti plasmid *vir* genes by acetoyringone and other wound-related compounds:
(1) acetosyringone released by plant cells at the wound site;
(2) acetosyringone activated Vir A protein located in the bacterial membrane; (3) activated Vir A modifies Vir G protein, possibly by phosphorylation; (4) modified Vir G binds to the operator;
(5) sequences of the *vir* genes resulting in their activation or up-regulation (Redrawn from Grierson, 1991)

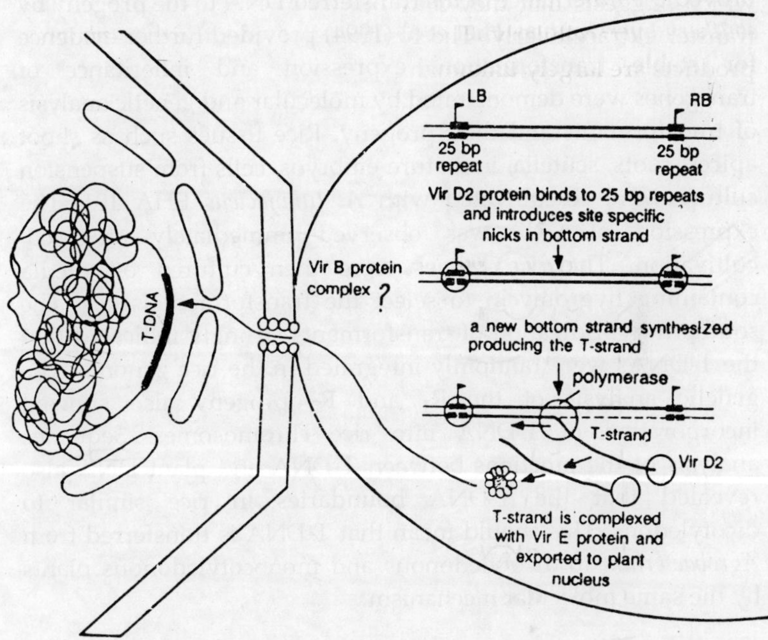

Figure 34
Molecular events in the transfer of T-DNA to plant cells
(Redrawn from Grierson, 1991)

Apart from the role of Ti plasmid, the gene located on *Agrobacterium* chromosome also help in virulence. These genes are involved in the synthesis and secretion of *glucons, cellulose, fibrils* and *cell surface proteins.* These loci are constitutively expressed and are also found in other soil bacteria associated with higher plants. Thus, these loci play a more general role in the virulence of *Agrobacterium*, and also in the *Agrobacterium* mediated gene transfer.

Transformation of cereal plants with *Agrobacterium* has been attempted in several laboratories (Raineri et al, 1990; Schlappi and Hohn, 1992). But none of these studies could provide an unequivocal evidence for stable transformation. A few transgenic plants of rice were obtained by Chan et al (1993) by inoculating immature embryos with *Agrobacterium.* They also proved the inheritance of the transferred DNA to the progeny by *southern hybridization.* Hiei et al (1994) provided further evidence for stable transformation, expression and inheritance of transgenes were demonstrated by molecular and genetic analysis of transformants and their progeny. Rice tissues such as shoot apices, roots, scutella, immature embryos, cells from suspension culture were co-cultivated with *A. tumifaciens* EHA 101. The expression of GUS was observed immediately after co-cultivation. The rice tissues were then cultured on media containing hygromycin to select the transformed tissues. The southern hybridization of transformants strongly indicated that the T-DNAs were randomly integrated in the rice genome. The genetic analysis of the R_1 and R_2 progeny also showed incorporation of T-DNA into rice chromosomes. Sequence analysis of the junctions between T-DNA and plant DNA also revealed that the T-DNA boundaries in rice similar to dicotyledons. This should mean that T-DNA is transferred from *Agrobacterium* to dicotyledonous and monocotyledonous plants by the same molecular mechanism.

Physical Delivery Methods or DNA Mediated Gene Transfer (DMGT)

Agrobacterium mediated gene transfer has been the most commonly used method of gene transfer, but the cereals are not amenable to this method. Further, in many crops including cereals and legumes, tissue culture techniques for regeneration are not very successful. These two limitations, gave rise to physical delivery of DNA or DNA mediated gene transfer. These include the following techniques:

Chemically Stimulated DNA Uptake by Protoplasts

Direct DNA uptake by protoplasts can be stimulated by chemicals like polyethylene glycol (PEG). The technique is so efficient that virtually every protoplast system has proven transformable. PEG is also used to stimulate the uptake of liposomes and to improve the efficiency of electroporation. PEG at high concentration (19-25%) will precipitate ionic macromolecules such as DNA and stimulate their uptake by endocytosis without any gross damage to protoplasts. This is followed by cell wall formation and initiation of cell division. These cells can now be plated at low density on selection medium.

Initial studies using the chemically stimulated DNA uptake by protoplasts were restricted to *Petunia* and *Nicotiana*. Other plants *Oryzae* and *Zea* were sucessfully used later. In these methods, PEG was used in combination with pure Ti plasmid, or calcium phosphate precipitated Ti plasmid mixed with a carrier DNA. Transformation frequencies upto 1 in 100 have been achieved by this method.

Protoplasts Mediated Transformation

Paszkowski et al (1984) developed transgenic fertile tobacco plants using direct DNA transfer into protoplats. Protoplasts were also used to transform the cells of *Lolium* (Potrykus et al, 1985). For the transformation, isolated protoplasts were suspended in a buffer solution containing plasmid DNA. The protoplast membrane is temporarily disrupted by electropolation or polyethyleneglycol treatment. This allows entry of DNA through the protoplast membrane. Protoplasts are then cultured in a medium containing a marker toxin to identify transformed lines expressing a marker gene conferred resistance. A number of selective agents and suitable resistance genes have been investigated for the efficient selection of transformed cells. The most widely used inhibitors are kanamycin, geneticin 418 and hygromycin (Table 37). These aminoglycoside antibiotics which interfere with the translational machinery of cells. These antibiotics can be inactivated by phosphorylaton reaction mediated by the neomycin phosphotransferase gene or from hygromycin B resistance gene (hpt) from *Escherichia coli* (Hauptmann et al, 1988; Dekeyser et al, 1989). Other selective agents used include bleomycin, methotrexate and phophinothricin. Phosphinothricin inhibits glutamine synthase which results in excessive accumulation of ammonia leading to cell death.

Table 37
Genes used in the production of transgenic cereals

Gene	*Gene product*	*Selection*	*Reference*
npt II (neo)	neomycin phospho transferase	Kanamycin, G 418, neomycin	Zhang *et al.* 1989 Davey et al. 1991 Lynch et al. 1992
hpt (aph IV)	hygromycin phospho transferase	hygromycin	Datta et al. 1990 Hayashimoto et al. 1990 Walters *et al.* 1992
bar	phosphino-thricin acetyl trans-ferase	Basta, I-phosphino-thricin Biala phos	Gordon - Kamme *et al.* 1990 Vasil *et al.* 1992 Nehra *et al.* 1994 Becker *et al.* 1994
gus A (uid A)	β-glucuronidase-		Rathore *et al.* 1993 Nehra *et al.* 1993 Nehra *et al.* 1994 Zhang & Wu 1988 Shimamoto 1989
EPSP synthase	5-enolpyruvyl shikimate synthase		Vasil *et al.* 1991
dhfr	dihydrofolate reduclase	methotrexale	Hauptmann, 1988

The bialophos resistance (bar) gene codes for phosphinothricin acetyl transferase, which converts phosphinothricin to acetylated form which is not toxic to the cell. The bar gene has been used in several more recent transformation experiments in crop plants (Weeks et al, 1993; Yoder and Goldsbrought, 1994). Cells of almost all the major cereals have been transformed using this method (Table 38). Transgenic plants of japonica and indica rice, maize and barley have also been produced. In the production of transgenic rice Zhang et al (1989) used the protoplasts of japonica rice variety Taipei 309 and electroporated the protoplasts with pCaMVNeo plasmid containing *npt* II gene. They reported that 400 kanomycin resistant protoplasts derived colonies produced 12 green plants and 2 albino plants. Out of the 6 plants selected randomly for analysis, all the plants showed 1.0 KbBamH1 fragment of pCaMVNeo carrying the npt II gene. Transformation of indica rice variety Chinsurah Boroll was reported by Datta et al (1990). They used microspore derived embryogenic suspension culture for protoplast isolation and used polyethylene glycol for gene delivery. The hygromycin phosphotransferase gene was detected in the regenerated plants by southern hybridization. The efficiency of stable transformation of cereal protoplasts may be high but the regeneration of transgenic plants from transgenic calli is rather low (Hayashimoto et al, 1990). Protoplasts have most successfully been used in the transformation of rice variety Taipei 309 and in some lines of maize (Omirulleh et al, 1993).

Table 38

Fertile transgenic plants produced using protoplasts in cereals

Cereal species	*References*
Oryza sativa	Toriyama *et al.* 1988
	Zhang & Wu 1988
	Zhang *et al.* 1989
	Shimamoto *et al.* 1989
	Datta *et al.* 1990a
	Davey *et al.* 1991
	Lynch *et al.* 1992
	Battraw & Hall, 1992
	Peng *et al.* 1995
	Rathore *et al.* 1993
Zea mays	Rhodes *et al.* 1988a
	Omirulleh *et al.* 1993
	Walter *et al.* 1992

Microinjection and Macroinjection

Plant regeneration from transformed protoplasts, still remains a problem. Therefore, cultured tissues, that encourage the continued development of immature structures, provide alternative cellular targets for transformation. These immature structures may include immature embryos, meristems, immature pollen, germinating pollen, isolated ovules, embryonic suspension cultured cells etc. The main disadvantage of this technique is the production of chimeric plants with only a part of the plant transformed. However, from these chimeric plants, transformed plants of single cell origin can be subsequently obtained.

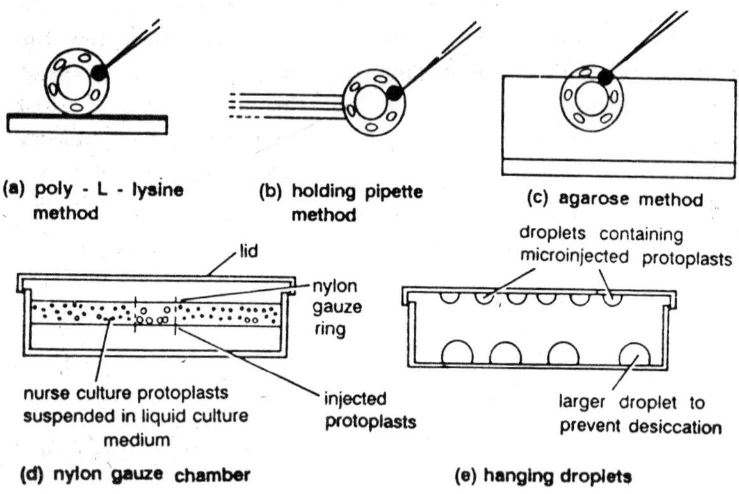

Figure 35
Methods for the microinjection and culture of plant protoplasts

When cells or protoplasts are used as targets, glass micropipettes with 0.5-10 μm diameter tip are used for transfer of macromolecules into the cytoplasm or the nucleus of a recipient cell or protoplasts. Recipient cells are immobilized on a solid support (coverslip or slide) or artificially bound to a substrate or held by a pipette (Figure 35) under suction. Often a specially designed micro-manipulator is employed for microinjecting the DNA. Although this technique gives high rate of success, the process is slow, expensive and requires highly skilled and experienced personnel.

DNA macroinjection employing needles with diameters greater than cell diameter has been tried in rye *(Secale cereale)*, where a marker gene was macroinjected into the stem below the immature floral meristem, so as to reach the sporogenous tissue leading to successful production of transgenic plants. Unfortunately, this technique could not be successfully repeated with any other cereal. Therefore, doubt has been expressed about the validity of earlier experiments (Potrykus, 1990).

There has been little success in stably transforming cereals by microinjection although the technique is relatively easy in achieving transient gene expression.

Microprojectiles (Biolistic or Particle Gun) for Gene Transfer

DNA delivery to plant cell is also possible when heavy microparticles (tungsten or gold) coated with DNA are accelerated to a very high initial velocity (1400 ft per sec). These microprojectiles, normally 1-3 μm in diameter, are carried by a macroprojectiles or the bullet, and are accelerated into living plant cells (target cells can be pollen, cultured cells, cells in differentiated tissues and meristems) so that they can penetrate cell walls of intact tissue. The acceleration is achieved either by an explosive charge *(cordite explosion)* or by using shock waves initiated by a high voltage electric discharge. The design of two particle guns used for acceleration of microprojectiles have been shown in Figure 36.

Transformed plants using this technique have been obtained in soybean, tobacco, maize, rice, wheat and onion. There is no other gene transfer approach, which has met with so much enthusiasm. The advantage of this technique include: (i) thousands of particles are accelerated at the same time, causing multiple hits resulting in transfer of genes into many cells simultaneously; (ii) since intact cells can be used, some of the difficulties encountered with the use of protoplasts are automatically eliminated; (iii) the method is universal in application, so that cell type, size and shape or the presence/ absence of cells walls do not significantly alter its effectiveness.

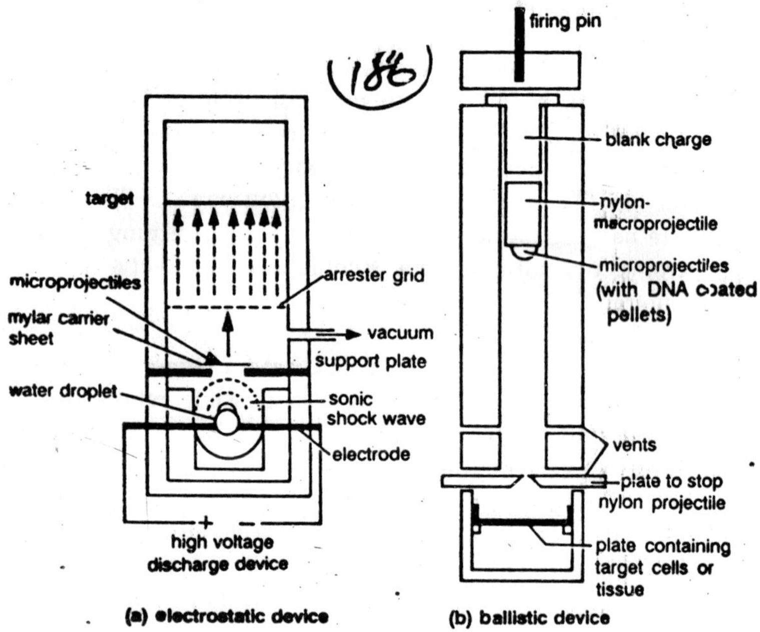

Figure 36
Microprojectile acceleration devices

Electroporation Technique for Gene Transfer

This technique is based on the use of short electrical impulses of high field strength. These impulses increase the permeability of protoplast membrane and facilitate entry of DNA molecules into the cells, if the DNA is in direct contact with the membrane.

Protoplasts in an ionic solution containing the vector DNA, are suspended between the electrodes. Electroporation pulse is generated by discharging a capacitor across the electrodes in a specially designed electroporation chamber (Figure 37). Either a high voltage (1.5 KV) ractangular wave-pulse of short duration or a low voltage (350 V) pulse of long duration is used (Figure 38).

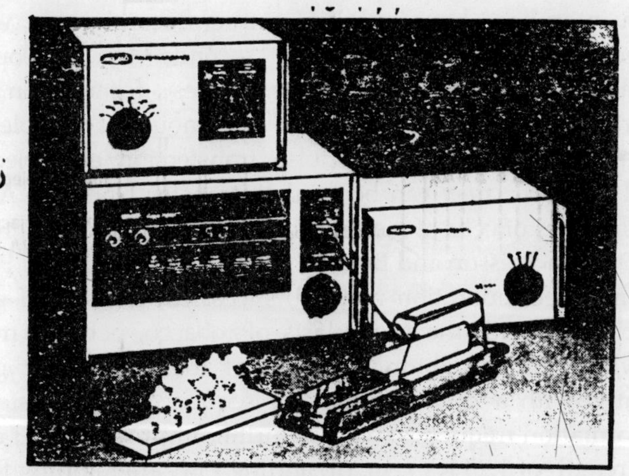

Figure 37
An electroporation system, suitable for transformation of bacteria, mammalian cells and plant cells

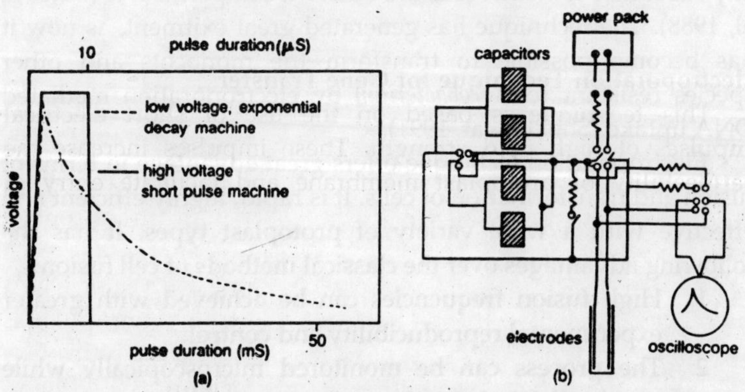

Figure 38
(a) Details of the output characteristics of the two principal types of electroporation devices in use, (b) circuit diagram of low voltage, exponential decay machine (Redrawn from Grierson, 1991)

The electrical pulse form transient pores in the plasma-membrane, without the induction of the fusion. The pores formed by a transient depolarization of the membrane lipid protein interphase have stability in microsecond range which can be increased upto several seconds by appropriate combination of Ca^{++}/phosphates. Pores formed in cell membranes can facilitate transfer of exogenous macromolecules into cells. The high voltage pulse transfection/electrojection, termed *electroportation* can be used to transfer DNA/RNA into different types of cells of mammals, plants and microbes (Fromm et al, 1986; Zachrisson and Borman, 1986; Riggs and Bates, 1986).

Using electroporation method, successful transfer of genes was achieved with the protoplasts of tobacco, petunia, maize, rice, wheat and sorghum. In most of these cases *cat* genes associated with a suitable promotor sequence was transferred. Transformation frequencies can be further improved by (i) using field strength of 1.25 KV/cm, (ii) adding PEG after adding DNA, (iii) heat shocking protoplasts at 45°C for 5 minutes before adding DNA and (iv) by using linear instead of circular DNA.

Electroporation uses high-voltage, electric field pulses to make cells permeable. This technique is effective with a wide variety of species and can be used to induce the uptake of almost any substance to which cells are otherwise impermeable (Bates et al, 1988). The technique has generated great exitment, as now it has become possible to transform the monocots and other species resistant to *Agrobacterium* by electroporation mediated DNA uptake (Shekhawat, 1991).

Electroporation technique offers great advantage in terms of fusion and transformation of cells. It is rapid, highly efficient and effective with a wide variety of protoplast types. It has the following advantages over the classical methods of cell fusion:

1. High fusion frequencies can be achieved with greater experimental reproducibility and control.
2. The process can be monitored microscopically while standardising the various parameters.

3. The fusion is synchronous and the number of protoplasts to be fused can be to some extent controlled.
4. Electrofusion does not require any severe chemical treatment of cells and it is not cytotoxic (Shekhawat, 1991).

Transformation of intact cells by electroporation has been reported by several workers (DeKeyser et al, 1989, Fennel and Hauptmann, 1992, Songstad et al, 1993; Chaudhary et al, 1994). First stable transformation was obtained by electroporating the immature maize embryo with *neo* genes (D'Halluin et al, 1992). They also confirmed transmission and Mendelian segregation of the transgene.

The particle bombardment technique developed in 1987 (Sanford et al, 1987; Klein et al, 1987) has overcome some of the problems of cereal transformation. Any regenerable tissue can be transformed by this method. The DNA transfer process is thus variety indepedent, giving the method great flexibility. Transient gene expression can be obtained if any plant species using particle gun mediated gene delivery is employed. If the targetted cells are regenerable, transgenic plants can also be obtained easily (Christou, 1992). Several factors, such as particle type, size and density, DNA preparation method, particle acceleration, speed can influence the efficiency of transformation. In the first report on cereal transformation using particle gun method, Klein et al (1987) used the suspension culture of maize and obtained stably transformed maize callus, plants were not regenerated. Later workers produced transgenic maize plants by bombarding embryogenic suspensions (Fromm et al, 1990; Gordom-Kamm et al, 1990) (Table 39). In wheat transgenic plants were first developed by Vasil et al (1992) by bombarding the embryogenic callus. They obtained fertile transgenic wheat plants resistant to the herbicide basta. The selective bar gene was delivered into type C cells of the embryogenic callus. The activity of bar gene was detected in all the 28 Ro plants analysed.

Table 39
Particle gun mediated transformation in cereals to produce transgenic plants

Species	*Reference*
Hordeum vulgare	Wan & Lemaux 1994
Avena sativa	Somers *et al.* 1992
Tritordium	Barcelo *et al.* 1994
Oryza sativa ssp. indica	Christou *et al.* 1991, 92
Oryza sativa ssp. japonica	Cao *et al.* 1992
	Christou *et al.* 1992
Triticum aestivum	Vasil *et al.* 1992
	Weeks *et al.* 1993
	Nehra *et al.* 1994
	Becker *et al.* 1994
Zea mays	Gordon-Kamm *et al.* 1990

In producing transgenic wheat Becker et al (1994) used scutellar tissues of immature embryos for particle bombardment. The bombardment was carried out using a PDS 1000/He gun (Bio-Rad). The plasmid *pDBI* containing the *gusA* gene under the conrol of the action promotor of rice and the marker gene bar under the control of *CaMV 355* promotor was used. They regenerated 59 plants from 1050 bombarded embryos. Both the marker genes were present in all the regenerated plants analysed through southern hybridization. Nehra et al (1994) also used immature embryos for particle gun mediated DNA delivery into wheat cells. They used three marker/reporter genes, namely *uidA bar* and *npt II* in their transformation experiments. It is highly desirable to produce transgenic plants from primary explants such as immature embryos, inflorescence etc, so that variations induced because of culturing can be excluded. Transgenic rice plants have thus been produced by bombardment of the immature embryo explants (Christou et al, 1991). In producing transgenic *Tritordium,* Barcelo et al (1994) used inflorescence tissue for micro-projectile bombardment. The method has been reported as highly efficient, produced 17 independent transformed *Tritordium* plants from 32 bombardments. The transformation efficiency was increased by preculture of the targeted tissues.

Silicon Carbide Fibres

In this method silicon carbide fibres are coated with DNA and then they are allowed to penetrate the cells. The DNA coated fibres and cells are mixed together and vortexed to effect cell penetration. Stable transformation of maize cells has been achieved by this method (Kaeppler, et al, 1990; 1992). This is a very simple technique and could replace all other available cell transformation methods if regenerating cells can be transformed by this method.

Liposome Mediated Gene Transfer

Liposomes are small lipid bags, in which large number of plasmids are enclosed. They can be induced to fuse with protoplasts using devices like PEG, and therefore have been used for gene transfer. In this technique, DNA enters the protoplasts due to endocytosis of liposomes, involving the following steps (Figure 39): (i) adhesion of the liposomes to the protoplast surface, (ii) fusion of liposomes at the site of adhesion, and (iii) release of plasmids inside the cell. This technique have been successfully used for tobacco, petunia, carrot etc.

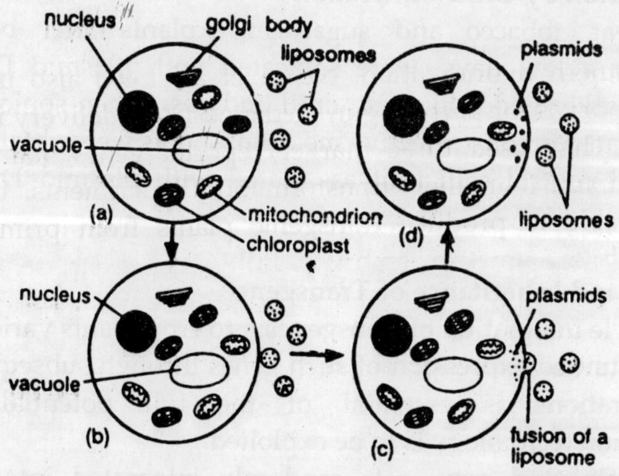

Figure 39

Fusion of plasmid filled liposomes with isolated protoplasts as a method of plasmid uptake

Calcium Phosphate Precipitation Method

Foreign DNA can be carried with the Ca^{++} ions, to be released inside the cell due to the precipitation of calcium in the form of calcium phosphate.

Pollen or Pollen Tube Method

Foreign DNA can be taken up by the germinating pollen and can either integrate into sperm nuclei or reach the zygote through the pollen tube pathway. Both these approaches have been tried and interesting phenotypic alterations suggesting gene transfer have been obtained. Transgenic plants have never been recovered using this approach and this method, though very attractive, seems to have little potential for gene transfer.

Incubation of Dry Seeds, Embryos, Tissues or Cells in DNA

Incubation of dry seeds, embryos, tissues or cells in known DNA has been tried in many cases. However, in no case proof of integrative transformation could be available. In all these cases plant cell wall not only works as efficient barrier, but are also efficient traps for DNA molecules.

Transformation by Ultrasonification

In wheat, tobacco and sugarbeet explants after being cultured for a few days, were sonicated with plasmid DNA (carrying marker genes like *Cat, nptII* and *gus.* When sonicated calli were transferred to selective medium, shoots were obtained, although all control calli (not associated with plasmid DNA) died.

Expression and Inheritance of Transgene

The stable integration of transgenes into crop plants varieties and the continued expression of such genes through subsequent seed generations is essential of the full potential of transformation technology is to be exploited.

The engineered gene gets randomly integrated into the recipient genome. Very often, multiple copies of the engineered foreign gene get integrated into host genome. The integration sites are generally at single or closely linked loci (Spencer et al, 1992), but integration of multiple loci has also been observed (Rathore et al, 1993). The inheritance of transgenes have been studied in rice, maize, wheat and barley. In maize, the transgenes usually segregate at single dominant allele (Spencer et al, 1992). In wheat inheritance of bar gene was shown to follow Mendelian segregation in R_1 and R_2 plants (Vasil et al, 1992; Nehra et al, 1994; Becker et al, 1994). Becker et al, (1994) reported that the uidA marker genes had a 1:1 ratio in pollen grains of all the transformed plants tested and the progeny from one of the transformed plants showed a 3:1 Mendelian segregation for uidA and bar genes. Inheritance of gusA and nptII genes have been investigated in three different R_o plants of rice (Peng et al, 1995). The two genes were co-transferred into rice protoplasts. The genes were residing on separate plasmids. The inheritance of transgenes was investigated by southern blot analysis and enzyme assays. The inheritance of gusA and neo genes was shown to be in Mendelian fashion in one plant and in an irregular fashion in other plant (Kothari and Chandra, 1995).

Herbicide Resistant Transgenics

Herbicides, widely used in modern agriculture, are chemical compounds that kill or inhibit the growth of plants. But though basically applied to control weeds, they also have deleterious effects on crop plants. Selective and rapid breakdown are not always obtained and leftover herbicides applied to weeds before a crop is planted persist in the soil and decrease crop yield. A promising alternative approach is the development of herbicide tolerant plants for use with broad spectrum or totally non-specific herbicides (Table 40). Three strategies have been adopted to obtain herbicide resistant plants (Botterman and Leemans, 1988; Mazur and Falco, 1989): (i) herbicide target modification, (ii) target enzyme over-production, and (iii) herbicide detoxification.

Table 40
Herbicide resistant transgenic plants

	Species modified	*Transgene source*	*Transgene product*
1.	*Target modification:*		
	Beta vulgaris	*Arabidopsis thaliana*	Acetolactate synthase
	Brassica napus.	*A. thaliana*	(as above)
	Festuca arundina-ceace.	*E. coli*	Hygromycin phosphotra-nsferase.
	Linum usitati-ssimum.	*A. thaliana*	Acetolactate synthase
	Nicotiana tabaccum	*A. thaliana*	(as above)
2.	*Enzyme Overproduction:*		
	Glycine max.	Plant and microbial genes	Analogue of EPSP synthase
	Linum usitati-ssimum.	(as above)	(as above)
3.	*Enzyme Detoxification:*		
	Agrostis palustris	*Streptomyces hygroscopicus*	Bialaphos
	Beta vulgaris	(as above) acetyltre-ansferase	Phosphino-thricin
	Brassica napus.	(as above)	(as above)
	B. oleracea.	(as above)	(as above)
	Festuca arundinaceae	(as above)	(as above)
	Gossypium hirsutum	*Alcaligenes eutrophus*	2, 4-D monooxy-genase
	Hordeum vulgare	(as above)	(as above)

Contd.

Species modified	*Transgene source*	*Transgene product*
Lycopersicon esculentum	*Streptomyces hygroscopicus*	Phosphinothricin acetyl transferase
Medicago sativa	(as above)	(as above)
Nicotiana tabaccum	*Alcaligenes eutrophus*	2, 4-D monooxygenase
Solanum tuberosum	*Streptomyces hygroscopicus*	Phosphinothricin acetyl transferase
Triticum aestivum	(as above)	(as above)

(i) Herbicide target modification

Herbicide targets are proteins and the action is two fold; it either inhibits photosynthesis or aminoacid biosynthesis. The most common herbicides used which inhibits photosynthesis are the triazines (atrazine and simazine). These herbicides block electron transport at PII by binding to the Qb protein present in the thlakoid membrane and enclosed by the psbA gene of chloroplast (cp) DNA (Zurawski et al, 1982). A single aminoacid substitution (Serine to glycine) at position 264 in the 32 kDa protein, results in decreased herbicide binding (Hirschberg and McIntosch, 1983). Triazine resistant mutants having one altered aminoacid have been identified from naturally occurring resistant weed biotypes or microbial species (Ohad and Hirchberg, 1990). Manipulation of the resistant chloroplast genome by transfer through sexual hybridization or by developing a chloroplast transformation system are the possible ways to obtain atrazine resistant plants. Beversdorf et al (1990) transferred atrazine resistant chloroplast from *Brassica campestris* to *Brassica napus* by back crossing the resistant plant to the female parent of *B. napus*. After eight backcross generations, the nuclear genome is almost isogenic. A faster approach was developed by Cheung et al (1988) in which tobacco cells were transformed by *Agrobacterium* vector harbouring a psbA gene encoding a triazine insensitive Qb protein, fused to the transit peptide of a nuclear encoded chloroplast protein. Transgenic plants showing increased tolerance to atrazine were produced.

Another group of herbicide—sulphonylureas and imidazolinones block and inhibit aminoacid biosynthesis. The target enzyme is acetolactate (ALS), (Ray, 1984; Shaver and Anderson, 1985). Mutant forms of ALS resistant to sulphobylureas and/or imidazolinones having one altered aminoacid from wild type ALS have been identified and isolated. Transgenic tobacco plants resistant to sulphonylurea through expression of a mutant ALS gene from *Arabidopsis* were obtained which could tolerate four times herbicide concentrations (Mazur and Falco, 1989; Haughn et al, 1988; Lee et al, 1988). Maize plants tolerant to imidazolines correlated with the presence of an altered ALS enzyme have also been produced (Shaver and Anderson, 1985).

(ii) Target enzyme overproduction

During the last decade, rapid progress has been made in developing herbicides which degrade rapidly and are non-toxic to animals. One of the most potent broad spectrum herbicide is glyphosate (N-phosphonomethyl glycine), marketed under the trade name of Roundup. It interferes aminoacid biosynthesis by inhibiting the enzyme 5-enolpyruvylshikimate-3-phosphate synthase (EPSP). High level herbicide tolerance has been obtained in plants by over production of EPSP in the chloroplast (Shah et al, 1988). An EPSP synthase cDNA isolated from a glyphosate tolerant *Petunia hybrida* cell line and joined to a CaMV 35 S promoter Ti nos (nopaline synthase) 3' polyadenylation signal was used for *Agrobacterium* mediated transformation of *Petunia* cells. Over production of EPSP synthase resulted in transgenics which could tolerate high doses enough to kill wild type *Petunia* plants (Steinrucken et al, 1986). In-vitro construction and transfer of a chimeric gene with plant EPSP synthase transit peptide joined to a bacterial EPSP coding region lad to transgenic tobacco (Comai et al, 1985; Comai and Stalker, 1986), and tomato due to the accumulation of stable glyphosate resistant enzyme in the chloroplasts. The combination of the chloroplast transit peptide sequence of *Petunia* cDNA clone and *E. coli* mutant resistant enzyme gave rise to fully resistant plants in tomato, potato, tobacco, soyabean, brassica and sugarbeet (Della et al, 1987; Kishore and Shah, 1988).

Phosphinothicine (PPT) is a non-selective herbicide and inhibits glutamine synthase (GS) (Leason et al, 1982). Inactivation of GS leads to accumulation of ammonia which is toxic to the cells. An alfalfa cell line resistant to PPT due to amplification of GS gene was reported by Donn et al (1984) suggesting that an over expression of the enzyme overcomes the toxic effect of the inhibitor. Insertion of the over expression GS gene from alfalfa lead to transgenic tobacco plants (Eckes et al, 1989; DeBlock et al, 1987).

(iii) Enzyme detoxification

Herbicide detoxifying enzymes counteract the effect of several herbicides by inactivating them before they are able to inhibit the target enzyme. The most successful example is the detoxification of phosphinothricine phosphate. Murakami et al (1986) isolated the bar gene from *Streptomyces hygroscopius* confering resistance to PPT by encoding an enzyme phosphinothricine acetyl transferase (PAT). This enzyme is able to inactivate PPT by acetylation of the free NH_2 group (Thompson, et al, 1987). The bar gene was inserted into an *Agrobacterium* vector and used for transforming several plants. Transgenic plants showing increased tolerance were obtained (DeGreef et al, 1989).

Other enzymes are glutathione-5 transferase (GST) which modifies triazineherbicides (Frear and Swanson, 1970) and 2, 4-D dichlorophenoyacetate mono-oxigenase involved in 2, 4-D degradable pathway. Transgenic tobacco plants were produced through genetic engineering of tfd gene of soil bacterium *Alcaligens eutrophus* which encodes the first 2, 4-D dichlorophenoxyacetate mono-oxigenase (Lyon et al, 1989; Streber and Willmitzer, 1989).

However, before commercialization of herbicide resistant crop plants, factors such as potential loss in vigour and yield, herbicide performance, crop and chemical registration costs, potential for outcrossing to weed species and potential for crop becoming a weed should be considered (Jain, 1993).

Table 41

Mechanism of action of different herbicides and the basis of achieving resistance against them in transgenic plants

Active principle of herbicide	*Inhibited pathway*	*Target product*	*Use*	*Basis of resistance*
1. Amino acid biosynthesis inhibitors				
1. Glyphosate (Roundup)	Aromatic amino acid biosynthsis	EPSPS	Broad spectrum	Over expression of EPSPS gene; bacterial *aroA* gene
2. Sulphonylurea and Imidazolinones	Branched chain amino acids	ALS	Selected crops	Mutant ALS gene
3. Phosphinothricin (Basta)	Glutamine biosynthesis	GS	Broad spectrum	Gene amplification, *bar* gene detoxification
II. Photosynthesis inhibitors				
4. Atrazine (Lasso)	Photosystem II	QB (32 kDa protein)	Selected crops	Mutant *PsbA* gene, GST gene; detoxification
5. Bromoxynil (Buctril)	Photosynt-hesis	—	Selected crops	*bxn* gene; detoxification

Table 42
Reporter genes used as scoreable markers and their enzyme assays

	Reporter gene for	*Substrate* and assay*	*Identification*
1.	Chloramphenicol acetyl-transferase (CAT)	(^{14}C) chloramphenicol + acetyl CoA; TLC separation	Detection of acetyl chloremphenicol by autoradigraphy
2.	Neomycin phosphotrans-ferase (NPT II)	Kanamycin+ (^{32}P) ATP (in situ assay)	Radioactivity detection (also dot blots)
3.	β-glucuronidase (GUS)	Glucuronides (PNPG, x-GLUC, NAG, REG	Fluorescence detection: colourimetric, fluorometric and histochemical
4.	β-galactosidase (lac Z)	β-galactoside (X-gal)	Colour of the colony
5.	Luciferase	(a) Decanal and $FMNH_2$ (b) ATP + O_2 + luciferin	Bioluminiscence (exposure of x-rays film)
6.	Octopine synthase (OCS)	Ariginine + pyruvate + NADH	
7.	Nopaline synthase (NOS)	Arginine + ketoglutaric acid + NDH	Electrophoresis

*PNPG = p-nitrophenyl glucuronide; x-GLUC = 5-bromo, 4-chloro, 3-indolyl glucuronide; NAG = napthol AS-BI glucuronide ; REG = resorufin glucuronide.

Insect Resistant Transgenics

Control of insect pests has been an integral part of the development of practices as crop damage caused by insects is a major economic factor in agriculture in tropics and temperate regions of the world. Modern high intensity agriculture which has been responsible for the tremendous increase in food production, is dependent on the use of chemical pesticides. But the drawbacks of this strategy are that pesticides are often highly toxic to non-target organisms, strong selection pressure on insects population imposed by insecticides causes rapid acquirement of resistance to such compounds and overuse of pesticides decrease the vigour of the crop and makes it more susceptible to an insect attack. Hence, attention was focussed on improving the inherent resistance of the crop plant to insect attack which resulted in the development of transgenic plants able to protect themselves against insects by expressing insecticidal proteins or a protinease inhibitor gene. These plants offered advantages:

(i) absence of non-proteinaceous residues in soil or ground water.

(ii) high specificity with respect to the target organism.

(iii) protection of plant parts such as roots which are difficult to reach by conventional methods.

Two methods of control have been developed in transgenic plants:

(i) Use of viral and bacterial toxin gene

Baculoviruses and *Bacillus thuringiensis* (Bt) provide an alternative to chemicals for controlling insect pests of agronomically important plant species. Hammock et al (1990) accomplished the production of genetically engineered baculovirus containing Juvenile Hormone Esterase (JHE) gene which resulted in reducing insect-feeding habit. Stewart et al (1991) constructed a baculovirus insecticide containing an insect-specific neurotoxin from the venom of the North African (Algerian) scorpion—*Androctonus australis*. In a similar report, a toxin (TxP-I) gene associated with the female mite *(Pyemotes tritici)* venum causes muscle-contractive paralysis when injected into insects. It was cloned, sequenced and expressed in baculovirus (Tomalski and Miller, 1991). Larvae infected with this recombinant baculovirus became paralysed during infection.

Table 43

Insect resistant transgenic plants (Kaul and Nirmala, 1995)

Plant species	*Insect pest*	*Gene source*	*Transgene product*
1. Resistance through bacterial toxin (Bt) gene :			
Lycopersicon esculentum.	*Manduca sexta, Keiteria lycopersicella*	*Bacillus thuriengiensis.* *(as above)*	Bt-insecticidal protein
Gossypium hirsutum	*Petinophora gossypiella*	*(as above)*	Bt-insecticidal protein ein
Zea mays	*Heliothis zea*	*(as above)*	(Bt-insecticidal protein.
2. Resistance through proteinase inhibitor gene :			
Nicotiana tabaccum	*Manduca sexta.*	*Vigna unguiculata*	Trypsin Inhibitor protein.

Gram positive bacteria *Bacillus thuringiensis* contain peptide toxins as crystals in their spores which when ingested by insects are cleaved by the proteases in the intestine resulting in the conversion of the protein into the active toxin. This toxin impaires digestion and midgut paralysis as a result of which the insect stops feeding and ultimately dies (Hofte and Whiteley, 1989). Successful transformations with *B. thuringiensis* have been reported (Vaeck et al, 1987; Fischhoff et al, 1987; Barton et al, 1987).

A chimeric gene with the structure CaMV 35 S promoter/*B. thuringiensis* toxin coding sequence/Tinos3′ termination sequence was constructed. This gene was placed in a Ti vector and tomato leaf disc cells were transformed by co-cultivation with *A. tumifaciens* (Fischhoff et al, 1987). The tomato plants were protected against feeding damage by larvae of the Lepidopterans specially *Manduca sexta.* Also significant control of tomato fruitworm *(Heliothis zea)* and the tomato pinworm *(Keiferia lycopersicela)* have been obtained (Perlak, et al, 1991; Brunke and Meeusen, 1991). Many other transgenic plants have also been produced (Table 43), which express the bacterial toxin in their vegetative and floral organs and are thereby effectively protected against attack by some insects. Bt toxin genes have been isolated from a number of different bacterial strains, but genetic engineering has been carried out only on strains active against Lepidopteran pests. However, as some insects strains show resistance to one type of endotoxin and are sensitive to another, e.g., *Plodia interpuctella* (Mc Gaughey, 1985), it is plausible to construct transgenic plants with two types of protein against the same insect. A further improvement in insect control by transgenic plants is increased endotoxin protein production which not only kills susceptible larva but also reduces the fertility of the mature insects, thus reducing their progeny.

(ii) Use of plant proteinase inhibitor gene

Proteinase inhibitors conferring endogenous resistance to insect attack are widespread among higher plants. In addition, they have anti-metabolic activity in a wide range of insects which provides an attractive strategy to make plants resistant to herbivorous insects by introducing genes for certain protease inhibitors.

Expression of a plant derived protenase inhibitor—cowpea trypsin was reported from transgenic tobacco. The cowpea trypsin inhibitor (CpTI) gene, identified from *Vigna unguiculata,* is an insecticidal component preventing development of the larvae of field and storage pests bruchid beetle *(Callosobruchus maculatus).* It has the advantage of insignificant toxicity as it is present in the cowpea seeds itself which is consumable. A library of cDNA clones was produced from mRNA isolated from developing cowpea seeds. A cauliflower mosaic virus (CaMV) promoter and a nopaline synthase gene from *Agrobacterium tumefaciens* providing a polyadenylation signal sequence and a transcription terminator was added to the cDNA to be incorporated into an *Agrobacterium* binary vector and was used to transform tobacco plants. The transgenic CpTI expressing plant showed a significant resistance to a wide spectrum of insects (McGaughey, 1985; Hilder et al, 1987), indicating the potential of such genes. The advantage of using these genes lies their broad spectrum of activity in many different insects and their nontoxicity, as such inhibitors are found in the food of humans and animals. The major disadvantage is the high level of protein needed to kill the insect larvae.

Viruses Resistant Transgenics

Of all plant pathogens, viruses are most intimately associated with their hosts, their genomes are relatively small and well-characterised. Significant progress has been made in protecting crops against a number of viral diseases through genetic engineering (Table 44), using the following: (i) coat protein mediated protection, (ii) protection by non-structural viral gene, (iii) protection via satellite RNA expression, (iv) protection via antisense RNA.

Table 44

Virus resistant transgenic plants (Kaul and Nirmala, 1995)

Host Plant	*Coat protein source*	*Protection against*
1. Viral Coat Protein Mediated Resistance:		
Lycopersicon esculentum	TMV, TMVI	TMV, TMVI
	AIMV	AIMV
Medicago sativa	*AIMV*	*AIMV*
Nicotiana tabaccum	*TMV*	*TMV*
	AIMV	AIMV
	CMV	CMV
	TSV	TSV
	TRV	TRV
	PVX	PVX
	SMV	TEV, PVY
Solanum tuberosum	*PVX or PVY*	*PVX or PVY*
	PLRV	PLRV
Carica papaya	PRV	PRV

Contd.

Host Plant	*Coat protein source*	*Protection against*
Cucumis sativus	*CMV*	CMN
2. Non-Structural Viral Gene Mediated Resistance:		
Nicotiana tabaccum	TMV-VI	TMV
3. Satellite RNA Gene Mediated Resistance:		
Nicotiana tabaccum	TRV, CMV	TRV, CMV
4. Antisense RNA Mediated Resistance:		
Nicotiana tabaccum	CMV, PVX, TMV	CMV, PVX, TMV

Abbreviations used: ALMV = Alfalafa Mosaic virus, CMV = Cucumber mosaic virus, PVX = Potato virus X, PVY = Potato virus Y, PRV = Papaya ringspot virus, PLRV = Potato leaf roll virus, PMV = Pepper mottle virus, SMV = soybean mosaic virus, TMV = Tobacco mosaic virus, TRV = Tobacco ringspot virus, TSV = Tobacco streak virus, TEV = Tobacco etch virus.

(i) Coat protein mediated protection

The most promising and successful way to produce viruses resistant plants is to insert and allow expression of a viral coat protein gene(s). The principle is based on the observation that infection of a plant with one viral strain protects against superinfection by another related strain. This phenomena is also referred to as cross-protection. Although the molecular mechanism involved are not as yet clear, it is assumed that the mild virus which infected the cell first, produces excessive amount of protein which remain in a free state. The unbound protein inhibit the uncoating of the RNA of the second aggressive virus subsequently delaying or preventing the RNA expression and replication (Hilder et al, 1990). Powel-Able et al, (1986) were the first to describe the expression of tobacco mosaic virus (TMV) coat protein when infected with TMV in transgenic tobacco. A cDNA coding for the TMV coat protein under the control of a CaMV35S-promoter was transferred to tobacco. The progeny of the engineered plants expressing the viral coat protein gene either showed delayed disease symptoms or failed to develop symptoms at all. Since then, coat protein mediated virus protection has also been introduced in other plants (Table 44). According to Beachy et al (1987; 1990) two steps are operative—one at early infection, and the other during the spread of infection.

An interesting feature of coal protein mediated protection is that in some cases the plant is not only protected against an infection of the virus from which the coat protein was received but also against other seriologically non-related viruses. Stark and Beachy (1989) demonstrated that expression of coat protein of SMV in transgenic tobacco led to resistance to two seriologically unrelated viruses PVY and TEV. The strategy of coat protein mediated viruses protection offers great potential to control many virus diseases and is being vigorously and successfully pursued by researchers. Sanders et al, (1992) demonstrated that under field conditions transgenic tomato plants expressed tomato mosaic virus (ToMV) coat protein showed high degree of resistance to MoMV.

(ii) Non-structural viral gene protection

Using a non-structural gene from TMV strain U1, Golemboski et al (1990) were able to produce highly virus resistant transgenic tobacco plants, the chimeric gene used contained a CaMV35S promoter and a coding region of open reading frame which encoded a 54 kD protein. Transgenic plants containing the chimeric 54kDaORF showed complete resistance even when inoculated with high concentrations of the virus. Analysis of the chimeric gene expression in these plants showed the presence of mRNA but not the 54 kDa protein which makes it unclear whether the resistance is due to the protein or its mRNA (Harrison et al, 1987).

(iii) Satellite RNA modification

Satellite RNAs are small molecules encapsulated by plant viruses and packaged together with a virus genome. Certain plant RNA viruses harbour these molecules. Such molecules are unable to replicate independently but require assistance from the virus. The ability of certain satellite RNA to modify and ameliorate plant viral disease symptoms aroused attention for using it as a means for controlling viral disease. This led Harrison et al (1987) and Gerlach et al (1987) to incorporate genomes of satellite RNA CMV and TRV under the control of constitutive promoters into tobacco plants. The transgenic plants which expressed the satellite RNA when infected with the corresponding virus exhibit significant delay in symptom development compared to the plants devoid of the gene coding RNAs. But there are potential risks for using satellite RNA for genetic engineering as (i) satellite RNA that are ameliorative in one species may be lethal to another (Waterworth et al, 1979), (ii) satellite RNA mutate and a single nucleotide change introduced in in-vitro mutagenesis can make ameliorative satellite necrogenic (Sleat and Palukaitis, 1990).

(iv) Antisense RNA resistance

An antisense RNA is an RNA complementary to the mRNA strand and contains base sequences complementary to the target (Sense) RNA transcripts. When present together they anneal to form duplex RNA molecules thereby blocking translation. Attempts have been made to produce virus resistant plants by using antisense RNA against viral coat protein gene (Cuozzo et al, 1988; Rezian et al, 1988). Plants showing some resistance to low inoculum have been produced for CMV, PVX and TMV. Experiments with CMV were not very optimistic as out of 12 lines, only one transgenic line showed some resistance to CMV while the others were as susceptible as control plants. A higher rate of resistance can be expected if the antisense RNAs expressed are complementary to key regions involved in the regulation of replication of gene expression.

Fungal Resistant Transgenics

Fungal diseases are a major problem in agriculture causing enormous world-wide economic losses. Though a variety of plants have a natural mechanism to resist attack of pathogenic fungi either through preformed barriers, such as cell walls and cuticle or through the production of defence enzymes, the pathogens have somehow also developed ways to evade these defence mechanisms (Bowles, 1990; Dickman et al, 1989; Murray et al, 1993). In order to combat the diseases, control by fungicides is essential but sometimes problematic. Therefore new approaches of introducing resistance genes to plants have been taken up. But in comparison to virus and insect resistance, methods of obtaining resistance against fungi and bacteria are less developed. The methods were developed taking into account, the natural defence related genes present in plants which was triggered by pathogen attack. Bowles (1990) grouped the defence genes into three classes: (i) genes synthesizing compounds involved in cell wall modification, (ii) genes encoding defence related proteins exhibiting anti-microbial activities or catalysed the synthesis of products that retard microbial activity, and (iii) genes encoding the pathogen related (PR) protein whose appearance was correlated with defence responses. Some of the fungal disease resistant transgenic plants have been shown in Table 45.

Table 45

Fungal disease resistant transgenic plants (Kaul and Nirmala, 1995)

Plant species transformed	*Transgene source*	*Transgene product*	*Resistance against*
Brassica napus	*Phaseolus vulgaris*	Bean endo chitinase	*Rhizoctonia solani*
Nicotiana tabaccum	(as above)	(as above)	(as above)
N. tabaccum	*Serratia marcescens*	Chitinase	*Alternaria longipipes*
N. tabaccum	*Hordeum vulgare*	Ribosome inhibiting or inactivating protein	*Rhizoctonia solani*

Bacteriacidal proteins (Lysozyme, Cecropins, and Attacins) have been identified in the pupae of the giant silk moth *(Hyalophora cecropia).* These lytic proteins have shown a potent in-vitro antifungal activity against several pathogenic fungi, including *Phytophthora infestans*. Transgenic tobacco plants that express a barley ribosome-inactivating protein (Stripe et al, 1992), exhibited hightened protection against agronomically deleterious fungus *Rhizoctonia solani* (Longemann et al, 1992). Similarly, tobacco pathogen related protein (PR), namely PR-S and osmotin, have been shown to be serologically related to zeamatin, an antifungal protein of maize (Vigers et al, 1992). Woloshuk et al, (1991) showed that osmotin and a related protein from tomato had antifungal activity against *Phytophthora infestans.*

Fungal Pathogen Related Proteins

An attack by a pathogen triggers a number of active responses in plants. One of the most important response is the synthesis of defence related proteins which include amylase proteinase inhibitors, thiomins, hydrolytic enzymes such as B-1, 3 glucanase, chitinase and enzymes involved in the synthesis of phytoalexins (Dickman et al, 1989; Murray et al, 1993).

(i) Phytoalexins

Phytoalexins, a secondary metabolism product, are low molecular weight antimicrobial compounds which act as broad spectrum antibiotics and play an important role in arresting the growth of fungal pathogens (Bailey, 1987; Bryngelsson and Collinge, 1991). Expression of phytoalexin genes may be due to elicitors produced by host and fungal cell wall breakdown or by abiotic agents such as methanical injury, ultraviolet irradiation and heavy metals. The first phytoalexin to be purified and identified was pisatin from pea (Perrin and Bottomley, 1962). Others include medicarpin, kievitone from legumes, and lubinin and rishitin from potato. Indirect proof for the role of phytoalexins in disease resistance has been supplied by genetic experiments utilizing pathogen strains that vary in their virulence and ability to degrade a particular plant phytoalexin (Deffardius and Gardaner, 1989; VonEthen et al, 1989). Hain et al, (1990) demonstrated that introduction of the key biosynthetic gene for a peanut stilbene phytoalexin into tobacco plants enables them to produce the peanut phytoalexin-resveratol. It is predicted that plants able to make large amounts of foreign phytoalexins would be resistant to pathogens that could detoxify the new chemical structures.

(ii) Chitinase

A common natural mechanism of plants to resist fungal attack is secretion of the chitinase which attacks the cell walls of the fungus. This enzyme is very stable, resistant to heat and inhibits fungal growth in-vitro. Strains of *Serratia marcescens* are effective in the biocontrol of a number of pathogenic fungi e.g., *Scerotium rolfsii* due to the secretion of chitinase. Introduction of a microbial chitinase gene into tobacco plants has shown promise for the control of certain fungal pathogens (Broglie et al, 1991). There are additional noval avenues suggested to develop disease resistant plants such as to introduce genes that detoxify pathogen toxins, inhibits essential pathogen enzymes and encode antimicrobial paptides. Such genes have been described from plants (Chrispeels and Raikhel, 1991; Vigers et al, 1991; Dimarcq et al, 1990; Lehrer et al, 1991).

(iii) Ribosome inhibiting proteins

Seeds of various cereals contain proteins that are toxic to some pathogens. For example, barley contains a ribosome inhibiting protein (RIP) which is a glysosylase that inhibits ribosome function by cleaving a glycosol from the 60 S sub-unit of ribosomes thus preventing peptide elongation. This protein while non-toxic to plants, inhibits the growth of a number of pathogenic fungi. Using a chimeric gene capable of expressing barley RIP in the stem and roots of tobacco, Longmann et al (1992) developed transgenic plants capable of resisting attack by *Rhizoctonia solani.*

Stress Tolerant Transgenics

Biological stress refers to any change in environmental conditions that might reduce or adversely change a plant's growth or development. Crop plants are subjected to a variety of environmental extremes such as drought and temperature stresses and breeders have long faced problems selecting for stability of performance over a range of environments, using extensive testing and an intricate biometrical approach. Environmental stress alters gene expression (Sachs and Ho, 1986) and permits of stress related genes. Improving resistance to environmental stress thus requires a combination of breeding, physiological and biotechnological approaches to understand (i) the structure and enzymatic functions of stress proteins, (ii) mechanisms regulating the stress genes, and (iii) identification, isolation and transfer of these genes by various transformation schemes. Although precise molecular basis of stress phenomena is poorly understood, monitoring of level of plant tolerance to cold, heat, drought and salts has been possible to some extent. Following are the common types of stresses the plant face:

(i) Water stress

Water stress is mainly considered in terms of drought stress. Survival of the plant depends upon its ability to function under water scarcity. This can be circumvented by drought-avoidance requiring a short growing season or dehydration-avoidance where the plant maintains sufficient tissue hydration for metabolic functioning. Abscisic acid (ABA) concentration effects yield under water stress conditions (Sachs and Ho, 1986; Milborrow, 1981). Thus selection for plants having high ABA accumulation should form an important selection criterion for drought resistance.

Table 47
Stress tolerant transgenic plants (Kaul and Nirmala, 1995)

Species	*Genetic modification*	*Transgene Source*	*Reference Product*
Lycopersicon esculentum	Frost protection	Fish, *Pseudo pleuronectes americana*	*Antifreeze* protein
Nicotiana tabaccum	Cadmium tolerance	Mouse	Metalloth ionein binding protein
N. tabaccum	Cold tolerance	*Arabidopsis thaliana*	Glycerol - 3 phosphate acyl transferase

(ii) Temperature stress

Temperature stress refers to any temperature outside the optimum for growth and development. It depends on growth stage, the most severe perturbation occurring at germination and fruit formation. Three different types of temperature stress are heat-stress, chilling-stress, and freezing-stress.

(a) Heat tolerance

Thermal tolerance is viewed in terms of stress, degree, exposure, duration and developmental stage. Of all biological processes, the reproductive-stage is most sensitive to heat leading to floral abscission, pollen sterility and poor fruit-set. All these lead to yield reductions. Heat shock response is characterised by (i) decrease in protein synthesis, (ii) production and accumulation of large amount of heat shock protein, (iii) gradual decline in heat shock protein synthesis and return to normal synthesis. Though plant breeders have been able to manipulate thermal tolerance as a heritable agronomic trait, the relationship between thermal tolerance in-vivo is ambiguous. As genetic resources for heat tolerance exist in rice, potatoes, soybean and tomato, for other crops, these need to be explored.

(b) Chilling Stress

It is the most severe environmental stress that reduces germination and growth rate, vegetative and reproductive growth and leads to deformed fruit formation and/or failure of fruit and/or seed set. Chilling tolerance becomes operational below 0°C. Genetic resources for chilling tolerance have been found in Maize (Mock and Eberhart, 1972). Such genes have been introduced in tomato from exotic germplasm (Patterson and Payne, 1983) and transgenic tomato resistant to chilling stress have been developed.

(c) Freezing stress

In crop plants, considerable genetic variability occurs in freezing tolerance below 0°C. Considered to be genetically conditioned, the inheritance pattern of freezing stress tolerance is scantily known in wheat and rice. Frost hardiness is considered a quantitatively inherited trait controlled by a complex interaction of several genes (Thomashaw, 1990). Plasma membrane is regarded as the target site for freezing because disruption of cellular membranes is the first symptom of freeze injury in plants. Steponkus et al (1988) regarded increased tolerance as related to change in plasma lipid composition. Series of biochemical alterations follow freezing viz., increase in proline and organic acids, sugar and soluble proteins. The expression of cold regulated genes parallels freezing tolerance in plants (Steponkus et al, 1988).

(iii) Salt stress

Efforts to develop salt resistant varieties have been unsuccessful due to multipartite nature of stress that makes difficult to predict the extent of stress in a saline environment as salinity causes: (i) water stress from the osmotic effects, (ii) mineral toxicity of the salt, and (iii) interruptions of the mineral nutrition of the plant. The saline fields are also inherently variable in their salt distribution and thus require an appropriate strategy to breed for high yields.

In-vitro selection and cell lines with enhanced salt resistance have been isolated (Tal, 1990), but this is rarely associated with resistance at the whole plant level (Orton, 1980; Warren and Gould, 1982).

Protein Modification in Transgenics

Seeds of higher plants contain large quantities of storage proteins which upon germination are hydrolysed providing nitrogen for growth. But such proteins are deficient in aminoacids that are essential for human and livestock nutrition (Nelson, 1979). Cereals are most limited in tryptophan, threonine and lysince and legumes in the sulphur containing aminoacids methionine and cysteine (Krebbers et al, 1993). Conventional plant breeding techniques to increase aminoacids in crops have met with low success (Poehlman and Sleper, 1995). Storage proteins are the products of multigene families comprising about 10 or more members, tightly linked at a given locus (Shotwel and Larkins, 1989). Plant genetic engineering aims at modifying seed storage proteins to improve nutritional properties of seeds. Two approaches followed in altering seed aminoacid composition by molecular means are:

(i) to find naturally occurring seed storage protein with high levels of the desired aminoacid, clone the corresponding gene and allow its expression.

(ii) to modify seed storage protein genes by recombinant DNA or in-vitro mutagenesis so that they encode proteins that are similar to wild type proteins, but contain higher levels of essential aminoacids.

Tissue specific, developmentally regulated expression of both dicot and monocot seed storage protein genes have been demonstrated in dicot transgenic plants. For instance in *Brassica napus,* both the chimeric and modified storage protein transgenes have expressed fully (Guerche et al, 1990; Van Dekerckhover, et al, 1989; DeClercq et al, 1990). Likewise, a chimeric phaseolin transgene was expressed in tobacco (Voelker et al, 1989). Modified storage protein genes exhibiting differential accumulation of four phaseolin glycoforms successfully expressed in transgenic tobacco (Buston, et al, 1991), and influensive role of the polypeptide glycan in post-translational processing and transport of barley lectin to vacuoles in transgenic tobacco (Wilkins et al, 1990) clearly indicate the decisive role and demonstrable expression of storage protein genes in transgenic plants.

Despite the modifications in seed storage protein genes in transgenic plants, the manipulation of seed quality through genetic engineering faces several obstacles. The addition of one gene to plant genomes may not be very effective in improving the plant phenotype due to the strong expression of the rest of the multigene family. Despite this, the multiple codons for lysine and tryptophan in a 19 kD Zein cDNA by site directed mutagenesis have been successfully incorporated (Ohtani et al, 1991). Although the engineered zein gene was correlated in its deficiency in essential aminoacids, the protein was not stably incorporated into its normal cell compartment in the endosperm of maize. Another major experimental constraint in modification of seed storage proteins was the time required to regenerate and obtain seeds from the transformed plants. In contrast to bacterial or cell culture systems in which modified proteins can be tested in a matter of days or weeks, in plants, it may approach a year or sometimes even more. Infact, expression of modified storage proteins in transgenic plants is still in its infancy and requires development of reliable systems for quick testing of modified storage proteins (Krebbers and Vandekerechkhoeve, 1990; Altenbach et al, 1990). No data are available about what effect over expressing proteins rich in a particular aminoacid will have on aminoacid pools or other physiological factors. DeLuman (1990) suggested that protein quality was not the only parameter of importance and the interaction of changes in protein quality and quantity with changes in oil or starch has to be closely monitored. All should be balanced in such a way as to provide a useful product accepted by the market (Table 48 and 49).

Table 48
Transgenic plants with improved quality (Kaul and Nirmala, 1995)

Plant species Reference	*Genetic modification*	*Transgene source*	*Transgene product*
A *Food processing quality:*			
Brassica napus	Increased stearic acid	*Brassica rapa*	Antisense stearoyl ACP
	Increased methionine	*Bertholletia excelsa.*	Seed storage protein
	Increased lysine	*Cornybacterium dap. E. coli*	Aspartokinase, dibydrodipico linic acid synthetase
Glycine max	Increased methionine	*Bertholletia excelsa*	Seed storage protein
	Increased lysine	*Cornybacterium* dap, *E. coli*	Aspartokinase, dihydrodipico-linic acid synthetase
Lycopersicon esculentum	Improved storage	*Lycopersicon esculentum*	Antisense poly-galacturonase
	Flavour enhanced	Artificially synthesised	Synthesised monellin

Contd.

Plant species Reference	*Genetic modification*	*Transgene source*	*Transgene product*
Medicago sativa	Improved protein quality	Chicken	Chicken ovalbumin
Solanum tuberosum	Increased starch content	*E. coli*	ADP-glucose pyrophosphorylase
Nicotiana tabaccum	Increased manitol	*E. coli*	Mannitol dehydrogenase
B *Speciality chemicals :*			
Arabidopsis thaliana	Biodegradable thermoplastic	*Alcaligenes eutrophus*	Polyhydroxy butyrate (PHB)
Brassica napus	Increased lauric acid	*Umbellularia californica*	Lauroyl-ACP thioesterase
	Enkephalins	Chimeric gene part from *Homo sapiens & Arabidopsis thaliana*	Leu-enkephalin
Petunia hybrida	Flower colour	*Zea mays* *Gerbera* spp	Dihydraflavonol 4-reductase (DFR)
Solanum tuberosum	Serum albumin	*Homo sapiens*	Human serum albumin
	Cyclodestrins	*Klebsiella pneumoniae*	Cyclodextrin glycosyltransferase

Table 49
Transgenic plants used for research investigations (Kaul and Nirmala, 1995)

Plant species	*Genetic modification*	*Transgene source*	*Transgene product*
Beta vulgaris	Plant persistence	*Streptomyces hygroscopicus.* & *E. coli.*	Phosphinothricin acetyltransferase & Neomycin phosphotrans ferase
Brassica napus.	Male sterility	*Bacillus amyloliquefaciens*	Ribonuclease & Ribonuclease inhibitor
	Gene expression	*E. coli*	Chloramphenicol acetyl transferase
	(as above)	(as above)	Neomycin phosphotransferase
	Pollen dispersal	*Streptomyces hygroscopicus*	Phosphinothricin acetyl transferase
	Plant persistence	(as above)	(as above) & Neomycin phosphotransferase
Gossypium hirsutum	Pollen dispersal	*E. coli*	(as above)
Lycopersicon esculentum	Gene regulation	(as above)	ADP glucose pyrophonsphorylase
		Zea mays	Sucrose phosphate synthase
Nicotiana tabaccum	Gene regulation	*E. coli*	Chloamphenicol acetyl transferase
Solanum tuberosum	Gene expression	(as above)	Neomycin phospho-transferase
	Gene regulation	(as above)	ADP glucose pyro-phosphorylase

Contd.

Plant species	*Genetic modification*	*Transgene source*	*Transgene product*
	Pollen dispersal Plant persistence	*Arabidopsis thaliana* *Streptomyces hygroscopicus*	Acetolactate synthase Phosphinoth-ricin acetyltransfe-rase & Neomycin
		E. coli	phosphotransfe-rase.

Table 50

Field released transgenic plants (Kaul and Nirmala, 1995)

Crop species	*Genetic modification achieved*
Beta vulgaris (Sugar beet)	Sulfonylurea (HR), Glufosinate (HR)
Brassica napus	Bt protein (IR), Glufosinate (HR), Glyphosate (HR),
(Rape seed)	Seed storage protein, Oil composition, Male sterility Bar marker gene, nptII marker gene
Brassica oleracea (Cauliflower)	Male sterility
Carica papaya (Papaya)	Papaya ring spot virus (VR)
Chrysanthemum	Flower colour
Cichorium intybus (Chicory)	Male sterility
Cucumis melo (Cantalope, melon)	Cucumber mosaic virus (VR)
Cucurbita pepo (Squash)	Cucumber mosaic virus (VR)
Glycine max (Soybean)	Glufosinate (HR), Glyphosate (HR), Soybean mosaic virus (VR), Seed storage protein
Gossypium hirsutum	Bt protein (IR), Bromoxynil (HR), Glyphosate (HR),
(Cotton)	Sulfonylurea (HR), npt II marker gene

Contd.

Crop species	*Genetic modification achieved*
Helianthus annuus (Sunflower)	Seed storage protein, male sterility
Juglans regia (Walnut)	Bt protein (IR)
Linum usitatiss-immum (Flax)	Glyphosate (HR), sulfonylurea (HR)
Lycopersicon esculentum. (Tomato)	Tobacco mosaic virus (VR), Tomato mosaic virus (VR), Bt protein (IR), Glyphosate (HR), Sulfonylurea (HR), Bromoxynil (HR), Glufosinate (HR), Fruit ripening Maize transposon AC/DS
Medicago sativa (Alfaalfa)	Alfalfa mosaic virus (VR), Glufosinate (HR), Lectin protein (IR)
Nicotiana tabaccum (Tobacco)	Tobacco mosaic virus (VR), Bt protein (IR), Tobacco etch virus (VR), Sulfonylurea (HR), Glufosinate (HR) Glyphosate (HR), Bromoxynil (HR), Heavey metal tolerance, CAT marker gene.
Oryza sativa (Rice)	Marker genes, Bt protein (IR), Seed protein storage genes, male sterility
Petunia hybrida (Petunia)	Flower colour pattern genes, male sterility
Populus (popular)	CAT marker gene
Prunus domestica (Prune-Plum)	Plum pox virus (VR)
Solanum tuberosum (Potato)	X and Y viruses (VR), Potato leafroll virus (VR) Bt protein (IR), Bromoxynil (HR), Glufosinate (HR), Increased starch content gene, Sulfonylurea (HR), npt II marker gene
Zea mays (Maize)	Bt protein (IR), European cornborer (IR), Glufosinate (HR), Bromoxynil (HR), sulfonylurea (HR), Glyphosate (HR), Modified protein gene, Male sterility.

HR = Herbicide Resistance, IR = Insect Resistance,
VR = Virus Resistance

(i) Production of high value protein

Pharmaceutically useful high value peptide-leuenkephalin was produced by inserting its gene sequence into a seed of *Arabidopsis thalians* (Van Dekerckhove et al, 1989). The seeds of this plant has the polupeptide in abundance. In fact tento several hundred grams/ha of this popypeptide was obtained in the seeds produced by transgenic rape seeds having the polypeptide transgene (Krebbers and Vandekerchkhoeve, 1990). The synthesis of human albumin by the tubers of transgenic potato and the production of immunoglobins in transgenic tobacco plant represents two more major exmples of production of high molecular weight useful proteins by transgenic plants.

Tasty and Long Lived Fruits and Vegetables in Transgenics

More than half of the fresh fruits and vegetables produced annually are lost due to spoilage caused by ethylene formation that triggers fruit ripening. To delay ripening, sequestrants of ethylene are used or fruits are harvested long before they are ripe. Whereas sequestering ethylene involves the use of chemicals and consequently results in price rise, harvesting ripened fruit produces unpleasant taste on use. One of the best examples of producing tasty and long lived transgenic fruits is tomato.

Tomatoes become mushy due to the production of a softening enzyme called polygalaturonase (PG) which causes pectin breakdown in the cell pulling apart the cells. All other changes associated with tomato ripening such as flavour and colour development are not affected by this enzyme. The Calgene scientists cloned complementary DNA to tomato PG and inserted into tomato, DNA in the antisense orientation. These transgenic tomatoes exhibit decreased PG level upto 99 per cent increasing thereby the shelf life of the tomato. Most interesting and important is that this transgenic tomato does not have a foreign gene but its own genome in reverse form. The transgenic tomato is more resistant to mechanical stress associated with handling, packaging, and transport without losing compressibility. This tomato is the first genetically engineered food crop released in the US market under the name *"Flaver-Savor"*.

Engineering Male Sterility in Transgenics

Male sterility obviates artificial and chemical emasculation, enhances possibility of outcrossing and ensures hybrid production (Kaul, 1988).

Commercial outcrossing of the usable male sterility is limited by its presalent high instability, delicately balanced genetic requirements and profound environmental sensitivity. Therefore, creation of usable stable male sterility by using gene technology is an asset for commercial hybrid production. This was achieved by Mariani et al, (1990).

A strategy to engineer male sterility in tobacco and oil rape seed was devised by Mariani et al (1990). This strategy takes advantage of the tapetal specific transcriptional activity of the tobacco TA 29 gene and an RNase/RNase inhibitor defence system utilized by bacteria—*Bacillus amyloliquifaciens.* The chimeric RNase gene TI and barnase gene containing the tobacco TA 29 gene promoter induced male sterility in both these plants. This TA 29 RNase gene selectively destroys the tapetal cell layer, prevents pollen germination and results in male sterility. The barstar is produced intracellularly and protects the bacteria from the lethal effects of barnase by forming a stable complex with barnase in the cytoplasm. Mariani et al, (1992) restored male fertility in the genetically engineered male sterile oil seed rape plants by introducing the barster gene in the male sterile plants. This introduction was done by conventional crossing using male steriles as females and those having barster gene as males. Thus, both the creation of male sterility and restoration of male fertility in male steriles, when fertility genes are not available or traceable, has been done by genetic engineering; the fertility restoration being done by a proteinaceous inhibitor of the RNase under the control of the same tapetum specific promoter and introduced in plants (Dilworth, 1991). This inhibitor suppresses the tapetum destroying RNase activity fully and pollen fertility gets restored.

Transgenic Plants as Bioreactors

Transgenic plants can be used as 'bioreactors' for obtaining virtually unlimited quantities of commercially useful protein, biologically active peptides and antibodies for large-scale production (Jain, 1992; Vasil, 1991). Hiatt et al, (1989) produced antibodies in tobacco transgenic plants by using cDNA derived from mouse hybridoma mRNA. Plants expressing single gamme or kappa immunoglobulin chains were crossed to yield progeny in which both chains were expressed simultaneously. A functional antibody accumulated 1.3% of the total leaf protein in plants expressing full-length cDNA containing leader sequences. A noval carbohydrate-cyclodextrin (cyclic oligosaccharide with potentially high commercial value)—has been produced in the tubers of transgenic potato plants, which are otherwise only produced by using bacterial enzyme cyclodextrin glycosyltransferase in a batch fermenter with starch (Oakes et al, 1991). Although the level of cyclodextrin production were low, this work represents the beginning of the use of genetic engineering for the 'molecular farming' of noval compounds. Pen et al, (1992) expressed *Bacillus licheniformis* α amylase in transgenic tobacco plants and applied seeds directly in starch liquifaction, the first step in starch processing. The resulting hydrolysis products were virtually identical to those obtained from degradation with α amylase from *B. techeniformis*.

Development of transgenic plants with altered low molecular weight (lipids, sugars, secondary metabolites) and high molecular weight compounds (proteins, carbohydrates, polymers, fibres) is in rapid progress. Nearly 30 per cent of the aminoacid of seed protein are sulphur aminoacids. The Brazil nut seed protein contributed upto 8% of the total seed protein in the transgenic tobacco plants resulting in a significant increase in the methionine content. If this gene is transferred to legumes, a major improvement in seed protein quality will be achieved and the biological value efficiency ratio and digestibility of the legume proteins will enhance dramatically.

A variation of plant transgenics and the support they need from industry and government have been emphasized by Dixon (1995) and Hoyle (1995), as the transgenic plants have not only been utilized in multifaceted investigations (Table 49), but are released out of necessity because of their immense utility (Table

50). In the developed world (Table 51), the major transformant being the vector *Agrobacterium* (Table 52).

Table 51
Total number of fieldtested transgenics in different countries upto 1995

Argentina 08 Australia 13 Belgium 52 Canada 63 Chile 08 *China 07 Costa Rica 05 Denmark 09 Finland 17 France 104 *Germany 19 Israel 14 Italy 09 *Japan 13 Mexico 09 New Zealand 17 Spain 16 Sweden 12 Switzerland 19 The Netherlands 28 United Kingdom 37 United States 193 Total 672

*Complete information not available due to government or people resistance, for the development or release of transgenics.

Table 52
Major successful transformation methods for obtaining transgenic plants

Transformation Method	*Species transformed**
Agrobacterium mediated gene transfer	*Actinidia deliciosa, Alocasarina verticillata, Apium graveolens, Arabidopsis thaliana, Arachis hypogea, Armoracia rusticana, Asparagus officinalis, Beta vulgaris, Brassica carinata, B. juncea, B. napus, B. oleracea, B. rapa, Carica papaya, Citrullus lanatus, Cucumis melo, C. sativus, Daucus carota, Dendrathema indicum, Dianthus caryophyllus, Frageria vesca, Gossypium hirsutum Glycine max Helianthus annuus, Ipomoea purpurea, Juglans regia Kalachoe lacinata, Lactuca sativa, Linum usitatissimum Lotus corniculatus, Lycopersicon esculentum, Medicago sativa, M. varia, Musa acuminatu, Nicotiana tabaccum, Passiflora edulis,* Phaseolus vulgaris
	Pisum sativum, Petunia hybrida, Poncirus trifoliata Populus nigra Prunus armeniaca, P. domestica, Pyrus malus, Solanum melongena, S.

Contd.

Transformation Method	*Species transformed**
	muricatum, S. tuberosum, Stylosanthes humilis, Synapsis alba, Vicia narbonensis, Vigna aconitifolia, Vitis rupestris, V. vinifera.
Direct DNA transfer to protoplast	*Agorostis alba, Brassica oleracea B. napus, Dactylis glomerata, Festuca arundinacea, Glycine max Lactuca sativa, Oryza sativa, Triticum aestivum, Zea mays.*
Biolistics	*Agrostis palustris, Avena sativa, Carica papaya, Glycine max Gossypium hirsutum, Hordeum vulgare, Musa sapientum Nicotiana tabaccum, Oryza sativa Phaseolus vulgaris Picea glauca Populus nigra, Saccharum Officinarum Secale cereale, Sorghum bicolor, Triticum aestivum, Zea mays*
Electroporation	*Asparagus officinalis, Oryza sativa, Phaseolus vulgaris Zea mays.*

It may be possible to grow instead of producing biodegradable plastics (termed as biopolymers) in transgenic plants (Pool, 1989). A common bacterium *(Alcaligenes eutrophus),* that turns out a polymer called polyhydroxybutyrate (PHB). A single molecule of PHB consists of thousands of hydroybutyrate molecules linked end to end, and in bulk, PHB is a stiff, brittle plastic that can be used for plastic soft-drink bottles.

Conclusion

Transgenics, the neospecies, are the organisms harbouring new genes within their genome. Included in these are also the organisms in which the resident genes have either been silenced or replaced by refined foeign genes. This foreign gene incorporation is done by direct or indirect methods. The vector mediated indirect method utilizes *Agrobacterium* whereas direct method uses many chemical or physical techniques for the gene transfer has been most widely used (Table 10). This paves an easy, reliable and useful way for commercial hybrid production and the consequent increased productivity. Currently, transgenic plants range from forest and fibre plants to cereals, fruits, ornamentals and vegetables.

Transgenic Potence

Despite numerous brilliant successes and break throughs in transgenic development technology, several obstacles exist in the transgenic cloning, transfer, expression and stability. Moreover, many transgenes stay silent either immediately after the transfer or get silenced after some generations of expression (Finnegan and McElroy, 1994). In addition, the transgenics pose grave ethical, economical, ecological and technological risks. Transgenes, especially those conferring resistance to pests, diseases, herbicides and stress may get transferred by cross pollination to sexually compatible wild weedy species, offering them a selective advantage over the cultivated ones. Moreover, repeated transformations of a genome, pyramiding of several transgenes following multiple rounds of transformations and elimination of ancillary sequences are the major issues facing global transgene marketing strategy. May be transgenics lead to proliferation of new viral, fungal and insect strains that gain resistance to transgenic resistant plants. This can have serious impacts on humans, birds, insects and other animals that feed over these plants. Overgrowth of transgenic plants inhabitats where indigenous relatives of these plants ordinarily grow will diminish indigenous species, genetic richness and consequently reduce biodiversity. This paves way for the species extinction from earth. Thus, transgenics will stand on trial, time and tedious test.

Transgenic Cotton for Improved Fibre Production

Cotton is the most important fibre crop being used in the textile industry. The production of lint per unit area is very low, and one of the most important reasons for low yield is the susceptibility of the crop to insect diseases and various environmental stresses including drought and cold. Conventional methods of breeding have not been adequate to cope with these problems because of the lack of sufficient genetic variability among cultivated types. Desirable genes for insect and disease resistance drought tolerance, and superior fibre qualities that are present in some wild *Gossypium* species cannot be transferred into cultivated types because of various incompatibility barriers. Therefore, there is a strong need to resort to unconventional biotechnology approach that can be helpful to increase genetic variability.

In the genus *Gossypium*, a number of wild species (Table 53) possess superior fibre qualities that are lacking in the cultivated species. If these traits could be incorporated in cultivated types, it would greatly enhance cotton production in the world.

Table 53
Some wild species of cotton possessing superior fibre quality

Species	*Genome*
G. anomalum Wawrex Wawre and Peyr	B_1
G. thurberi Tod	D_1
G. tomentosum Nutt ex Seem	AD

The ultimate index of quality in cotton is determined by its spinning performance. Among those fibre properties that contribute most to the spinning value are fiber length, fineness and strength. Fiber uniformity and maturity are also important. Depeding upon fiber length, cotton can be classified into short 19 mm or below), medium (20-21.5 mm), superior medium (22-24 mm), long (24-26 mm), and superior long (27 mm or above). Fiber fineness is usually expressed in terms of weight per unit length of fiber.

Cotton cross-pollinates naturally. The conventional cotton breeding usually consists of introduction, selection, and hybridization. The prime requisite for the success of any crop improvement program is the presence of sufficient genetic variability for the traits that are to be improved. When the required variability is not present in a species, then it has to be induced by mutations.

In-vitro techniques form an important component of biotechnology, and have the potential not only to improve present cultivars, but also for the synthesis of noval plants and early release of high yielding resistant to various diseases, pests, stresses, temperature and such.

It has been demonstrated that cell and tissue culture methods can be manipulated for the introduction of genetic variability in cotton. Embryo/ovule culture methods have been used to produce hybrids between different species. Irradiation of seeds and pollen in segregating generations may increase chances of transferring a few genes of the wild species into the cultivated types. In-vitro environment may also be suitable for enhancing chromosome breakage and reunion events in the cultured cells. A tissue culture cycle of the hybrid material can enhance the frequency of genetic exchange between chromosomes from different species. The hybrid callus will then be a rich source of genetic variability. Once the problem of regeneration is solved, somatic hybrid cells and cellular genetic variants can be evaluated among regenerated plants, and may be included in breeding programs (Bajaj and Gill).

Chapter 6

Transgenic Animals

The introduction of genes into the germ line of animals is one of the recent advances in biology. Transgenic animals have been instrumental in providing new insights into mechanisms of development and developmental gene regulation, into the action of oncogenes, and into the intricate cell interactions within the immune system. Furthermore, the transgenic technology offers exciting possibilities for generating precise animal models for human genetic diseases and for producing large quantities of economically important proteins by means of genetically engineered farm animals.

The first animals carrying experimentally introduced foreign genes were derived by microinjection of simian virus 40 (SV 40) DNA into the blastocyst cavity (Jaenisch and Mintz, 1974). However, integration of the viral DNA into the germ line was not demonstrated in these early experiments. It was later suggested that some of the viral SV 40 DNA remained episomal in somatic tissues (Willison et al, 1983). Germline transmission of foreign DNA was detected in subsequent studies when mouse embryos were exposed to infectious Moloney Leukemia Retrovirus (M-MuLV), which resulted in the generation of the first transgenic mouse strain (Jaenisch, 1977).

Since then many transgenic animals—cattle, sheep, goats, pigs, rabbits, chickens and fish have been produced (Table 54) and will be utilized in future for a variety of purposes including:

(a) their efficiency in utilizing feed;

(b) ability to give leaner meat;

(c) ability to grow to marketable size sooner;
(d) resistance to certain diseases.

Table 54

Transgenic animals produced and the promoter, enhancer and structural transgenes involved in each case

Animal	*Genes transferred (promoter or enhancer/structural genes)*
1. Mouse	*mMT/rGH, mMT/bGH, mMT/oGH, mMT/hGH, mMT./hGRF, mMT/hFIX*
2. Chicken	*ALV, REV*
3. Cow	*BPV lactoferrin*
4. Fish	*hGH, mMT/hGH, mMT/bGal, ed-crystallin SV/hygro, AFP*
5. Pig	*mMT/hGH, mMT/bGH, hMT/pGH, MLV/rGH, bPRL/bGH*
6. Rabbit	*mMT/hGH hMT/hGH, rbEu/rb, e-mye*
7. Sheep	*mMT/hGH, mMT/TK, mMT/bGH, mMT/hGRF, oBLG/hFIX, oBLG/a. 1 AT, oMT/oGH,*
8. Goats	A variant of *tPA* gene (*LAtPA)*

Note:- The list of genes; promoter/or enhancer is separated from structural gene by a slash *(l)*. The source of the gene is given by a lower case preceding the gene: b = bovine *(ox): c* = chicken; h = human ; m = mouse: o = ovine; p = porcine: r = rat; rb = rabbit. Gene abbreviations— *ALV* = avian leukosis virus: *a1AT* = *a1* anti-trypsin; *BPV* = bovine papilloma virus; *Eu* = immunoglobulin heavy chain; *FIX* = factor IX; *GH* = growth hormoe; GRF = growth releasing factor; βGal = β galctosidase; *hygro* = hygromycin: *BLG* = β-lactoglobulin; *MT* = metallothionein; *MLV* = moloney murine leukemia virus; *c-myc* = a cellular proto-oncogene; *REV* = reticuloendotheliosis; *PRL* = prolactin ; *SV* = SV 40; *TK* = thymidine kinase; *AFP* = antifreeze protein; *tPA* = human tissue - type plasminogen activator.

More than this, efforts are being made to use transgenic animals as living bioreactors. Transgenic animals produced for this purpose will secrete valuable recombinant proteins and pharmaceuticals into their milk, blood and urine which can be used for extraction of these drugs. This new possibility of manufacturing drugs through transgenic animals is often described as *'molecular farming'*.

Methods for Introducing Genes into animals

Genetic material from one organism to another can be transferred by (a) microinjection of DNA into pronucleus, (b) retrovirus infection, and (c) embryonic stem cells with a vector carrying the gene of interest.

(a) Microinjection of DNA into pronucleus

Microinjection of cloned DNA directly into a pronucleus of a fertilized mouse egg has been the most widely used method for generating transgenic mice (Figure 40). The DNA is microinjected into a fertilized egg before the fusion of male and female nuclei. The egg is first immobilized by applying mild suction to the large blunt end of a narrow glass microneedle. Through recombination at the DNA level, the inserted gene may get integrated into the host genome and inherited in a Mendelian manner. This method has been successfully utilized for production of transgenic mice, chicken, cows, fish, pigs, rabbits and sheep.

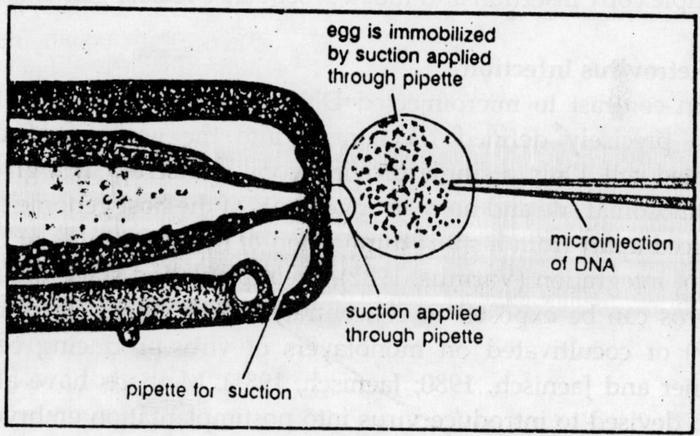

Figure 40
Microinjection of foreign DNA into a fertilized egg for production of transgenic animals

Typically, multiple DNA molecules arranged in head-to tail array integrate stably into the host genome. It is thought that the injected DNA molecules associate by homologoes recombination before integration and in most cases insert subsequently at a single chromosomal site. It has been proposed that random chromosome breads, possibly caused by repair enzymes that are induced by the free ends of the injected DNA molecules, may serve as integration sites of the foreign DNA (Brinster et al, 1985). Frequently rearrangements, deletions, duplications (Mark et al, 1985), or translocations (Mahon et al, 1988) of the host sequences occur at the insertion sites. However, the injected DNA does not always integrate into the host genome.

The principal advantage of direct microinjection of recombinant DNA into the pronucleus is the efficiency of generating transgenic lines that express most genes in a predictable manner. However, one disadvantage of this method is that it cannot be used to introduce genes into cells at later developmental stages. Moreover, the cloning of the chromosomal insertion site may be difficult because of the multiple copy insertion and the host sequence rearrangements.

(b) Retrovirus infection

In contrast to microinjected DNA, retroviruses integration by a precisely defined mechanism into the genome of the infected cell. Only a single proviral copy is inserted at a given chromosomal site and no rearrangements of the host genome are induced apart from a short duplication of host sequences at the site of integration (Varmus, 1982). Preimplantation stage mouse embros can be exposed to concentrated virus stocks (Jaenisch, 1977) or cocultivated on monolayers of virus-producing cells (Jahner and Jaenisch, 1980; Jaenisch, 1981). Methods have also been devised to introduce virus into postimplantation embryos. While this allows infection of cells from many somatic tissues, germ cells are infected with a low frequency (Soriano and Jaenish, 1986).

The main advantage of the use of retroviruses or retroviral vector for gene transfer into animals is the technical ease of introducing virus into the embryos at various developmental stages. Furthermore, it has proved much easier to isolate the flanking host sequences of a proviral insert than those flanking a DNA insert derived from pronuclear injection. This is of considerable advantage when attempting to identify the host gene disrupted by insertion of proviral DNA. The main disadvantage of the use of retrovirus gene transfer are the size limitations for transduced DNA and the unresolved problems of reproducibly expressing the transduced gene in the animals.

(c) Gene targeting using embryonic stem (ES) cells

Embryonic stem cells are established in-vitro from explanted blastocysts and retain their normal karyotype in nature. These stem cells are injected with a vector containing the desirable gene into the host blastocyst. This allows targetting of gene to a particular site by homologous recombination. The rare targetted cells are isolated, multiplied and used for injection of explanted blastocysts, which are then implanted into *surrogate mothers* to allow the development to be completed. The different steps of the technique have been shown in Figure 41.

By means of this technique, mice have been generated from cell clones that were selected in-vitro. This opens exciting possibilities for deriving mouse strains carrying specific mutations.

The last few years have witnessed an extraordinary increase in the use of transgenic animals. Each of the three methods for generating transgenic animals has distinct advantages for some and disadvantages for other applications.

It is likely that rapid advances will occur in future in the following areas: (i) It will be important to isolate and characterize chromosomal regulatory elements controlling developmental gene activation over large distances. Inclusion of such elements in gene constructs should guarantee predictable and efficient expression independent of the chromosomal integration site. This will be particularly important for genetic engineering of large farm animals where cost constraints limit the number of transgenic lines that can be generated and evaluated; (ii) The various possibilities of marking early embryonic cells or ablating specific lineages give experimental access to stages of mammalian development as yet not amenable to easy experimental manipulation. This undoubtedly will accelerate our understanding of the complex cell interactions in mammalian development; (iii) The prospect for generating recessive or dominant mutations in preselected genes not only will permit the derivation of precise animal models for human hereditary diseases but also will mark the beginning of a systematic genetic dissection of developmental processes that will radically change the future of experimental mammalian genetics.

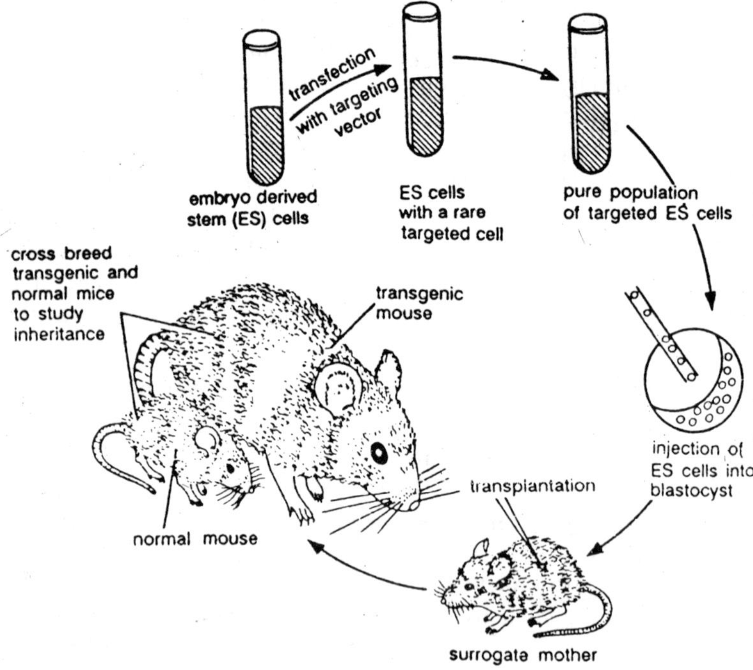

Figure 41
Embryonic stem cell transfection with a vector carrying the gene of interest, to produce transgenic mice

Improved Wool Production in Sheep Through Genetic Engineering (Transgenic Sheep)

Wool is a natural fibre popular in the manufacture of outer clothing and is a vital source of income for several countries. Wool has increased in popularity for the manufacture of quality fabrics since textile research overcame problems such as shrinkage and felting. Present day fabrics exhibit the characteristics of pure wool (such as the thermal insulation even when moist) and have fine texture, resilience, excellent drape and durability.

Australia heads the list of wool trading countries, producing one-third of the world's wool, followed by South Africa, Argentina, Urguay, and New Zealand. In United Kingdom there are about thirty different sheep breeds, whereas in Australia the main breeds are high-yielding fine wool sheep based on the Merino and developed since the first Spanish Merinos arrived in 1797 (Ryder and Stephenson, 1968). Selective breeding has produced sheep with high fleece densities, the product of a total body population of 10^8 hair follicles, each of which grows a fibre that is highly uniform in diameter (20 um or less). The most important consequence of this breeding is that wool growth is limited by the competition between the follicles for available nutrient supply (Black and Reis, 1979; Black, 1987). Central to this limitation are the supplies of energy for cell proliferation and of aminoacids for the synthesis of the keratins of the wool fibre. Each breed of sheep has a genetically determined follicle, the actual rate of wool growth in an individual sheep can vary 4-fold depending on the composition and the amount of the aminoacid supply (Reiz, 1989). Gluconeogenesis, which uses aminoacids, adds to the burden on the sheeps aminoacid supply (Black, 1987). The microflora of the rumen can also contribute to the aminoacid supply.

The digestive pathway in the sheep and the ways in which nutrients are used for wool production are summerised in Figure 42. The use of recombinant DNA techniques can potentially intervene at some point to enable sheep to each their maximum capability of wool growth and increase the efficiency with which they use feeds.

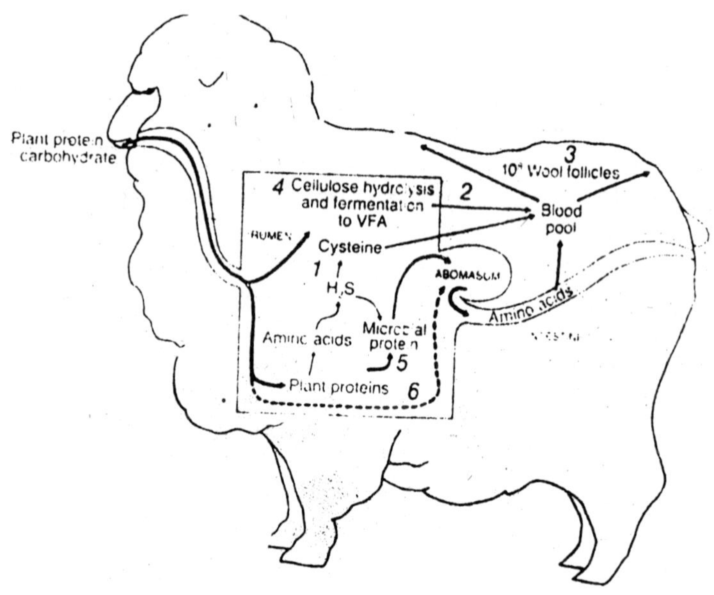

Figure 42
Ruminal/abdominal digestion in relation to wool growth in sheep

There are three principal approaches for genetic manipulation: (a) to introduce new genes into the sheep's genome, (b) to introduce new genes into one of the ruminal micro-organism; or (c) to engineer forage plants which make proteins which can withstand ruminal breakdown.

(a) Insertion of new genes into the sheeps genome

One way to produce transgenic sheep is to insert bacterial genes for cysteine synthesis. Two genes were isolated from *Salmonella typhimurium,* separately fused to the SV 40 late promoter and polyadenylation signal in a plasmid vector and expressed in sheep cells in culture, producing active serine acetyltransferase (SAT) and O-acetylserine sulphydrylase (OAS). The two genes have been linked in a plasmid construct so that they can be transferred together. With both genes separately under the control of the promoter derived from the long terminal repeat of Rous sarcoma virus (RSV-LTR). Constitutive expression has been observed in cultured sheep cells. Microinjection of the double gene construct into sheep yielded one transgenic animal per four lambs born.

The genes from *E. coli* encoding SAT and OAS have been fused with the propoter from the sheep metallothionein gene (Ia (MT-Ia).

A pathway involved in promoting glucose synthesis in the liver or in the wool follicles is a possible target for genetic engineering. The sheep obtains 70-80 per cent of its energy from acetate and other volatile fatty acids produced by carbohydrates fermentation in the rumen. Most of the glucose is produced from propionate and aminoacids through gluconeogenesis, especially when sheep are on a poor quality diet. The demand for glucose reduces the availability of aminoacids for wool growth. A way to supply glucose would be to by-pass part of the tricarboxylic acid (TCA) cycle by introducing two genes for two enzymes of the glyoxalate pathway, which functions in bacteria but not in mammals. The first step in the pathway is the conversion of isocitrate to succinate and glyoxalate. In the second step, glyoxalate is converted to malate (Figure 43). The overall result is that two molecules of acetate yield succinate; this could increase the amount of glucose produced via gluconeogenesis.

Genes essential for synthesis of some important aminoacids found in the Keratin protein of wool, have been cloned and introduced in embryos to produce transgenic sheep. For instance, genes (Cys E and cys M) for two enzymes (Serine acetyltransferase=SAT and O-acetylserine sulphydrylase=GAS), involved in cysteine biosynthesis, were isolated from bacteria and cloned in a vector. These genes were introduced in sheep cells, ultimately leading to the production of transgenic sheep, where these genes are expressed.

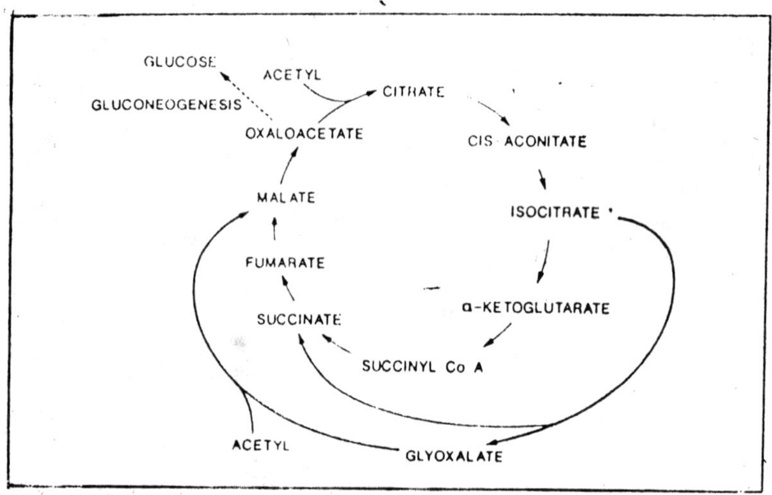

Figure 43
Tricarboxylic acid cycle with glyoxalate pathway by-pass in *E. coli*

The two genes for the glyoxalate pathway, aceA and aceB, have been isolated form *E. coli* and inserted into a plasmid under the control of promoters that function in animal cells. Glucose produced in this way would not only reduce the use of aminoacids but would also supply the energy needed for wool growth (Chapman and Ward, 1979).

Growth hormone (GH) genes have also been introduced into transgenic sheep, for basic studies of harmonal regulation of growth, body growth in lambs, to increase the efficiency of food utilization, and to increase wool growth rate. This is some what surprising because injected GH causes a decrease in wool growth rate, followed by an increase when discontinued.

(b) Insertion of new genes into rumen bacteria

It is the bacteria rather than the protozoa and anaerobic fungi which are essential for the digestion and fermentation of the cellulosic pasture feeds consumed by sheep. Most of the rumen bacteria (about 10^{10} cells per ml rumen contents) are obligate anaerobes, belonging to the genera *Bacteroides, Ruminococcus,* and *Butyrivibrio.* Cellulases produced by these bacteria degrade celluloses and hemicelluloses to sugars, which are then fermented to produce the VFAs, mainly acetate, propionate and butyrate. In the ruminal breakdown of the plant fibres, the plant proteins are hydrolysed, and the aminoacids degraded and then resynthesized into microbial proteins. The sheep obtains its supply of aminoacids from the rumen bacteria.

The manipulation of ruminal bacteria with noval genes aims either to improve either the efficiency of cellulose breakdown, or the availability of essential aminoacids. Our knowledge of ruminal plasmids and bacteriophages, which could be the basis of stable transformations of ruminal bacteria is still primitive.

(c) Insertion of new genes into forage plants

The genetic manipulation of forage plants is another way in which higher amount of aminoacids might be supplied for wool growth.

A low molecular weight seed protein (pea albumin 1 or PA_1) rich in cystine was found to be resistant to rumen fluid. The gene for PA_1, and another pea seed-protein, vicilin, have been introduced into tobacco test plants *(Nicotiana tobaccum)* and into lucerne *(Medicago sativa)* utilizing the binary vector system derived from the Ti plasmid of *Agrobacterium tumefaciens.* The PA_1 and vicilin were both expressed in the leaves of tobacco and lucerne.

There are several criteria for genetic manipulation of forage plants with their utility for increased wool production by sheep. First, a gene which encodes a protein with the appropriate aminoacid composition is required. Second, the protein must be relatively resistant to ruminal degradation. Third, the gene must be highly expressed in the leaves and stems of pastures species such as lucerne and subterranean clover.

Chapter 7

Biotechnology for Pollution Abatement

We are so much obsessed with and possessed of industrial growth that our ecosystem has undergone many ramifications and diversions; its pristine glory and vigour, vitality and utility has been completely lost. Therefore, it is the boundain duty of us all to be aware of what is happening around us and how this dither-down can be halted. In the initial stage of industrialization, more importance was attached to stepping-up of production—regardless to the need to install pollution free technologies. While some form of environmental pollution has always existed and will continue to exist, it is only in the last two decades or so that we have found the concern for environmental pollution becoming more and more pervasive. The time clearly is now to have a look at our technological capabilities to protect our environment (Agarwal, 1996).

So far we have relied more upon physical and chemical methods of pollution control. Today, biotechnology is being considered as an emerging technology in environmental protection. It involves the use of micro-organisms, the oldest inhabitants of the earth, which are likely to prove as more suitable for pollution control due to their versatility and adaptability to changing environments.

The application of biotechnology in the field of air pollution abatement and water pollution abatement shall be discussed here in more details.

Biotechnology for Air Pollution Abatement

Sulphur dioxide, Nitrogen oxides, Volatile organic Compounds and Particulates are the four major components of air pollution and are responsible for environmental hazards. Sulphur dioxide and Nitrogen oxides gases are just two of the four main pollutants known to be important determinants of local air quality. Addition of Volatile Organic Compounds and fine Particulates rounds out what Rhodes and Middleton (1983) have termed the *"gand of four"* air pollutants.

Increase in environmental awareness has resulted in more attention of people to air pollution. Pollution is sensed by people by offensive odour just before their receiving the damage. Waste gases with an offensive odour may be generated during the production process or they may generate from open water—water treatment plants and garbage composting plants.

Three types of biological waste gas purification systems are in operation. These are: (1) Bioscrubbers, (2) Biofilters, and (3) Biotrickling filters.

(1) Bioscrubbers

A typical bioscrubber consists of an absorption column and one or more bioreactor (Figure 44 and 45). Biological oxidation tales place in these bioreactors. The reaction tank are aerated and supplied with a nutrient solution. The microbial mass mainly remains in the circulating liquor which passes through the absorption column. Circulation rate is fast and not much of biofilm will develop in the absorption column. The biofilm is removed from time to time.

(2) Biofilters

Many micro-organisms have the ability to oxidise volatile organic compounds. Biofilters may be of open type or closed type (Figure 46). Soil, compost, peat, heather, bark etc are used in combination in biobeds. Uniformity, permeability of biobeds will decide the proper gas treatment and bypassing or chocking should not occur. Biobeds require a lot of space, they are turned 2-3 times in a year. Proper drainage is essential at the bottom. Height of packing of bed is 1 m and flow rate of gas is 130 m^3/

hm^2. Micro-organisms used in biofilters are mesophilic. Temperature 15-40°C, moisture 40-60% and gas contact time 10-30 seconds are maintained for biofilters.

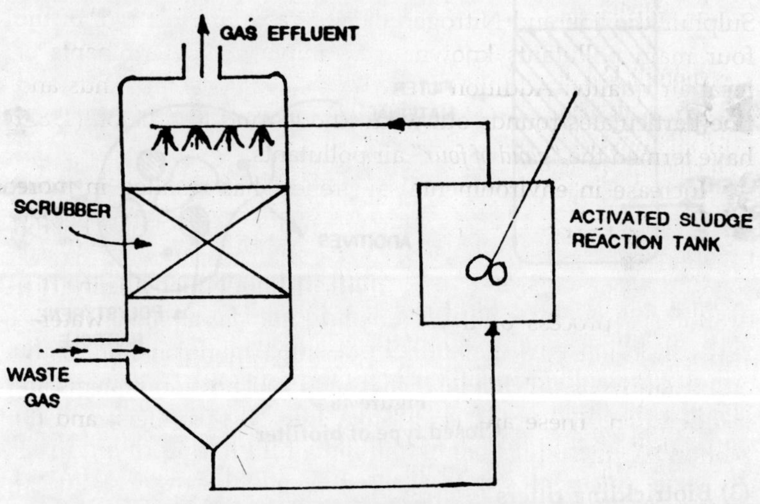

Figure 44
Activated sludge bioscrubber

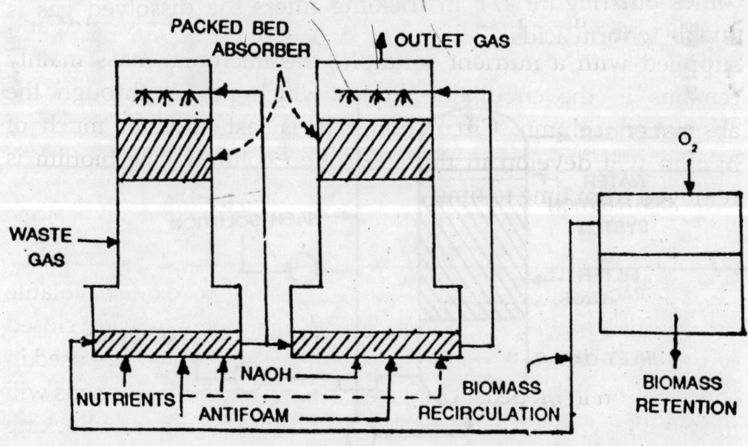

Figure 45
Venturi scrubber

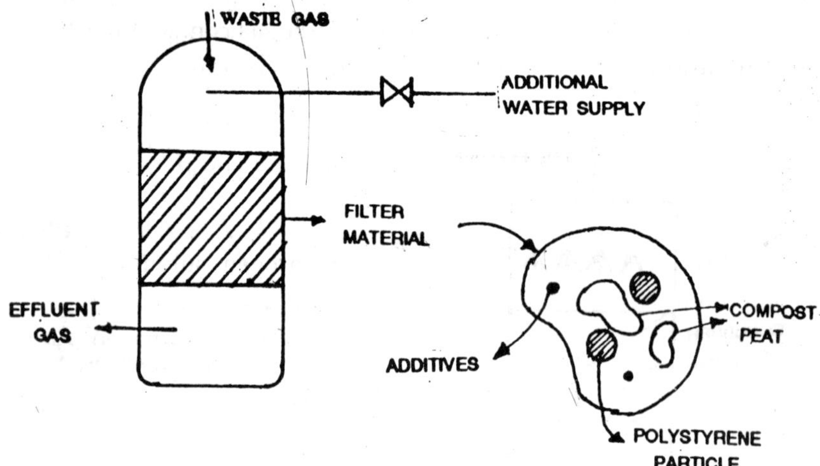

Figure 46
Closed type of biofilter

(3) Biotrickling filters

They possess a biofilter packing through which gases are passed, water trickles down dissolving the gas, and the pure air comes out (Figure 47). In trickling filters the dissolved gas is unable to form acids.

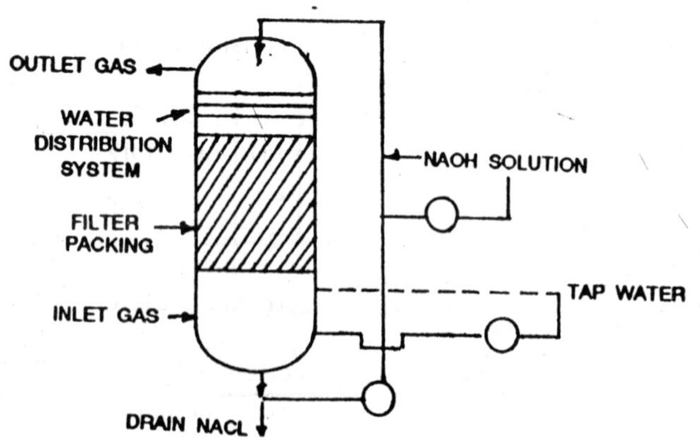

Figure 47
Biotrickling filter

Biosorption of polluted air have been increasing in recent past. A large number of micro-organisms have been observed to degrade specific air pollutants (Table 55). Pure cultre of *Pseudomonas* sp degrades Trichloroethylene (TCE). Genetically engineered *E. coli* have been found to be efficient TCE degrader. Volatile organic compounds are degraded and end products are carbon dioxide, water, biomass and inorganic salts. The process is cheaper than thermal and catalytic oxidation. *Chiobacillus ferroxidans* is used for treatment of Hydrogen sulphide and Sulphur dioxide, and gives solid sulphur.

Biotechnology for Water Pollution Abatement

The undesirable waste characteristics of polluted water include the following: (i) Suspended solid and soluble organic compounds, which undergo progressive decomposition and thus result in oxygen depletion and production of noxious gases; (ii) Heavy metals, cyanides and other toxic organics which are deleterious to aquatic life; (iii) Undesirable levels of nitrogen and phosphorus, which enhances eutripjication (excessive plant growth, which kills animals due to deprivation of oxygen) and stimulate undesirable algal growth; (iv) Non-biodegradable chemical and volatile materials like Hydrogen sulphide and Sulphur dioxide.

The objective of biological treatment of waste water are to coagulate and remove the non-settleable colloidal solids and to stabilize the organic matter. They are considered better because they not only remove colour, but help in removal of carbonaceous organic matter in waste water which is measured as BOD (Biological Oxygen Demand), or COD (Chemical Oxygen Demand), or TOC (Total Organic Carbon), nitrification, denitrification and stabilization are the purposes of biological treatment.

Biological treatment can be carried out if the effluents are rich in unstable organic matter. The microbes break-up these unstable organic pollutants into stable products like carbon dioxide, carbon monoxide, ammonia, methane, hydrogen sulphide etc.

Table 55
Microbial degradation of air pollutants with intense odours

	Substrate	*Micro-organism*	*Degradation Products*	*Degradation Pathway*
(1)	Methanol, formaldehyde	*Pseudomonas* AMI	CO_2, water	Assimilation via serine, transhy-Broxy methylase
(2)	Lower alcohols and fatty acids	Many bacteria and fungi	acetyl COA	β oxidation
(3)	Dimethylamines	*Pseudomonas aminovorans*	Methylamine and formaldehyde	Hydroxylase
(4)	n-propylamine	*Mycobacterium convolutum*	Propionate	Amine dehydrogenase
(5)	Phenol	*Pseudomonas putida*	Acetaldehyde and pyruvate	Meta cleavage
(6)	Phenol	*Trichosporon cutaneum*	Acetyl COA and succinate	Ortho cleavage
(7)	p-cresol	*Pseudomonas fluorescence*	Protocatechuic acid	Ortho cleavage
(8)	m-cresol	*Pseudomonas* Spp.	Fumerate and pyruvate	Gentisic acid pathway
(9)	Benzaldehyde	*Acetobacter ascendens*	Benzyl alcohol & benzoic acid	Dismutation
(10)	Aniline	*Nocardia* Spp. *Pseudomonas* Spp.	Pyrocatechol	Dioxygenation
(11)	Indole	*Chromobacterium violaceum*	Pyrocatechol	—
(12)	Indole	*Neurospora crassa*	Tryptophan	—
(13)	Camphor	*Pseudomonas putida*	Lactonic acid	—
(14)	Dimethyl sulphide	*Hypomicrobium* Spp.	—	—

Industrial effluents vary in load, concentration of pollutants, toxic materials and are often nutritionally unbalanced. They may contain biodegradable or non-biodegradable or both types of pollutants. They show seasonal variations related with production.

Biological treatment of waste waters is a relatively neglected area and is also less properly understood and controlled. Many people do not give importance to choosing of right biological treatment method. Many people handle biological treatment in the most crude manner. Ignorance about the composition of effluents just cannot be excused.

Effluent treatment systems broadly fall into two categories: (i) Aerobic, and (ii) Anaerobic (Table 56).

Aerobic Biological Treatment

They are based on the system of sewage treatment and are meant to handle easily biodegradable organic matter. When applying to industrial effluents, a careful treatability study should be done to determine the design and working parameters. The basic reaction in aerobic treatment plant is:

$$\text{Organic material} + O_2 \xrightarrow[\text{Other Nutrients}]{\text{Cells}} CO_2 + H_2O + \text{New Cells}$$

Microbial cells undergo progressive auto-oxidation of the cell mass:

$$\text{Cells} + O_2 \rightarrow CO_2 + H_2O + NH_3$$

Lagoons and low rate biological filters have limited industrial applications but activated sludge and fixed film systems of different types are widely used for industrial effluents. Advanced activated sludge systems use pure oxygen instead of air and can operate at higher biomass concentration.

Activated sludge process

It operates as a homogeneous continuous culture and is aerated. Biosorption and floculation removes the organic matter rapidly while oxidation and biosynthesis proceed at a lower rate. Subsequently flocs settle into the next stage of secondary

Table 56

Major points of comparison between aerobic and anaerobic treatment of industrial effluents (Jogdand, 1995)

Aerobic Treatment	*Anaerobic Treatment*
(1) Range of waste-waters can be treated	(1) Limitation in treatment applications.
(2) Process stability and control.	(2) Less process stability and control.
(3) Require more power input.	(3) Require less power input.
(4) Produce more sludge.	(4) Produce less sludge.
(5) Percentage BOD removed is more.	(5) Percentage BOD removed is less.
(6) Nitrogen removal is better.	(6) Nitrogen removal poor.
(7) Phosphorus removal is better.	(7) Phosphorus removal poor.
(8) Can cope up with low substrate level wastes.	(8) Advantageous for high substrate level wastes.
(9) Works in meophilic range of temperature.	(9) Advantageous at higher temperatures.
(10) —	(10) Reductive dechlorination occurs.
(11) Nitrification, denitrification phosphorus accumulation, ligninase activity occur.	(11) —

sedimentation tank. A portion of this floc may be returned as inoculum. BOD and suspended solids are reduced by 85-95 per cent. Organisms in activated sludge are similar to that in percolating filters. In the activated sludge, conditions are not suitable for macro-invertibrates grazer population. Resultantly, activated sludge plants do not suffer from nuisance by flies. Also fungi are less dominant. So less sludge bulking. Protozoans are abundant in activated sludge. Nematodes and rotifers are small in number.

The contents of the reaction vessel are referred to as Mixed Liquor Suspended Solids (MLSS) or Mixed Liquor Volatile Suspended Solids (MLVSS) and consists mostly of micro-organisms and inert and non-biodegradable suspended matter. The original activated sludge process was introduced in 1914. It has certain drawbacks: (1) High running costs; (2) Difficult to operate and maintain; (3) Produce large surplus biomass. There are various modifications of activated sludge process:

(A) *Tapered aeration:* Here aeration capacity is related to demand and it is less at the outlet than at the inlet.

(B) *Step aeration:* Here feeding as well as aeration is done at steps in the system throughout length of the tank (Figure 48).

(C) *Contact stabilization:* Here returned sludge is aerated to encourage organisms to utilize any stored nutrients. More wastes digested. Sludge volume is reduced through aerobic digester stage. It is similar in principle to the extended aeration treatment. For extended aeration treatment, aeration and mixing of sludge and effluent is done in the same unit which is conducted separate in contact stabilization.

(D) *Advanced activated sludge process:* Most of them operate with pure oxygen. So they can operate at a higher biomass concentration. This reduces residence time and bulking (i.e. excessive growth of filamentous bacteria and fungi which may inhibit sludge setting) is inhibited.

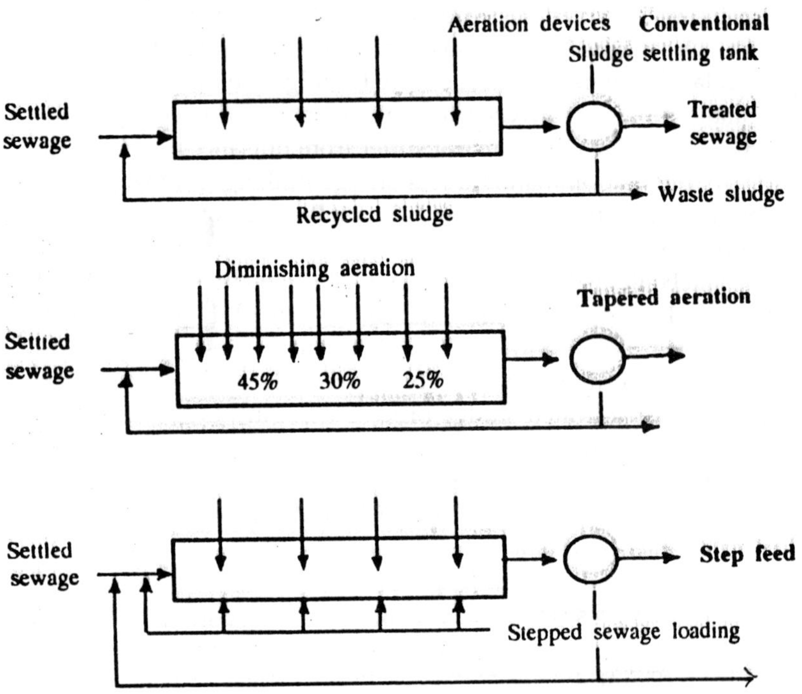

Figure 48
Modifications of activated sludge process

Powdered activated carbon (PAC) may be added as additive in the activated sludge process. This absorbs organic compounds which cannot be degraded. It also reduces colour and metal in the effluent. It reduces effective level of inhibitors. PAC can also overcome toxicity to nitrification in activated sludge plant treating coking plant effluents.

Activated sludge is still considered as a cost-effective technology. The common problems of activated sludge are nutritional deficiency, pH our of 6-8.5, temperature beyond mesophilic range, poor flocculation of biomass due to perhaps very few different degradable substrates and few species of bacteria.

Carrier Activated Sludge Process

Activated sludge systems modified by the addition of inert suspended solid particles (carriers) are widely used. These CASP systems combine the characteristics of fluidized bed systems and conventional activated sludge systems. Here biomass concentration may vary from 8000 to 30,000 mg/l as MLVSS. Disadvantages experienced in fludized bed systems like ineffective suspended solid removal and operational difficulties are overcome here. CASP requires minor modifications for existing process installations conversion. Biomass is attached as well as in flocs form. Surface area is considerably increased. Biomass concentration is much more than conventional activated sludge process. Carriers may include sand, plastics, glass, powdered activated carbon, clay etc. Microcarriers if used help flocs formation in addition to providing of attached growth support. Biological flocs capture particulate matter and hence effectively remove fine suspended solids.

Advanced Activated Sludge Process

(High Purity Oxygen Systems)

Closed tank systems	Open tanks systems
(1) Monitoring and Maintenance is difficult.	(1) Ease of access for monitoring and maintenance.
(2) Less treatment capacity.	(2) Treatment capacity is increased.
(3) CO_2 accumulates in Head space. Hazardous.	(3) Allows escapes of CO_2 from mixed liquor to atmosphere.
(4) Examples — (i) UNOX system (ii) OASES system (iii) Forced Free-full 'F^3O' system (iv) Marox system	(4) Examples — (i) Vitox system (ii) Megox system (iii) Primox system (iv) Simplox system
(5) It prevents loss of unutilised oxygen and permits recirculation.	(5) Unutilised oxygen is lost.
(6) It prevents odour and spray nuisance which are normally	(6) Odour and spray nuisance exists.

associated with activated sludge processes.	
(7) Combustible materials like oil, fat, grease burn vigorously in O_2 rich atmosphere. Hydrocarbons should not be allowed to accumulate in head space by doing purging and also oxygen should not come in contact with hydrocarbon lubricants. This will reduce possible hazards. Chances of hazards are also low as system is operated at ambient temperature and low pressure and gas phase being saturated with water vapour.	(7) No possibility of hazard as no accumulation of hydrocarbons in headspace.

Biological filters—Fixed film systems

Micro-organisms are attached to an inert supporting medium which is packed into a tower or tank. There are some variations like rotating biological contactors which basically are aerobic systems with fixed film of micro-organisms growing on discs on a rotating shaft. Even distribution of effluent is done in the reactor and air is introduced from the bottom vents and passes it across the media bed as effluent percolates or distributes.

Microbial slime develops on media support by using organic matter and oxygen. When the thickness of the slime increases, extra biomass sloughs off. This sludge is collected by gravity in a sedimentation tank. Sludge is later on treated and disposed off the same way as for activated sludge.

The first trickling filter was put in operation in England in 1893. Filter media used in trickling filter normally consists of rocks varying in size from 25-100 mm in diameter. The depth of the rock varies with design and ranges from 3 to 8 ft. The use of plastic medium in trickling filters is relatively a recent innovation and tanks are square-shaped with depths of 30-40 ft. Rock filter beds are circular with a rotating arm distributing liquid effluent at the top of the bed. Filters have an underdrain system to collect sludge and treated effluents. Organic matter is decomposed by micro-organisms grown as film on inert support. Development of slime, metabolic activities of micro-organisms in slime, increase in thickness of slime, detachment of slime when micro-organisms near to medium reach endogenous phase (lose the ability to cling), formation of new slime is a continuous process occurring in a cyclic manner. Facultative bacteria like *Achromobacter, Flavobacterium, Pseudomonas, Adcaligenes,* filamentous forms like *Sphaerotilus natans, Baggiatoa* and at lower levels of bed nitrifying bacteria like *Nitrosomonas, Nitrobacter* are common. Also few fungi, algae and protozoa are present.

Percolating liquid washes the slime off the medium. Sloughing off depends on organic and hydraulic loading of the filter. Hydraulic loading accounts for shear velocities and organic loading accounts for the rate of metabolism in the slime layer. On the basis of hydraulic and organic loading rates, filters are divided into two classes: low rate and high rate.

Rotating Biological Contactors (RBC)

This is one of the principle types of fixed film-moving medium systems used as a digester. Biological growth is established on disc surface made of polystyrene, polyethylene, polypropylene, stainless steel, cement, aluminium, glass, PVC, rubber, teflon, wood, wire screens etc.

Discs are 2-3 meter in diameter, 10-20 mm wide and mounted on a horizontal shaft. The distance between adjacent discs is 20 mm. Discs are partly submerged (40% area in the medium). Discs are rotated at 1-7 revolutions per minute. Air is sparged to reduce the risk of anaerobiosis in multiple unit systems. Retention time of medium is comparable with percolating time. Biomass on discs is 200 gm dry weight per square metre of disc surface. Biological growth is 2.4 mm thick (Figure 49).

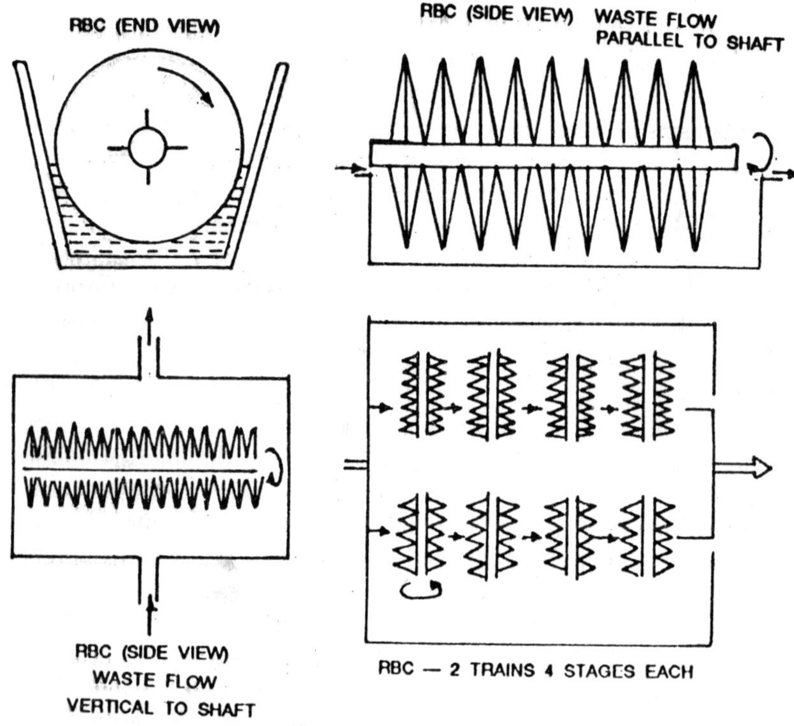

Figure 49
Rotating biological contactors (RBC)

Advantages of RBC

(1) They are simple to operate, have a low maintenance (only lubrication is required).
(2) It accommodates shock loadings.
(3) It does not have the channeling problem as in the percolators.
(4) It has reduced power costs (no sludge recycling) and less operational and maintenance costs, too.
(5) It requires less space than activated sludge plant.
(6) Fast startup, efficient mixing, little sloughing off of biomass.

(7) Effluent quality achieved is as good as after tertiary treatment.
(8) Foaming, aerosol, airstripping is reduced.
(9) Problems faced in trickling filters (percolators) like clogging, ponding, filter flies are eliminated.
(10) High waste-water temperature, oxygenation may be easier with RBC than with the activated sludge system.
(11) Lower head loss compared to the trickling filters.
(12) Process is self-regulating with respect to cell retention time and stability of process is a plus point as biomass neatly attaches to media support.

Disadvantages/Problems with RBC

(1) Lack of operational control due to oversimplicity of the process.
(2) Process is still new and people have less experience.
(3) There are difficulties in kinetic evaluation due to the complexity of interactions between biomass and solid, liquid, gaseous phase.
(4) Undesirable heavy growth of *Baggiatoa, Thiothrix* may occur. More sludge production.
(5) High total Vs. Soluble BOD in RBC effluents.
(6) Dissolved oxygen may remain limited.
(7) Odour problem.
(8) Enclosures required for RBC to protect it from sun, cold and heavy precipitation.

RBC systems find applications for effluent treatment for the following purposes: (a) secondary treatment, (b) nitrification, (c) landfill leachate and runoff, (d) pretreatment of water supplies, (e) denitrification, and (f) phosphate removal etc.

Fluidised Bed Reactors (FBR)

It is combination of attached growth (percolating filter) and suspended growth (activated sludge) systems. Biological slime film is developed and maintained on a solid support medium consisting of particles small enough to be maintained in suspension by the upward flow of liquid being treated. Support medium particles neither sink nor outflow. Reactors are generally cylindrical with perforated distribution plates and tapered or conical entry sections. Fluidization of support particles is allowed but clumping prevented (Figure 50).

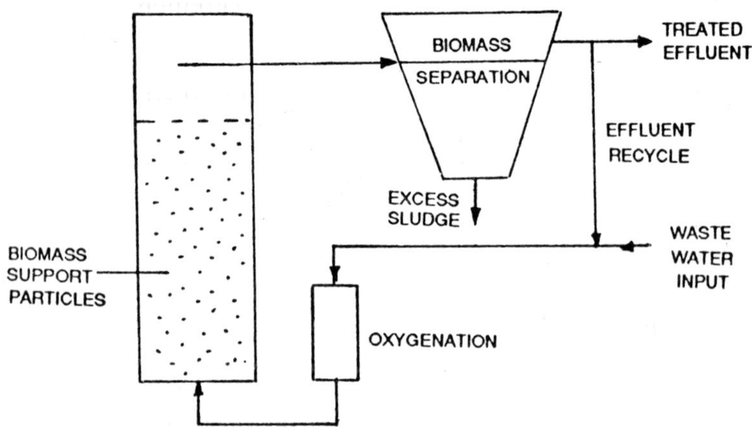

Figure 50
Fluidised bed reactor

Support media used may be sand, carbon, flyash, anthracite, glass calcinated clay with particle size 0.2 to 0.3 mm. Factors important in the selection of support particles are particle density, particle size, size distribution, surface area, sphericity, composition of material, hardness, bulk density, cost. These mineral media support growth only on the outer surface. Smaller the size, larger is the surface of the film action but lower is the setting rate is reactor. There are also fabricated media developed which allow biomass to grow within the porous internal structure.

Advantages of FBRs

(1) Compact small reactors; so space saved.
(2) Easy to install, easy for shipment. Only piping to be done on site.
(3) It is a quick, economical expansion alternative in treatment plant.

(4) Modules can be easily added to the existing system.
(5) It is suitable for high-strengh industrial waste-water with >1000 COD.
(6) Low capital and operating costs.
(7) Higher reactor biomass concentration than other suspended growth or fixed film systems.
(8) Highly efficient as compared to other systems.
(9) Secondary clarifiers can be possibly eliminated for solid separation and recycling.
(10) High exposed liquid/biofilm surface area.
(11) Sludge transport and disposal costs are lower.
(12) Low biomass sludge production.

Inverse fluidised bed biofilm reactor (IFBBR)

Fluidised bed reactors that are in use today are all operating with upflow systems, consisting of gas-solid, liquid-solid or gas-liquid-solid phases. Density of bioparticles is higher than the density of the medium. Upflow biofluidization is inconvenient for certain applications carrying aerobic processes. Collisions between the bioparticles and shear stress affects biofilm formation and then biofilm thickness. Inverse fluidization (Figure 51) removes the drawbacks of normal upflow FBR and gives a higher performance. Three advantages of inverse biofluidization are: (a) effective and simple control of biofilm thickness, (b) large specific support surface area, and (c) fast biofilm formation.

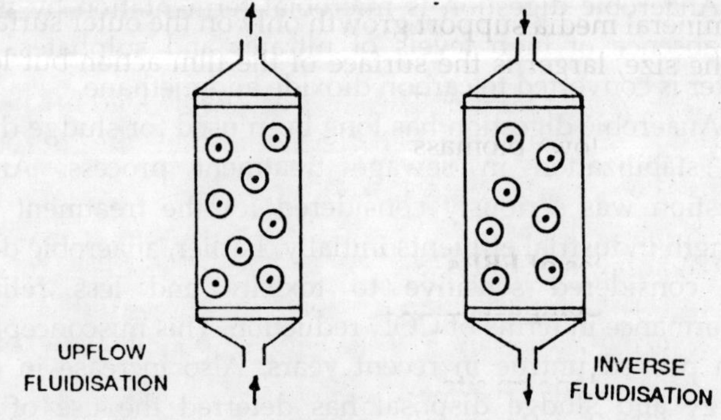

Figure 51: Upflow and inverse fluidisation bed biofilm reactor

Expanded bed reactor (EBR)

The basic mode of operation is similar to the packed filters and fluidized bed processes. Fluidized processes were developed earlier and have diverse applications. The expanded bed system uses the operating mode of fluidized bed reactor. For biological applications, an expanded bed is a different process than fluidized bed in certain aspects. These are: (a) velocities or maintain the delicate attached living film, (b) separation, retardation and biocoagulation of fine suspended solids, (c) achievement of maximum biomass concentration (Figure 52).

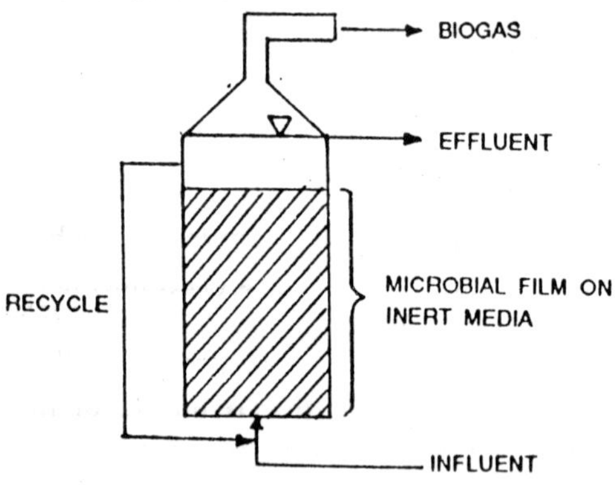

Figure 52
Expanded bed reactor

Anaerobic Biological Treatment

Anaerobic digestion is microbial fermentation by which in the absence of high levels of nitrates and sulphates, organic matter is converted to carbon dioxide and methane.

Anaerobic digestion has long been used for sludge digestion and stabilization in sewage treatment process. Anaerobic digestion was seriously considered for the treatment of high strength industrial effluents initially. Earlier, anaerobic digestion was considered sensitive to toxicity and less reliable in performance in terms of COD reduction. This misconception has been proved untrue in recent years. Also increase in costs of energy and sludge disposal has deterred the use of aerobic biological treatment procedures which are energy consuming sludge producing and acting towards net loss of material. As against this anaerobic digestion of effluents has several advantages over aerobic digestion. These are:

(1) very low sludge production.
(2) Lower consumption of energy.
(3) Production of methane which has high calorific value.
(4) Process can operate at high organic loading rate.
(5) No environmental nuisance of odour, aerosol as is the case with aerobic treatments.
(6) Micro-organisms can remain dormant for several months and become operational within a week of start up. This is suitable for seasonally produced waste-waters.

Anaerobic contact digesters

This is an aerobic equivalent of activated sludge process. It consists of a stirred tank and a tank under anaerobic conditions. The output of completely stirred tank (digester) is settled under anaerobic conditions and a part of settled sludge in returned to the digester. This results in the concentration of the sludge and longer retention time. This enables retaining of methanogenic organisms over a wide range of loading. Separation of bacteria is hindered by gassing of the effluent, hence the effluent is usually degassed before settling of the biomass.

Anaerobic contact digesters (Figure 53) are currently used for treating effluents from sugar processing, distilleries, citric acid and yeast production, industries producing saurkraut, canned vegetables, pectin, starch and meat products, farm slurries. Contact digesters are not much affected by the suspended solids in the feed as it happens in retained biomass types of digesters.

One of the main problems with contact digesters is the poor settlement of solids because of attachment of product gas to solid particles.

Packed bed reactors (Packed column reactors)

They are simple in design, easy to construct and operate (Figure 54). Organisms are contained within the packing medium in an enclosed vessel and liquid wastes pass upwards. Organisms do not form slime on packing. Regular backwashing will prevent clogging. This will also prevent high concentration of suspended solids sloughing continuously into the effluent. The treatment (removal) rate is directly related to surface area of packing.

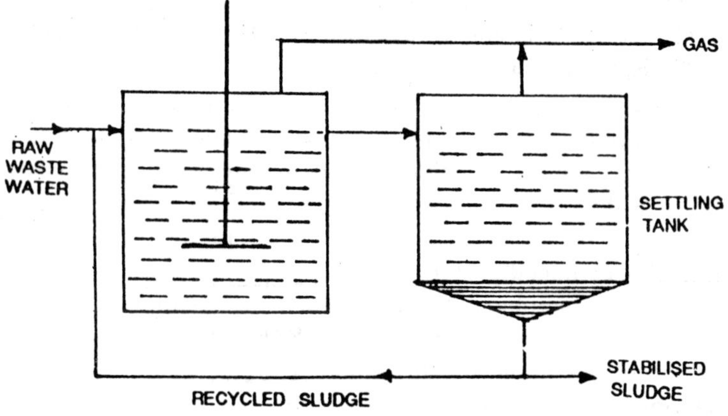

Figure 53
Anaerobic contact digester

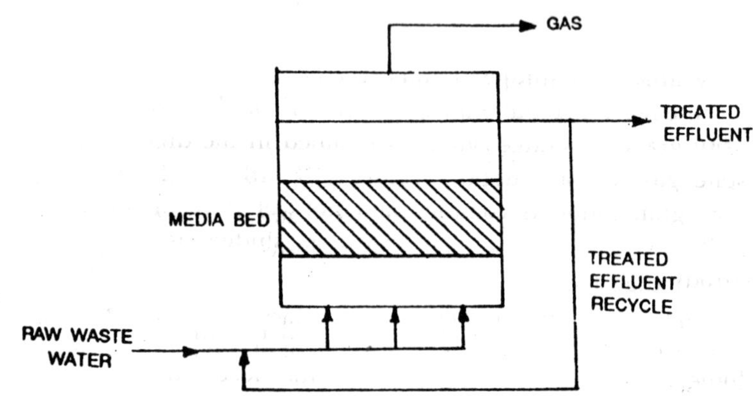

Figure 54
Anaerobic packed bed reactor

Polystyrene spheres in anaerobic filter or Neptune-Microfloc filter containing sand, silica, anthracite coal are used.

Anaerobic baffled digesters

These reactors have walls across the tank built from the top to the bottom, so that effluent flowing along the tank has to go alternately under and over the baffles. The baffles tend to keep the bacteria in the tank and also help to prevent problems with the floating solids (Figure 55).

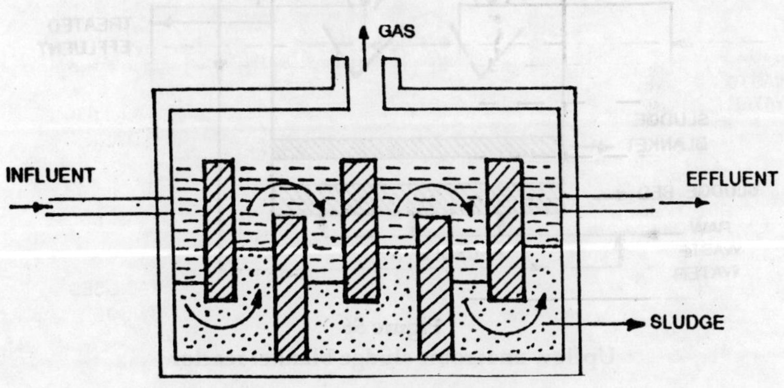

Figure 55
Anaerobic baffled digesters

Upflow anaerobic sludge blanket reactor

These reactors require the active bacteria in the form of high-density granular sludge which is retained in the digester tank despite gassing and upflow velocity of effluent (Figure 56). Sludge granulation is complex and not yet fully understood. Initially 10-15% inocula of granular sludge is required. Hydrodynamics of the digester created by feed distribution and the shape of digester are important in retaining the correct granular form. Well-adapted sludge may be sufficient in 1% volume only. The key element in the feed substrate for successful granule formation are calcium, phosphorus, magnesium, ammonia, aluminium, silicon. A large population of filamentous micro-organisms (e.g., *Methanothrix* spp) is also essential. Granular sludge develops with the dissolved wastes.

Baffles in the digester, promote gas solid separation along with the shape of digester and upward liquid velocity. Baffles provide the surface area which reduces the upwards flow velocity and promotes biomass flocculation. Baffles are corrosion-resistant reinforced plastic. Long chain fatty acids prove toxic to methanogenic bacteria and addition of calcium chloride can reduce this toxicity and can help better granulation.

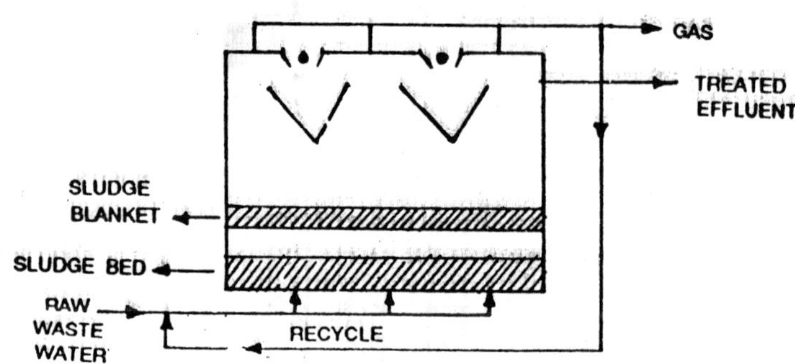

Figure 56
Upflow anaerobic sludge blanket reactor

Periodic biological reactors

One of the most recent advances in biological treatment processes is the use of Periodic biological reactors (Figure 57). These reactors markedly reduce the problems associated with effluent variations and eliminate sludge recycling.

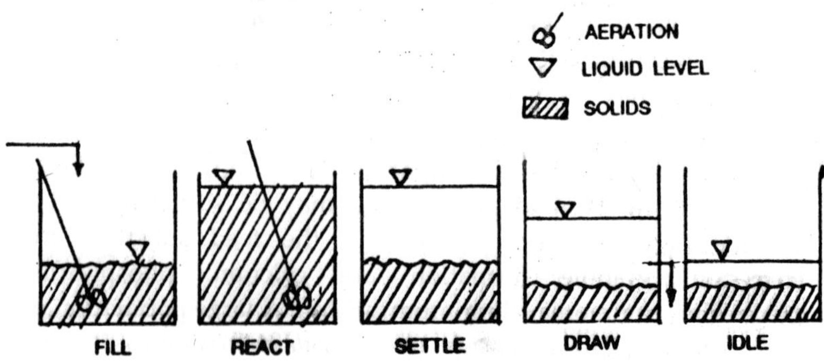

Figure 57
Periodic biological reactors

Membrane bioreactors

Membrane separation is useful to remove, the inhibitory effects of biodegradable pollutants. Membrane bioreactor for volatile organic compounds (Figure 58) consists effluent inlet chamber, from where the effluents pass through membrane and reach over biofilm. Degradation occurs in biofilm. Aerating gas and waste water do not come in direct contact. The use of specialised cultures in conjunction with bioreactors dedicated to treatment of 'point source' discharges of particular waste water is advocated.

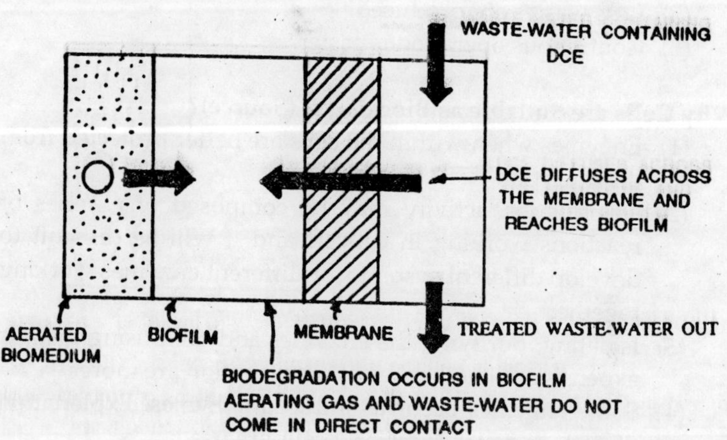

Figure 58
Membrane bioreactor for Volatile Organic Compounds

Use of immobilised enzymes or microbial cells for effluent treatment

Suspended growth and fixed film systems are the two most common types of treatment used for effluents. Both are based on microbial growth, metabolic activities of micro-organisms, eventual death of organisms sludge production and bioconversions occurring during the whole process. Immobilisation of biocatalyst (Figure 59) cells or enzymes can add to the advantages of biological treatment system.

An immobilised biocatalyst can be defined as a biocatalyst for which movement in space in completely or severally restricted to form a distinct phase within the bulk phase in which substrate, effector, inhibitor molecules are dispersed and their exchange is possible.

Why Enzymes are Suitable as a Biocatalyst?

(1) Enzymes require low energy inputs.
(2) Enzymes have low heat of reaction.
(3) Enzymes are heat labile, so it is easy to terminate reaction.
(4) Enzymes require much less complex media.
(5) Less wastes are produced.
(6) Continuous operation is easier.

Why Cells are Suitable as Biocatalyst (Source)?

(1) Enzymes when within the cells are better protected from denaturation.
(2) Degradation activity may be composed of a series of reactions working in concert and it will be difficult to develop different systems for different enzymes working together.
(3) Isolating, purifying the enzymes and then using them is expensive. Also chances of denaturation are more.
(4) Some enzymes occur as induced enzymes. Exploitation of such enzymes is easier if cells are used.
(5) Degradation rates and resistance to toxic pollutants will be great when cells as monoculture or mixed culture are used rather than using isolated enzymes.

Advantages of Immobilisation of a Biocatalyst

(1) Reuse of continuous use of cells enhances overall efficiency.
(2) Biocatalyst (cells/enzymes) does not contaminate the product.
(3) Cells are more evently dispersed by immobilisation, so diffusional restrictions are minimised.

(4) Concentration of biocatalyst possibly is more, so smaller size reactors can be used.
(5) Immobilised cells are used with more ease to exploit the kinetic features of continuously-stirred and packed bed reactors.

Methods of Immobilisation of a Biocatalyst

(1) Entrapment in polymer matrix.
(2) Adsorption of charged biocatalyst on oppositely-charged support material.
(3) Covalent attachment to chemically-activated supports.
(4) Encapsulation inside semipermeable membrane.
(5) Aggregation of biocatalyst into flocs.
(6) Biospecific attachment to supports by means of lectins etc.

It will be out of place to discuss the details of various methods of immobilisation and their advantages and disadvantages. But the general advantages of immobilisation are already mentioned and methods of immobilisation are illustrated in Fig. 3.25.

Applications of Immobilised Cells and Enzymes in Waste-water Treatment Falls into Four Categories:

(a) BOD/COD reduction.
(b) Specific pollutant detoxification.
(c) As biosensors.
(d) For bioconversion of waste to get specific products.

Flocculation of cells that occurs in activated sludge process or cells that are trapped on a slime layer in trickling filter beds may be considered as crude or preliminary way of applying/using immobilized bio-catalyst. Improvements in these immobiliation techniques and purposeful procedures of immobilisation have proved useful in waste-water treatment. Natural flocculation of cells in suspended growth systems can be improved by the addition of synthetic polyelectrolytes.

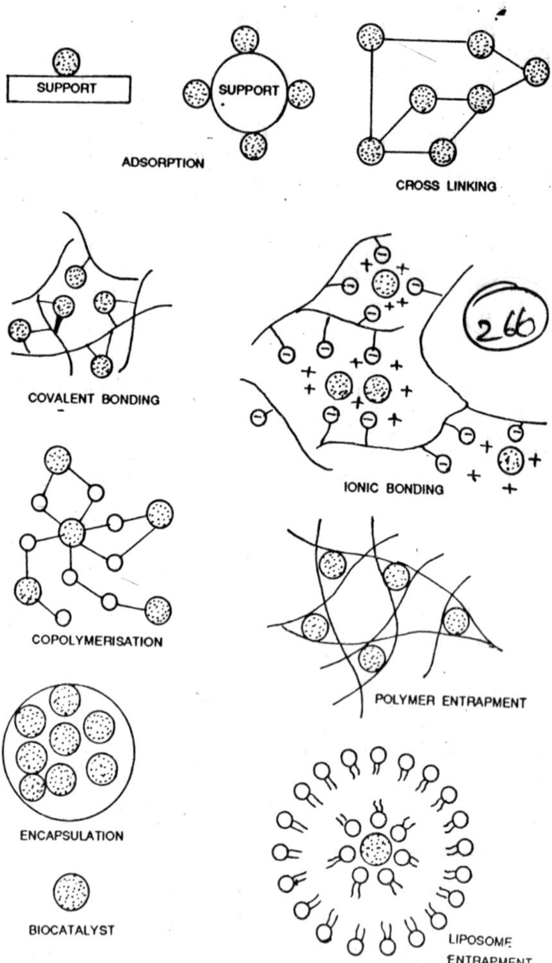

Figure 59
Methods of immobilisation of biocatalyst

Chapter 8

Bioremediation

'Bioremediation' encompasses biological methods for clean-up of contaminated soil and water. It means giving nature a helping hand. It involves establishing the conditions in contaminated environment so that appropriate micro-organisms flourish and carry out the metabolic activities to detoxify the contaminants. During bioremediation, micro-organisms may use the contaminants as nutrients or energy source or it may be degraded by co-metabolism.

Establishing suitable conditions for bioremediation may mean adding nutrients to promote the growth of a particular organism, adding terminal electron acceptor (O_2 or NO_2), adjusting moisture conditions or raising the temperature etc. The basic concept is to provide critical environmental requirements which may be adverse in particular site. Unavailability of critical environmental factors for microbial activity may be responsible for the persistance of the otherwise degradable substances in many places.

Bioremediation operations may be made either on-site or off-site, in-situ or ex-situ. Bioremediation may also mean promoting the growth of microflora that are indigenous to the site or it may mean the addition of component consortia of micro-organisms with the specific desired activities and characteristics. Seeding contaminated sites with competent microflora produced in fermenters and used to speed up bioremediation is the approach known as 'bioaugmentation'.

Bioremediation can be defined as stimulation of micro-organisms to rapidly degrade hazardous organic contaminants to environmentally safe levels in soils, subsurface materials, water, sludge, residues. Bioremediation has a vast potential to treat soil and ground water contaminated by a variety of hazardous chemicals, including refractory organics, oils, benzene, styrene, vinyl chloride, pentachlorophenols, polyaromatic hydrocarbons, toluene, xylene, phenols etc.

Bioreactors for Bioremediation

For bioremediation of polluted soils, sediments or groundwaters, two approaches are normally used:

(a) In-situ operation—polluted zone or compartment is subjected to restoration.

(b) Ex-situ operation—contaminated isolated volume of material is subjected to treatment for restoration on site.

The hetrogenicity of micro-organisms exists in both cases. On-site bioremediation of polluted soil uses three types of bioreactors:

(1) Static beds and heaps of soil that can be intermittently irrigated, aerated and mixed.

(2) Rotating cylindrical and horizontal screw mixed moving bed systems that are continuously aerated and mixed and intermittently irrigated.

(3) High and low aspect ratio bioreactors in which soil undergoing treatment is present as an aqueous slurry subjected to aeration and agitation.

The conversion of pollutant to any hazardous or toxic form that can migrate in soil is not permitted. Agitation enhances the process rate and avoids segregation: (1) Process economics, (2) System reliability, (3) Safety, (4) Operational flexibility, (5) Acceptable residual concentration of pollutant after treatment are the factors to be considered in the assessment of the right process.

Types of Bioremediation

The bioremediation may be natural or intervention process. The intervention process may be solid phase process or slurry process.

Natural bioremediation

In this process emphasis is given on the use of indigenous microflora for the desired biodegradation. Indigenous microflora should be adopted for the purpose. Adaptation may be done by:

(a) An increase in the contaminant-degrading micro-organisms by adding nutrients or removal of toxins or elimination of predators;

(b) Induction of enzymes responsible for contaminant degradation;

(c) Depletion of preferentially metabolized substrate before metabolism of targeted substrate.

Polluted sites contain a mixture of pollutants and mixed cultures do degradation but are slow. Indigenous microflora use the bulk of the ingredients of mixed pollutants but produce toxic intermediates, and therefore, overall toxicity of the site is not reduced.

Natural micro-organisms have not evolved genetic competence to utilize synthetic compounds. Natural micro-organisms have capacities to degrade pollutants but are not fast. Application of genetic selection and gene manipulation techniques, however, can make the process of capacity development fast.

Emphasis is on natural adaptation due to the fear of release of gene manipulated organisms in the environment which may start new problems in future. Also regulatory constraints deter the use of genetically engineered special cultures for remediation purposes.

Solid phase bioremediation

This depends on the assimilative capacity of soil to degrade and immobilize the constituents in waste. The indigenous or added micro-organisms carry it. Volatilization, leaching, water runoff are prevented from the area under treatment. Nutrients and energy sources may be added to promote microbial activity. Hydrolysis, photolysis, chemical degradation may occur along with biodegradation. *Land-farming* involves spreading of contaminated soil in thin layers (15-30 cm) on plot and supplemented by nutrients, micro-organisms and proper pH.

Tilling is done periodically for aeration and mixing. This technique requires land area and the process is slow. Two years are required to achieve 90 per cent reduction in fuel oil in the soil under this process. *Soil banking* involves construction of lined facility in which waste is mixed with clean soil and contained. It has a drainage and leachate collection system. The leachate may be recycled on the heaps or treated separately in bioreactors. The treatment time to degrade hydrocarbons is often have in soil banking in comparison to land farming.

Aeration of soil and large land area are two main requirements of these processes, but it requires least management as compared to other bioremediation techniques. However, the long time requirement is a major drawback. Restrictions on application of hazardous wastes to land for treatment purposes has reduced the use of land treatment methods for hazardous wastes. However, pretreated hazardous wastes can be processed by land treatment.

Slurry phase bioremediation

Hazardous wastes is liquid-solid contact are treated in a closed reactor or open pit or lagoon. Waste is taken in a reactor in the form of slurry and mixed with micro-organisms, organic and inorganic nutrients. Oxygen is supplied by aerators. Aerobic bioremediation is faster, single batch or sequence batch reactors may be used. The reactor may be mobile above the ground tank or lined in-situe lagoons or pits. Neutralizing agents, surfactants, dispersants are added to hasten degradation through pH adjustment, toxicity reduction, area of action increase, availability increase. After treatment, soils are allowed to settle, approximately treated or disposed off. The liquids are discharged in environmentally safe manner.

Materials most often treated by this method are wood-preserving and oil refinery wastes containing phenol, benzene, toluene, naphthalene, polycyclic aromatic hydrocarbons, asphalt, tar, material with high oil and grease content (more than 30%) cannot be treated by the slurry process.

Bioventing

It is also called soil venting or soil vacuum extraction. It is used for the removal of oily phase contaminants above the water table. A well is bored near the point of contamination but above the water table, a vacuum is applied from this extraction well and volatile emissions safely vented. This causes oxygenated air coming in contact with the undissolved contaminated subsurface material which gets biodegraded. Appearance of carbon dioxide in the extraction well is the indication of biodegradation activity. When the oxygen content of soil are above 1% hydrocarbon degradation by microbes is independent of hydrocarbon concentration. Rates differ from 2-20 mg hydrocarbon/kg acquifer material/day. The slow removal of air and maintaining 5 per cent oxygen in subsurface is the recent practice. This increases biotransformation. Bioventing is cheaper as compared to the use of nitrates or hydrogen peroxide as the source of electron acceptors.

Application of Bioremediation

Pollution prevention frequently is viewed in terms of hierarchy of management options, the principle ones of which are: reduction of wastes at source; recycling of wastes on-site within a manufacturing or other process; reuse of waste as a secondary feestock for other (bio) technological processes; treatment of wastes either in-line, end-of-pipe or off-site when their prevention or recycle cannot be avoided; and, failing all other options, disposal (Bull, 1992). Early preoccupations with environmental problems tended to focus on the *symptoms* of such pollution rather than its *causes*, a perspective which provides the driving force for developing treatment technology while helping to direct attention away from long-term strategies required to bring about waste minimization. Throughout the past decade, the focus has gradually shifted to such that the preferred options are prevention and recycling. More specific priorities for biotreatment research should address:

(a) The treatment of micro-pollutants and dispersed pollutants. These problems will be significantly more difficult and expensive to resolve than those contained pollutants that can be controlled by end-of-pipe treatment. The dispersion of heavy metals in sewage, sludge, estuarine and marine sediments, chlorinated solvents in ground water, and persistant estrogenic chemicals, and mimics such as p-nonyl-phenols are illustrative of these classes of problems.

(b) The development of robust (bio) sensors for rapid on-line and in-situ measurement and monitoring of environmental chemicals.

(c) The strengthening of ecotoxicology research with a sharper focus on the identification and measurement of damage-impact on ecosystems, and effects on human populations.

(d) Improved understanding and predictive behaviour of the physiology of detoxifying micro-organisms particularly when growing in communities and attached to surfaces. Moreover in vary many cases the identity of organisms constituting competent biodegradative communities—whether operating in biotreatment reactors or in-situ—simple are not known or their presence even recognised; under these circumstances rational design of remediation processes will be severely compromised. The study of xenobiotic-degrading B-Proteobacteria by Busse et al (1992) is illustrative of the type of taxonomic work that needs to be done in this field; base-line data of this sort are so few that considerable investment of research effort will be necessary to enable unequivocal analysis of population dynamics and thence process control in biodegradation operations.

Further developments, such as the construction of taxon-specific, fluorochrome-tagged DNA probes, are opening up possibilities for identifying the physiological significance of different components of biodegradative communities. This latter approach is going to be critically important for assessing the ecological significance of so-called 'viable but non-culturable' bacteria. Thus, the analysis of activated sludge bacteria with DNA probes for α–, β-, or r-Probacteria has revealed that the β-group is discriminated against by current culture methodology and hence its contribution to ecosystem function should be seriously under estimated (Wagner et al, 1993). While such DNA probing facilitates studies of population dynamics, even of non-culturable organisms, an additional challenge for the bacteriologists is to devise culture media for the enumeration of the β-group.

(e) Assessment of the efficacy, need and desirability of developing genetically modified organisms (GMOs) for contained and in-situ biotreatment. Development in pathway engineering notwithstanding complementary research directed at the detection and isolation of natural micro-organisms capable of metabolising recalcitrant xenobiotic chemicals remains the most important priority for biotreatment technology (Bull, 1996).

Biodegradation

Genes coding for some enzymes essential for biodegradation of many organic compounds are plasmid borne. Plasmids can be exchanged among organisms by gene manipulation technique. Thus, many packaged micro-organisms have been produced, which possess plasmids derived from other organisms. These packaged micro-organisms have broad range biodegradation capacities.

Biological detoxification may be carried out using pure culture or mixed culture of micro-organisms. Mixed culture have a potential advantage over pure culture in the degradation of toxic compounds in hazardous wastes. Mixed cultures are particularly useful when complete degradation of toxic organics to carbon dioxide, methane, hydrogen sulphide, nitrogen etc is required. Pure cultures, on the contrary, produce toxic intermediates which may accumulate.

Biostimulation and *bioaugmentation* are the two main ways of initiating biological treatment. Biostimulation makes use of existing micro-organisms by providing favourable conditions for their action by adjustment of nutrients, pH, Temperature, growth factors etc. On the other hand, bioaugmentation involves externally introduced culture with specific degradation capacities. In bioaugmentation, the introduced micro-organism should be able to remain viable, should compete with the existing micro-organism.

Biodegradation of Xenobiotics

Industrial revolution has resulted in many materials which are man-made and resembling to natural ones. The term 'xenobiotics' (Strange to life) is derived from the Greek word 'xenon'—a stranger and 'bios'—life. For environmentalist, xenobiotic usually implies foreign to the biosphere. These materials may be products, intermediates or wastes of our industrial production activities. After their entry into nature, they cannot become a part of cycles of matter.

Major sources of xenobiotic compounds are agrochemical and petrochemical industries. These compounds have their half-lives which indicate the extent of persistence. It has been estimated that 150×10^6 tonnes of xenobiotic organic chemicals are produced annually which find their way into the soil, surface water and ground water as constituents of industrial effluents. Aromatic compounds and pesticides are the most significant among them.

Natural biodegradation of xenobiotics is brought about by micro-organisms. However, it is often a slow process and depends upon the complexity of waste, concentration of individual pollutants, toxic intermediates produced, pH, temperature, oxygen availability, redox potential, types of micro-organisms and their number, competition by easily degradable substrates etc.

In most cases, xenobiotic degradation is brought about by mixed consortia in a sequential mechanism rather than by individual species alone. The existence of mixture of species enhances the degradative process by allowing better adaptation, more tolerance and by supplementing nutrient and cofactor needs of each other in the community. Inducible enzymes of many types are better produced when a community of different species acts on a mixture of pollutants. Many micro-organisms have biodegradation potential for xenobiotics (Table 57).

Table 57
Micro-organisms with biodegradation potential for xenobiotics (Jogdand, 1995)

	Organism	*Toxic Chemicals*
(1)	*Pseudomonas spp*	Benzene, Anthracene, hydrocarbons Naphthalene, PCBs, 4 alkylbenzoates, Alkylamine oxides, p-xylene, p-cymene, phenylureas, polycyclic aromatics phenanthrene, toluenes, phenolics, phenoxyacetates, organophosphates, melathion, parathion, rubber etc.
(2)	*Alcaligenes* spp	Halogenated hydrocarbons, linear alkylbenzene sulfonates, polycyclic aromatics, PCBs.
(3)	*Arthrobacter* spp	Benzene, hydrocarbons, pentachlorophenol, phenoxyacetate, polycyclic aromatic.
(4)	*Bacillus* spp	Aromatics, long chain alkanes, phenyl ureas, salicylates, phenol, cresol.
(5)	*Corynebacterium* spp	Halogenated hydrocarbons, phenoxyacetates.
(6)	*Azotobacter* spp	Aromatics.
(7)	*Flavobacterium* spp	Aromatics.
(8)	*Rhodococcus* spp	Naphthalene, Biphenyl.
(9)	*Mycobacterium* spp	Aromatics, Branched hydrocarbons Benzene, cycloparaffins.
(10)	*Nocardia* spp	Hydrocarbons, alkylbenzenes, naphthalenes phenoxyacetate, polycyclic aromatics, phenols, substituted phenols.
(11)	*Methosinus trichosporium*	Benzene, Toluene, Cresols, Ethylbenzene, Phenylacetate etc.
(12)	*Methanogens*	Aromatics.
(13)	*Xanthomonas* spp	Hydrocarbons, Polycyclic hydrocarbons.
(14)	*Streptomyces* spp	Phenoxyacetate, halogenated hydrocarbon Diazinon.
(15)	*Candida tropicalis*	PCBs, formaldehyde.

Contd.

	Organism	*Toxic Chemicals*
(16)	*Cunniughamela elegans*	PCBs, polycyclic aromatics Biphenyls.
(17)	*Fusarium solani*	Propanil.
(18)	*Trichosporon cutaneum*	Phenol.
(19)	*Lipomyces*	Herbicide paraquet.
(20)	*Rhodotorula*	Benzaldehyde.

Methanogens are important in xenobiotic degradation. *Methancoccus thermolithotrophicus, M. delte, M. thermoautotrophicum* could degrade bromo-and chloro-ethanes, trichloroethylene, common industrial solvents and pollutants. Hydrolysis, dehydrohalogenation and reductive dehalogenation were proposed mechanisms in the production of ethane, ethylene and acetylene from parent compounds. Breakdown of carbon tetrachloride, chloroform and bromoform occur by reductive dechlorination. The actual mechanism by which methanogens detoxify such chemicals is not known.

Biodegradation of Hazardous Wastes

Wastes are generated in natural processes and also by human commercial-industrial activities. It is difficult to define 'waste' and it is further difficult to define 'hazardous wastes'. Anything which is discarded or otherwise dealt with as if it were waste is presumed to be waste unless the contrary is proved. The problem with hazardous wastes is that there is no internationally accepted definition of a hazardous waste. However, technical criteria such as toxicity, flammability, corrosivity, ignitability and reactivity have been proposed and used to some extent to identify and designate the waste as hazardous (Agarwal, 1996).

The minimum or trace amount of hazardous wastes are generated from metallurgical, iron and steel, fertilizer, and thermal power production sector. The maximum amount of hazardous waste is generated from chloro-alkali, dye and pigment, organic chemical and pesticide industrial sector. The inorganic chemical sector on the other hand, generate low volume high toxic wastes. The synthetic drug manufacture units in drugs and pharma sector generates maximum quantity of hazardous wastes (Kumar, 1986).

United States generates about 60 million tonnes of hazardous wastes every year, and the European Economic Community produces about half of that. Roughly 10 to 20 per cent of all the industrial manufacturing waste is potentially hazardous.

Biodegradation of hazardous wastes is still considered as an emerging technology. Evaluation and demonstration of biodegradation process is done in the laboratory, in field and in specific situations, but the commercial potential is not established in most cases. The fundamental studies of the biodegradation is well advanced for most of the hazardous wastes.

Biological cyanide detoxification is brought about by fungi and bacteria. Fungi can degrade simple cyanides but are less successful with metallocyanides. Bacteria are more efficient with metallocyanides. Among fungi *Fusarium moniliforme, Mycoleptodiscus terrestris, Helmonthosporium sorghicola, Gloeocercospora sorghi* and *Stemphylium loti* the first one is the most efficient in cyanide detoxification. Biological cyanide detoxification is technically competitive to traditional alternatives but it does not give any cyanide recovery. Biological cyanide degradation is today limited to relatively dilute solutions of sodium cyanide, hydrocyanic acid and thiocyanate.

Pseudomonas sp detoxify sodium oxalate (a hazardous product of Aluminium extraction from bauxite process). *P. oxalaticus* degrade byproduct sodium oxalate with 95 per cent efficiency.

Spores of *Fusarium flocciferum* can be used to degrade phenol following immobilization on diatomic particles. Cell loading of upto 50 per cent dry weight have been obtained and the immobilized system can efficiently degrade a high concentration of substrate and is active for over two months.

Biological nitrification is brought about by *Nitrosomonas* and *Nitrobacter,* for wate water with intermediate or low levels of ammonia nitrogen. High pH is often harmful to nitrification as gaseous ammonia produced is detrimental to the nitrification process. Low strength waste-waters can be easily used for biological NH_4-N removal.

Biological phosphorus removal is possible by the exposure of micro-organisms to alternating aerobic and anaerobic conditions. Phosphorous removed is incorporated into the cells. Stress on micro-organisms increases their phosphorous uptake.

The polar biodegradale oils and greases are decomposed by biological treatment and effluent is left with only 2-8 mg/l. Non-polar oils, however, are not degraded but are recovered in a primary separator.

Living organisms or dead biomass both find use in the removal of heavy metals from waste waters. Blooms of algae and Cyanobacteria reduce levels of Copper, Cadmium, Lead, Nickel, Zinc, Mercury and Iron in contaminated waste water. *Potomogeton, Typha, Eichhornia* trap heavy metals (Agarwal, 1993).

Biodegradation of TNT Wastes

The waste water from manufacture of the high explosive TNT was red in colour, acidic in nature, and had high organic load (COD around 44000 mg/L) and contained TNT, trinitrobenzene, dinitrotoluene, mononitrotoluene, sulphuric acid and nitric acid. Micro-organisms were isolated from soil on the bank of the river. *Pseudomonas mendocina, Citrobacter freundi, Bacillus polymyxa* and *Micrococcus cryophilus* were the prominant species having the ability to utilize TNT as the sole source of carbon and nitrogen. *Pseudomonas trinitrotoluenophila* mineralized TNT to carbon dioxide through an aromatic amino acid, di-dinitro-phenyl-Lysine and an organic acid namely succinic acid indicating that end product of microbial degradation are non-toxic. These species could degrade the waste containing TNT to the extent of 50 per cent in terms of COD and TNT content within 7 days under stationary culture conditions (Kanekar and Godbole, 1983 and 1984).

Trinitrotoluene (TNT), hexahydro 1, 3, 5-trinitro 1, 3, 5 triazine (RDX) and Octahydro 1, 3, 5, 7 tetranitro 1, 3, 5, 7 tetraazocine (HMK) are the explosives which are significantly transformed by the composting process over 153 days. The scientists from the Argonne National Laboratory and the University of Notre Dame have found that four species of *Pseudomonas* can cometabolise 2, 4, 6-trinitrotoluene (TNT) 100 per cent in contaminated soil. *P. acidovorens, P. fluorescens, P. mendocina* and *P. aeruginosa* are the active species isolated from TNT contaminated soil near an army ammunition plant in Illinois. The two intermediates produced by TNT metabolism are 4-amino-2, 6-dinitrotoluene and 2-amino-4, 6-dinitrotoluene and the third intermediate is perhaps 2-amino-6-nitrotoluene. The Ligninase enzyme system of *Phaneochate chrysosporium* (white rot fungus) has high affinity for TNT. Intech, one of the eighty Corporations Company and Mycotech Company are commercializing the use of *P. chrysosporium* PhC technology. The existing method of degradation for explosive contaminated soil in incineration which is expensive and energy intensive. Biodegradation of the explosives like TNT, RDX, and HMX may provide safe and cost effective approach.

Biodegradation of Dyestuff Wastes

The waste water arising from the manufacture of dyes namely Methul violet, Rhodamine B, Chrysoidine and Nigrosine contained phenol, aniline and traces of dyes imparting purplish red colour (Optical density 0.936 at 580 nm) and exerted organic load in terms of BOD 575 mg/L, COD 5000 to 8000 mg/L, TOC of 400 to 1200 mg/L and phenol content 100 to 1100 mg/L, pH of the waste was 8.9 to 9.2. The waste was found to be amenable to microbial degradation and *Pseudomonas alcaligenes, P. mendocina,* and *Micrococcus* sp were predominant bacterial species from cattle dung and soil, degrading the waste (Kanekar, 1990).

Biodegradation of Pesticides

Environmental damage due to pesticides is a serious problem. The pesticides can enter into the environment from: (a) manufacturing wastes, (b) user wastes, and (c) after application wastes. The complex chemistry of pesticides, persistance in the environment, toxicity to animals and human beings, bioaccumulation risks make pesticide pollution a citical problem.

Safe and economic disposal of pesticides is a current problem of significant magnitude.

The majority of pesticides can be degraded by a fungi and bacteria. The transportation of parent pesticide molecule to, less complex products is often achieved by concerted efforts of microbial communities. The extreme toxicity is often lost in the first transformation step and hence potential exists for the development of simple microbial treatment for pesticide detoxification.

Pseudomonas cepeciae degrade 2, 4, 5-T to clean up soil containing 20,000 ppm of 2, 4, 5-T. The herbicide was reduced to a point that lecttuce plants could grow in the treated soil.

Flavobacterium sp produces Parathion hydrolase enzyme, which can rapidly hydrolyse the organophosphate insecticide 'Coumaphos to' less toxic products.

Pseudomonas spp can remove 94-98 per cent residual Parathion from pesticide container in 16 hours at the substrate level of 1% (W/v). The degradation is brought about by parathion hydrolase enzyme contained in them.

Fusarium solani can be used to act on herbicide Propanil (3, 4-dichloro-propionanilide) to form 3, 4-dichloroaniline. The enzyme involved is acylamidase in immobilized form.

Trichoderma viridae produces enzyme that can hydrolyse the insecticide Malathion.

Geotrichum candidum produces enzyme that can degrade 4-methylphenol.

Phanerochaete chrysosprium when grown in nitrogen free medium containing a suitable carbon source, can degrade pesticide such as DDT and Lindane.

Chrysoporium lingorum, Trametes versicolor, Phanerochaete chrysosporium and *Sternum hirsutum* can mineralize, 3, 4-dichloroaniline, Dieldrin and Phenathrene.

A mixture of *Brevibacterium, Azotomonas* and *Xanthomonas* can tolerate upto 10000 ppm of Parathion and degrade it in 36 days.

Pseudomonas putida, Candida tropicalis, Fusarium flocciferum, Auroobasidium pullulans, Aspergillus niger and *A. japonicus* can degrade herbicide of Chlorobenzoate class.

Pseudomonas putida, P. aeruginosa, P. diminuta, P. cepeciae, P. flurescens etc possess genes responsible for degradation of 3-chlorobenzoic acid and other closely related compounds 4-chloro-and 3, 5-dichlorobenzoic acids. For every compound, one separate plasmid is required. It is not like that one plasmid can degrade all the toxic compounds of different groups. Other bacteria from which pollutant degrading plasmids have been isolated are *Acnetobactor* sp, *Alcoligenes* sp, *Beijerinkia* sp etc (Chakraborty et al, 1982; 1983 a, b).

Degradation of Oil-Spills

Oil-spills from oil tankers and from distant oil spills, have been recognised as a major environmental hazard. The spilled oil is believed to destroy the habitat of seabirds, marine mammals and fish. The thick and gummy crude oil discharges can cause immediate harm to fish and wildlife, degrade oceans and coastal habitats, and over time, even threaten human health. Therefore, remediation efforts in this direction have been made, using chemical dispersants. These chemicals are believed to cause major pollution problems in shallow waters due to their toxic nature and persistance in the environment. Bioremediation of oil has been found useful for this problem. Initially, oleophilic fertilizers were used for bioremediation of oil spilled fouled beaches. The nutrients present in the fertilizers allows micro-organisms to utilize the hydrocarbons in oil as a food source—in effect, to remove the oil by eating it. A single application of fertilizer made the microbial consumption of oil two or three times faster than on untreated shoreline. No environmental hazard have been found to be associated with using fertilizer to degrade oil. In recent years, however, using genetic engineering, oil utilizing micro-organisms have been successfully produced, which would grow rapidly on oil. A strain of *Pseudomonas aeruginosa* have been developed, which produces a glycolpid emulsifier, that reduces the surface tension of an oil-water interface and thus helps in removal oil from water. This microbial emulsifier is non-toxic and biodegradable, thus being ecologically sound.

It was demonstrated in laboratory experiments that this *biosurfactant* releases oil to an extent of 2-3 times that of water alone. Thus it has the potential for future use, whenever its release is allowed. The oil-spill *Superbug* was a *Pseudomonas* strain carrying together four plasmids CAM-Camphor, OCT-octane, SAL-salicylate, NAH-naphthalens. These plasmids code for enzymes required for attack on respective compounds suggested with their names. The multiplasmid strains, however, have to compete with other organisms and still face shortage of nitrogen and phosphorous. Hence they are useful in closed systems (in sewage treatment plants) but for open ocean oil slicks their effectivity in questionable.

The use of genetically engineered microbes for cleaning spilled oil is facilitated by mixing the microbes with straw, which can be stored. When needed, the straw with mixed microbes can be scattered over the spilled oil so that the straw will first soak oily water and then the microbes will breakup the oil into harmless, non-polluting materials, rendering the water and the site harmless.

Biodegradation of Other Pollutants

Some 40 bacterial strains and species have been isolated that can degrade many of the pollutants of the greatest hazard to the environment and to health. The prospects for constructing or finding in nature catabolic pathways, that is, the pathways used by micro-organisms to breakdown the wastes are encouraging. There are many sources of catabolic genes, including in-vitro genetic engineering, laboratory selection techniques, and naturally occurring organisms.

By manipulating a recombinant gene in the DNA of soil bacteria of the genus *Rhodococus,* scientists have created an organism that "eats and metabolites sulphur in the substrate". The "newly reprogrammed" bacteria are mixed with water and added to fossil fuel. Once the bacteria eat the sulphur, they are filtered out of the coal and reused. Thus, the occurrence of acid-rain due to coal burning for power generation can be avoided.

Biotechnology has proved the way for removing phosphorous, a problematic and expensive pollutant in sewage sludge. The traditional method to remove phosphorous using chemicals to precipitate into metals is expensive. On the otherhand, phosphorous removed by microbes is being used to produce fertilizers.

Four strains of *Pseudomonas* can co-metabolize 2, 4, 6-Trinitrotoluene (TNT) 100% in contaminated soil. *P. acidovorans, P. fluerescens, P. mendocina* and *P. aeruginosa* are the active species isolated from TNT contaminated soil near an army ammunition plant in Illinois, USA. The two major intermediates produced by TNT metabolism are 4-amino-2, 6-dinitrotoluene and 2-amino-4, 6-dinitrotoluene. One of the four species could degrade 100 ppm concentration of TNT in soil completely within four days. The existing method of remediation for explosive contaminated soil is incineration which is expensive and energy-intensive and leaves behind ash as residue. Bioremediation of TNT may provide safe and cost-effective approach.

Ligninase enzyme system of *Phanerochate chrysosporium* (white rot fungus) is non specific and has a high affinity for a wide range of pollutants like PCB, DDT, TNT, PCP, Creosote, Coaltar and heavy fuels. Wood chips are added to promote the growth of this fungus while using it for bioremeidation of these pollutants in soil. 85-90% reduction have been observed in PCP levels which were 10000 ppm and PCP was converted to carbon dioxide and non toxic organic matter.

A number of plasmids ranging in size from about 30 to 500 Kb coding for biodegradation of a variety of organic compounds including chlorinated aliphatic and aromatic compounds have been detected in bacteria belonging mainly to the genera *Pseudomonas* and *Alcaligesns* (table 58).

Table 58

Pollutant degradative plasmids (Kunhi, 1991)

Plasmids	*Degradative pathway*	*Size*
CAM	Camphor	500 kb
OCT	n-octane	500 kb
SAL	Salicylate	85 kb
NAH	Napthalene	83 kb

Contd.

Plasmids	*Degradative pathway*	*Size*
TOL	Xylene, toluene	117 kb
XYL-K	Xylene, toluene	90 MD
2 - HP	2-Hydroxypyridine	N.D.
NIC	Nicotine, Nicotinate	N.D.
pRA500	3, 5-xylenol	500 kb
pCITI	Aniline	100 kb
pEG	Styrene	37 kb
pCS1	Parathion hydrolysis	68 kb
pWR1	3-chlorobenzoic acid	111 kb
pAC8 (RP4-TOL)	Xylene, toluene	76 MD
PAC21	p-chlorobiphenyl	65 MD
PAC25	3-chlorobenzoate	117 kb
PAC27	3- and 4-chlorobenzoate	59 MD
PJP1	2, 4-dichlorophenoxy acetic acid	58 MD
PJP2	-do-	58 kb
PJP4		83 kb

Most of these pasmids posses a wide range of hosts and are capable of conjugal transfer from one strain to another strain. They carry a number of genes useful in the biodegradation of pollutants. Recombinational exchange of these genes may lead to the construction of pathways of rapid breakdown and recycling of most of the organic molecules entering the soil ecosystem.

The first report on plasmid borne degradative pathway genes was that for degradation of Camphor which is an active ingredient in mothballs (insect repellant) by a strain of *Pseudomonas putida*. This was followed by the isolation of a naththalene degradative plasmid (NAH) in 1973. In the following years a number of other pollutant degradative plasmids were detected which degrade such diverse compounds as xylene and toluene (YOL), the herbicide 2, 4-

dichlorophenoxyacetic acid (2, 4-D), 3-chlorobenzoate, 3-and 4-chlorobenzoate etc. An interesting feature of the several 2, 4-D plasmids so far isolated from strains of *Alcaligenes eutrophus* and *A. paradoxus* is that they have a broad host range. One particular plasmid, pJP4 can be transferred to such diverse gram-negative bacteria as the photosynthetic bacterium *Rhodopseudomonas sphaeroides,* the symbiotic nitrogen fixer *Rhizobium,* the plant pathogenic *Agrobacterium tumifaciens,* the human gut inhabitant *E. coli* and the soil saprophytic *Pseudomonas putida.*

It is quite obvious that much of the data on biodegradation of pollutants so far obtained point to the absence of existence of an interated system in nature for effective and reasonably rapid removal of unwanted pollutants from the environment. The construction of such a system should be possible with the help of tools and systems of bacterial molecular genetics. First on the heels of *E. coli* molecular genetics many other sytems, particurally those in *Pseudomonas* have been developed that should be of value in evolving genetically tailored organisms for successful elimination of pollutants.

Caution for Using Bioremediation

A word of caution is called for before a micro-organism can be tested outside a contained environment. An evaluation is needed of what happens after it has done its work. Predictive ecological research is essential to ensure that biodegradation of wastes does not unintentionally create new and potentially serious environmental problems. The most pertinent theme of the criticism concerning the release of genetically engineered micro-organisms in an open environment, has been the fear of adverse ecological and epidemiological consequences that might stem from the accidental or deliberate release of self-propagating genetically engineered organisms into the biosphere. Further, tests for safety in the genetic engineering revolution are yet to be conceived, since how the genetically modified life-forms interact with other organisms is totally unknown.

The appliction of new genetic methods to constructing noval microbial strains that have improved capacities for degrading various synthetic chemicals is currently an active area of research. However, before such strains are constructed the researcher must answer the question "Why are the toxic chemicals not being degraded normally?" This requires a detailed knowledge of the pathways for degrading structurally related compounds, including an understanding of the specificities of the enzymes involved in their regulations.

Microbes may not be capable of accomplishing the complete degradation of a particular compound for a number of reasons. The organism may not have all the enzymes needed, or the enzymes may be present but too slowly acting. Such problems could be solved by introducing into the organism new genes that code for enzymes with the desired specifications. Alternatively, mutations could be introduced into the appropriate gene of the microbe to change the enzyme specificity to allow recognition of the desired substrate. If the problem is the slowness of the degradation pathway, the solution could be to increase the amount of enzymes produced by cloning the corresponding gene(s) in a plasmid that would be present in multiple copies in the cell.

Another possible explanation for lack of degradation of a chemical is its failure to enter the microbial cell. This might be overcome by introducing genetic mutations into an existing transport system to widen its specificity range to include the chemical in question. Finally, the synthesis of degradative enzymes may be closely regulated so that they are normally turned-off and are not turned-on by the chemical in question. In that event, the solution would be to select for regulatory mutants that produce the desired enzymes without the need for specific activation of the genes.

It is now possible to construct total pathways for chemical degradation by combining genetic capacities of bacterial strains. The genes required for degradation of synthetic chemicals are normally carried by plasmids, rather than in the chromosomal DNA. Plasmids can be readily transferred between species of *Pseudomonas.*

Genetically manipulated microbes offer potential as inocula for detoxification, unfortunately lack of knowledge on the fate and the effect of engineered microbes in the environment is hindering further progress. The stability of engineered strains to compete or co-exist with indigenous microflora, their transport from original application site, effect on ecosystem function and the ability of the strain in mixed wastes, are but a few of the parameters that are important and needs consideration.

Although bacteria have a great potential for degrading contaminants in the environment, the reaction carried out by the organisms do not always produce less toxic substances. Sometimes the products are more dangerous than the parent compound. For example, the DDT (dichloro diphenyl trichloro ethane) is converted by the bacteria in soil and aquatic environments to DDD (dichloro diphenyl dichloro ethane), which is more potent insecticide than the original compound and also persists longer in the environment, and is more toxic to higher animals.

Biosorption

Metabolism independent binding or absorption of heavy metals to living cells or dead cells, extracellular polysacchrides, capsules and slime layers are referred to as 'biosorption'. Cell walls and envelops of bacteria, yeasts, algae and fungi are very efficient in biosorption due to the charges groups present in them. Biosorption is influenced by environmental factors such as pH and ion composition and some metabolism aspects which can affect the micro-environment of cells. Variations in chemical behaviour of metal species, composition of microbial cell wall and extracellular material also affect the biosorption capacity. Gold and silver crystallize due to the reduction while uranium and thorium crystallize due to the formation of hydrolytic products. Metals may deposit around cells in the form of phosphates, sulphides or oxides.

The microbial removal of toxic and/or valuable metals can be done from industrial effluents. This can result in detoxification and safe environmental discharge.

Use of bacteria in biosorption

Pseudomonas fluorescens immobilised on polyvinyl chloride granules was used primarily to remove nitrate but was found to be capable of removing metals such as lead and zinc. *P. aeruginosa* immobilised on an oxygen plasma treated polypropylene web showed plutonium removal efficiency of 98-99 per cent with a useful life of 2 weeks. This system was not used on an industrial scale and no attempt was made to recover plutonium. It was suggested the plutonium loaded biosorbant could be packed in drums and dumped.

Citrobacter sp 'resting cells' accumulate heavy metals by surface located phosphatase enzyme that releases HPO_4^{-2} from supplied substrate and precipitates divalent cations as $MHPO_4$ at the cell surface. The process is non-specific and depends on the fact that certain metal phosphats are insoluble. Cadmium, Lead, Copper can be precipitated singly or in combination. *Citrobacter* process is capable of long-term use, functional with a high metal load, possibility of metal recovery and biomass regeneration are the advantages.

Pseudomonas sp, *Micrococcus leutus, Streptomyces phaceromogenes* are capable of acting as adsorbants. One gram of bacteria can recover as much as 180 mg of gold. Recovery occurs in 5-10 minutes.

Sphaerotilus natans accumulate metals in mucilaginous layer outside sheath. It accumulates Iron, Copper, Magnesium, Cobalt, Cadmium when present as sulphates but growth is inhibited if these metals are present as chlorides.

Uranium can be recovered by *Pseudomonas aerugenosa. Streptomyces virido-chromogens* immobilized in poly-acrylamide gels have been shown to remove uranium. The bound uranium could be released in sodium carbonate.

Use of fungi in biosorption

Active research is going on where removal of heavy metals or radio nuclides from effluents is done with live of dead fungal mycelium. This can help in pollution control and precious metal recovery. Fungal biomass can be used to adsorb metal particulates, elemental sulphur, insoluble sulphides, clays, charcoal, magnetite etc. Particulate adsorption by fungal cell wall occurs independent of cellular metabolism. Young actively growing mycelial tips are more effective. Technology of biosorption of metals with fungal biomass allows 500-2000 tonnes of metals to be recovered without regenerating the biomass and several times more if regeneration is used. Cell walls of fungi contain many anionic sites such as amines, carboxyl, hydroxyl and phosphate groups. Adsorption of metallic cations takes place on these sites.

Rhizopus arrhizus can take up 170 mg radionuclides per gram dry weight and 30-130 mg Cadmium per gram dry weight. *Aspergillus niger* mycelium can adsorb silver, however, silver cannot be desorbed easily. *Aspergillus niger, Penicillium chrysogenum, Claviceps paspali* mixture can adsorb Zinc, which can be recovered in HCl. *Rhizomucor meihei* or *Mucor mucedo* in fibrous form treated with alkali is used for recovery of metals. *Trichoderma viride* is used to remove Copper from effluents. *Aspergillus oryzae* is used to remove Cadmium with 99 per cent efficiency. *Penicillium chrysogenum* is usedfor removal of uranium and radium.

Use of algae in biosorption

Algal biomass is capable of binding considerable amount of heavy metals such as Zinc, Copper, Cadmium, Lead, Gold, Uranium from mixtures. Cells of *Chlorella vulgaris* immobilised in polyacrylamide gel have been used to remove uranium from seawater. *C. vulgaris* binds Gold, Silver, Mercury at low pH (2-3). Gold binds 98500 mcg/gm dry-weight biomass from a solution of 19.7 mcg/ml. Zinc and Copper can be released by an acetate buffer at pH 5 and 3.5 respectively. Mercury can be released by same reagent at pH 2. Thus, simple and cheap recovery of metals is possible with the help of algae.

Ascophylum nodosum and *Sargassum natans* can adsorb Cadmium and Lead from the effluents. *Fucus vesiculosus* can adsorb Lead, while *Halmeda opuntia* can adsorb Chromium.

Chapter 9

Biomineralization

Ores are chemical compounds of different metals occurring in nature and are known as minerals. These minerals exist either in the form of oxides or sulphides. Conventional mining for metallic minerals involved: (i) crushing of vast quantities of ore from the potential sites and its grinding; (ii) smelting of ore at high temperature. This process was highly polluting and energy intensive approach. Pollution of air due to release of particulates in the form of dust during excavation, digging, shoveling, blasting or processing of minerals; use of acids, alkalies or other chemicals in benefication of ore leading to formation of waste water; blasting, drilling and use of compressed air in metallurgical operations creating noise pollution were only a few significant areas out of several others which demand immediate action. Moreover, when rocks are having high grade ores, then only conventional process is economically worthwhile. However, if high grade rocks are in short supply, and a large amount of metal is available in the low grade ore, the conventional process of metal extraction is not economical. In view of this, biometallurgy, involving the use of specialized groups of microbes to extract and concentrate metals in a cheaper and more efficient way, is an approach which is environment friendly. Application of biometallurgy in ore processing for metal recovery incorporates three technological components—biohydrometallurgy or bacterial metal leaching, biobenefication and biosorption. The biohydrometallurgy (bioleaching) utilizes biochemical oxidation reduction process, catalysed by living organisms for extracting metals in soluble form from ores. Biobenefication or enrichment of ores, is concerned with removal of other metals and elements mixed with the ore microbiologically, for enriching the main ore body. Biosorption of metals from leached solution is the strategy to extract the metals from solution when in very low concentration.

The biometallurgy is of significance due to the following advantages:

— Its wide application for recovery of valuable metals from low grade ores.
— The relative absence of air, water, land and noise pollution.
— Low energy requirements and operating costs as compared to conventional processes for recovery of metals from low grade ores.
— Degradation of a variety of mineral forms.
— Selective leaching possible, one metal solubilized while the other remains insoluble.
— Application most useful is desulphurization of coal for burning is free of sulphur and further pollution problems avoidable.
— Suitable for less developed countries as it eliminates the need for some costly imported heavy mining equipment.

Biomineralization process has certain limitations:

— Process is slow.
— in-situ application still under development.
— Control difficult.
— More possible for acid-producing minerals and not acid-consuming minerals.

Recovery of metals using biotechnological process for mining of ores (with the help of microbes) can be done at two places: (i) applying microbial leaching for low grade ores; (ii) concentration of metal from waste water (scavenging) or from dilute metal solutions. Recovery by the first way is secondary recovery in mining, while by the second way, it is a recycling process and a means of avoiding heavy metal pollution disasters.

Micro-organisms Involved in Bioleaching of Ores

Micro-organisms identified to be useful in bioleaching of different ores for the recovery of various metals have been listed in Table 59. The *Thiobacillus* group of bacteria has been well recognised to be associated in metal leaching. These are chemolithotrophic micro-organisms which are known to utilize inorganic sulphur and ferrous iron compounds in acidic solution. These organisms are rod shaped bacteria about 0.5 x 1.0 μm in size, can utilize carbon dioxide as carbon source (Bryner and Anderson, 1957), and require supplementation of additional nitrogen, phosphorus and some trace elements which are not usually present in their environment. This group of organisms requires different temperature with respect to various strains and substrates, but the optimum is considered to be within the range of 25°C to 35°C. Thermophilic iron oxidising *Thiobacillus* bacteria have also been reported to be isolated (Brierley, 1976; Brierley and Lockwood, 1977; Brierley et al, 1978). Guay et al (1975) discussed the role of different *Thiobacillus* in pyrite oxidation when cultured alone and in mixed cultures. The *Thiobacillus* group of organisms have been successfully investigated for bioleaching of Aresenic (As), Bismuth (Bi), Cadmium (Cd), Cobalt (Co), Copper (Cu), Gallium (Ga), Germanium (Ge), Lanthanum (La), Lead (Pb), Nickel (Ni), zinc (Zn) and Iron (Fe) from their respective ores (Brierley and Lockwood, 1977). *Thiobacillus* group of organisms are generally used for bioleaching of above metals from their ores containing sulphides namely Pyrite, Zinc sulphide, Cobalt sulphide, Chalcopyrite, Chalcocite, Covellite and Galena (Guay et al, 1975; Brierley and Brierley, 1976). Micro-organisms belonging to thermophilic groups can grow and survive at temperatures as high as 95°C (Brierley, 1977). Acidophilic bacteria of *Sulfolobus* sp and *Acidonus* sp are spherical in shape and are found to be active between pH values of 1 and 6 with an optima between pH values 2 and 3 (Brierley, 1977; Trudinger, 1971). The leaching of the Molybdenite (MoS_2) by *Sulfolobus* was described more thoroughly by Brierley in 1974 (Karaivko and Mnoshniakova, 1977; Duncan and Walden, 1972). Several copper sulphide ores and copper concentrates have been leached in stationary batch reactors using *Sulfolobus* (Brierley, 1978). The ability to leach these substrates is dependent on supplements added to the media, such as ferrous iron and yeast extract and mineralogy of samples. *Sulpholobus* has also been successfully used for leaching of copper from chalcopyrite ore and concentrate of chalcopyrite with an efficiency of 51 per cent in 60 days at 60°C (with initial copper 29%), having ore particle size less than 212 μm (Brierley, 1978). *Sulpholobus* has also been reported for leaching of zinc and nickel from their sulphide ores. However, these organisms do not indicate positive results in leaching of lead from lead sulphide. Application in the case of leaching of uranium ore indicates that uranium is not readily solubilized by the organism, and in fact, conditions manifested by the uranium are not conducive to the viability of organism (Duncan and Bruynesteyn, 1971).

Table 59
Micro-organisms for bioleaching of ores

Group		*Species*
Bacteria	—	*Thiobacillus ferro-oxidans, T. thiooxidans, T. novellus, T. neopolitanus, T. perometabolie, T. acidophilus, Sulpholobus sps, Arhbacter sps, Ferrobacillus ferrooxidans, Leptospirillum, Pseudomonas aeruginosa, Leptothrix discophora, Beijerinckia sp, Desulfovibrio sps.*
Yeast	—	*Candida lipolytica.*
Fungi	—	*Aspergillus niger.*
Algae	—	Chlorella pyrenoidosa.

A number of other groups of organisms—*Acromonas* sps, *Agrobacterium* sp, *Bacillus* sp, *Pseudomonas* sp etc have also been reported for bioleaching of various metals from different ores. The *Bejerinckia, Pacillomyces, Aspergillus* sps and *Zooglea* have also been used for bioleaching of copper from chalcopyrite. Similarly, leaching of iron from high grade silica used in the manufacture of glass by citric acid producing *Aspergillus* sps and removal of manganese from manganese containing silver ore for which a bacterium and fungus, naturally associated with ore are used. Hence, recovery of manganese which is present as an impurity in the ore enables easy silver recovery. Removal of silica from magnesite and bauxite for aluminium production has been reported using *Aspergillus* sp, *Bacillus* sp and *Panicillium* sps. Further bioleaching of aluminium from its ore has also been carried out by *Penicillium* sps and *Scopulaniosis* sps (Brierley, 1978; Silverman and Ehrlichm 1964).

Microbial leaching can be done in three ways: (i) vat leaching, (ii) in-situ leaching, and (iii) heap leaching.

Vat leaching is generally applied for the extraction of uranium, gold, silver and copper from their ores. It is applied for ores with high metal concentrations.

In-situ leaching is done by injecting a leach solution (chemical oxidants like H_2O_2 oxidase and solubilize uranium) into unfractured uranium ore body via injection well and collecting the leached metal from the production well.

Heap leaching or dump leaching is the one which is mostly used. It is mainly employed with wash material from conventional mining and extraction from discarded residues. Dumps upto 1200 ft high and 4 billion ton in weight are piled from waste residues after primary mining. Water is sprayed. *Thiobacillus* organisms are ubiquitous and so are present in dumps and no inoculation is required. Area on which heaps are built is usually covered with clay or asphalt so that metal rich liquids are collected in pools.

Mechanism for Bioleaching

The mechanism of bicobial leaching is not clearly understood. Different metals can be dissolved from insoluble minerals either directly by the metabolism of micro-organisms or indirectly by the products of their metabolism. The biological reactions involved in extractive metallurgy are primarily oxidation of sulphur or mineral sulphides. Generalized reaction often used to express the biological oxidation of mineral sulphide is:

$$MS + 2O_2 \rightarrow MS + O_2 \rightarrow MSO_2$$

Where 'M' is a bivalent metal. The mineral sulphide serves as energy source in the presence of other nutrients.

Direct Mechanisms

Some of the chemolithotrophis micro-organisms are capable of direct oxidative attack on mineral sulphides. It has been proved by electron beaming studies that the bacteria attach themselves to the surface of sulphide mineral in solutions supplemented with nutrients (Brierley, 1978; Silverman and Ehrlich, 1964). Results indicate that bacteria dissolve sulphide surface of the crystal by means of cell contact. Tato and Lundgren (1978) have discussed the nature of the cell surface of a mineral sulphide oxidising bacterium and have developed a model for illustration of oxidation of iron and sulphide. Further, the cells of living organisms have also been found to attach with hydroxides of metals in solution. Insoluble sulphide minerals can be biotransformed by micro-organisms in absence of ferric iron under conditions that preclude any likely involvement of ferrous-ferric cycle (Murr et al, 1978; Mizoguchi et al, 1976). Present evidence from research using soluble sources of sulphides and reduced iron substrates supports the idea that iron and sulphide can be simultaneously oxidised by bacteria and both the oxidative processes contribute metal leaching (Torma and Sakaguchi, 1978; Braithwaite and Wadsworth, 1976). Duncan and Bruynesteyn (1971) have demonstrated that direct mechanism for bioleaching of zinc from zinc sulphide ore, rich with iron and without iron. Authors concluded that the leaching of zinc is not mediated by ferric iron in presence of *Thiobacillus ferroxidans.* Experiments of Duncan and Walden (1972) on bioleaching of chalcopyrite, chalcocite and marmatic zinc sulphide have concluded that these minerals contain traces of iron preclude stating that iron is not involved in the metals. Alternatively, these experiments do not rule out *T. ferroxidans* directly attacking the mineral structure (Duncan and Walden, 1972).

Although many aspects of the direct attack by bacteria on mineral sulphides remain unknown, it is apparent that specific iron and sulphide oxidizers might be playing an important role. Microbial involvement is influenced by the chemical nature of both aqueous and solid crystal phase (Murr et al, 1978). Bacteria appear to attach specifically to the sulphide moiety of mineral rock surfaces, which are the regions that contain the energy supply for the bacteria. Attachment and sulphide oxidation results in pitting of the mineral surface, the extent of surface corrosion varies from crystal to crystal and is related to orientation of the mineral (Brierley and Brierley, 1973; Murr et al, 1978; Brierley, 1978). The metabolite that aids the dissolution of mineral sulphide may also be produced. This substance has been postulated to act either by (a) Oxidation of Fe^{2+}; (b) Stabilization of molecular sulphur on the surface of crystals; or (c) Direct attack on crystallized mineral sulphide surfaces.

Indirect Mechanism

The role of ferric sulphate and oxygen has been cited in the oxidation of mineral sulphides. Minerals such as galena (PbS), Chalcopyrite ($CuFeS_2$), bornite ($Cu_5Fe\ S_4$) and sphalerite (ZnS) are oxidised more rapidly than pyrite and marcasite (FeS_2), where as covellite (CuS), chalcocite (Cu_2S) and molybdenite (MoS_2) are oxidised more slowly (Silverman and Ehrlich, 1964; Lundgren and Dean, 1979). Ferric iron, either alone or in combination, is the most important chemical species involved in the indirect attack on the sulphide minerals. Thus, in short, stage I involves bacterial generation of Fe^{3+} from Fe^{2+}, where as stage II involves chemical leaching of ore by Fe^{3+} in acid solution. The reaction involved in indirect mechanism are:

$$2\ Fe^{2+} \xrightarrow{\text{Bacteria}} 2\ Fe^{3+}$$

$$MeS + 2\,Fe^{3+} + H_2O + 2O_2 \rightarrow Me^{2+} + 2\,Fe^{2} + SO_4 + 2H^{+}$$

In the presence of iron oxidizing bacteria, the ferrous ion produced by these reactions can be oxidized to ferric ion, thereby establishing a cyclic process.

In aerobic process, the ferric sulphate generates ferrous sulphate and elemental sulphur from pyrite:

$$F_2\,(SO_4)_3 + 2\,Fe\,S_2 \rightarrow 3\,FeSO_4 + 2\,S$$

Similarly, Lovely and Phillips (1962) showed a possibility for use of sulphate reducing bacteria *Desulphovibrio desulfuricans* for conversion of uranium hexavalent U (VI) to U (IV) which results in extra cellular precipitation of U (IV) from mineral uranite.

Biomineralization results from interaction of extracted metabolic products (iron, gases, polypeptides, electrons) with extraneous metal ions in the surrounding environment. Some of the processes of biomineralization have been shown in Table 60.

Biochemical Reactions Involved in Bioleaching

The general physiology and metabolism of iron oxidising bacteria—*Thiobacillus* is well documented in literature as compared to others and the organisms belonging to this group have more potential for commercial application. Further, the specific growth conditions for the organis, especially low pH, avoid the cross-contamination. The bioleaching of metals is related with mode of metabolism which is governed with physiological process of the organism. This metabolism incorporates the oxidation of suitable substrate and is carried out at the expense of certain nutrients taken into the bacteria in support of the growth. The energy source of *T. ferrooxidans* is reduced iron; which is stable in acid solution. The following biochemical reactions occur for the control of physiology of the organism:

$$2\,Fe^{2+} \rightarrow 2\,Fe^{3+} + 2\,e$$

The characteristic reducing potential is available for the reduction of molecular oxygen through the following mechanism:

$$1/2\,O_2 + 2\,e + 2\,H^+ \rightarrow H_2O$$

The formation of ATP through oxidative phosphorylation is governed by the reaction:

$$ADP + iP \rightarrow ATP$$

The pair of electrons transferred yields enough energy to ensure ATP formation.

Thus, the chemical energy is converted by oxidative phosphorylation—the biochemical equivalent of respiration to ATP, that is universally recognised as the form of metabolic energy which the cell utilizes for transporting substrate and nutrients into the cell and metabolic products out of the cell, vibration and locomotion, and for biosynthesis. ATP is

Table 60
Bacterial biomineralization processes

Process	*Mechanism*	*Mineral*	*Examples*
Biologically induced (biotransfer) (extracellular)	Soluble biopolymers	Mn/FeOOH	*Leptothrix* *Pedomicrobium*
	Spore coats	MnOOH	*Bacillus*
	Gas/ion exchange		
	H_2S	Fe/Cus	*Desulfovibrio*
	CO_2/pH	$CaCo_3$	*Calothrix*
	pH	$MgNH_4PO_4$	*Proteus mirabilis*
	Proteolipid/ion transport	$Ca_5(OH)(PO_4)_3$	*Streptococcus*
	Phosphatase activity	$(UO_2)_3(PO_4)_2$	*Citrobacter*
	Electron transfer	Fe_3O_4	GS—15
		UO_2	GS—15
	Nucleation proteins	H_2O(ice)	*Pseudomonas syringae*
	S-layer templates	FeOOH	*Leptothrix*
Biologically controlled (biosynthesis) (intracellular)		Fe_3O_4	*Aquaspirillum magnetotacticum*
		Fe_3S_4/FeS_2	Wild types
	Ferritin micelles	$FeO(OH)/(PO_4)$	*P. aeruginosa*

hydrolysed to ADP and inorganic phosphate and releases the energy.

Two other important reactions for autotrophic growth, the formation of reducing power and fixation of carbon dioxide, which are dependent upon the energy generating processes of metabolism. The reduction of pyridine (NADH, NADPH) provides the organism with reducing power and is generated from the oxidation of ferrous ion. Energy is also required to move the electrons against a redox potential gradient of + 38 V (cytochrome) to 34 V for the NAD/NADH:

$$NAD^+ + 2e + 2H + 2ATP \rightarrow NADH + H^+ + 2ADP$$
$$NADH + H^+ + NADP^+ \rightarrow NAD^+ + NADPH + H^+$$

The carbon dioxide fixation reaction in *T. ferooxidans* is linked with production of energy from Fe^{2+} oxidation. The pathway for carbon dioxide assimilation is a reductive pentose phosphate cycle that includes two characteristic enzymatic reactions: (a) Phosphorylation of ribulose-5-phosphate, catalyzed by phosphoribulose-kinase, to yield ribulose 1, 5-diphosphate; and (b) Ribulose-biphosphate-carboxylase which catalyzes a reaction between one molecule of carbon dioxide and one molecule of ribulose-1, 5-diphosphate with a concomitant conversion to yield two molecules of 3-phosphoglyceric acid (Torma, 1977; George et al, 1967).

Laboratory studies have also indicated that the iron grown cultures can be adopted to use organic substrates for energy and carbon under specific conditions. However, little is known about the capacity of organisms for heterotrophy and/or mixotrophy in nature.

Like the biochemistry of iron oxidation, the biochemistry and physiology of inorganic sulphur oxidation by *Thiobacillus* is well known. The mechanism of sulphur oxidation is similar to that of other *Thiobacilli.* Elemental sulphur, tetrathionate, thiosulphate and sulphide (mostly of various non-iron minerals) support the growth of the organisms. In all the cases, sulphite (SO_3) is the key intermediate molecule. Energy is produced from oxidation by the use of the enzyme sulphite oxidase, ADP sulphurylase, APS reductase and Adenylate kinase. An alternative oxidation mechanism for sulphur has been proposed in which elemental sulphur is oxidised to sulphide and then to sulphate by *T. ferrooxidans* through ferric iron system (Murr et al, 1978).

Bioleaching of Some Ores for Metal Recovery

A number of ores can be bioleached for recover of metals by direct or indirect mechanisms using micro-organisms.

Iron

Attachment of bacteria to surfaces of pyrite (FeS_2) leads to biological reaction:

$$FeS_2 + 30_2 + H_2O \xrightarrow{\text{Bacteria}} FeSO_4 + H_2SO_4$$

$$2\ FeSO_4 + * / 2\ O_2 + H_2SO_4 \xrightarrow{\text{Bacteria}} Fe_2(SO_4)_3 + H_2O$$

$$FeS_2 + Fe(SO_4)_3 \xrightarrow{\text{Bacteria}} 2\ FeSO_4 + 2S$$

$$2S + 30_2 + 2H_2O \xrightarrow{\text{Bacteria}} 2\ H_2SO_4$$

Copper

Biometallurgy of copper involves the following steps (Figure 60): (i) the low grade or and tailings, left from conventional mining, are piled up in an area, where the ground has been made permeable; (ii) the pile is sprayed with a leaching solution that contains iron in the form of Fe^{3+} ion, sulphate ions (SO_4^{2-}) and *T. ferrooxidans;* the sulphate ion in the leaching solution form sulphuric acid giving acidic solution essential for growth and activity of bacteria; (iii) copper released due to catalytic chemical (facilitated by *T. ferrooxidans)* is drained in the form of a solution; (iv) copper is then removed from the solution with the help of another solvent; (v) the remaining, leaching solution flows into an open pond, where *T. ferrooxidans* catalyses a reaction that oxidises Fe^{2+} into Fe^{3+} ions and sulphide into sulphate ions, so that the leaching solution in new recharged; (vi) the recharged leaching solution is pumped back to the top of the pile for the cycle to begin again; (vii) the copper is extracted from the solution as sheets through electrowinning process by passing electricity, so that copper is collected on negative terminals. This process is costly and non-biological, but 'biosorption filters' such as algae will be used in future to make the process entirely biological.

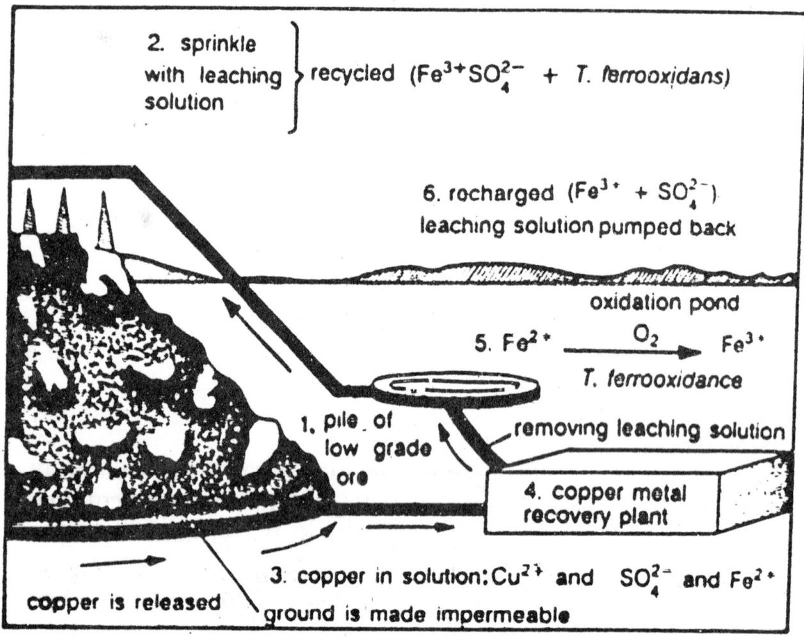

Figure 60
Different steps involved in mining of copper with bacteria

The copper minerals—chalcopyrite, chalcocite and covellite can be processed biologically for leaching of copper as follows:

Chalcopyrite: This is iron containing mineral and both mechanisms of bioleaching operate simultaneously and are difficult to separate. The following sequence of reactions results on the oxidation of chalcopyrite in presence of biocatalysts:

$$48\ CuFeS_2 + 204\ O_2 + 24\ H_2SO_4 \rightarrow 48\ CuSO_4 + 24\ Fe_2\ (SO_4)3 + 24\ H_2O$$

$$12\ CuFeS_2 + 24\ Fe_2\ (SO_4)_3 \rightarrow 12\ CuSO_4 + 60\ FeSO_4 + 24\ S°$$

$$60\ FeSO_4 + 30\ H_2SO_4 + 15\ O_2 \rightarrow 30\ Fe_2\ (SO_4)_3 + 30\ H_2O$$

$$24\ S° + 24\ H_2O + 36\ O_2 \rightarrow 24\ H_2SO_4$$

$$30\ Fe_2(SO_4)_3 + 120\ H_2O \rightarrow 20\ (HFe(SO_4)_3.2Fe\ (OH)_3)) + 50\ H_2\ SO_4$$

$$60\ CuFeS_2 + 2550_2 + 90\ H_2O \rightarrow 60\ CuSO_4 + 20\ (H(Fe(SO_4)_2.\ 2Fe\ (OH)_3)) + 20\ H_2SO_4$$

The early studies of chalcopyrite bioleaching using *Thiobacillus* in single stage column on bench scale resulted in 25 per cent yield of metal in 60 days (Bryner et al, 1954; Razzel, 1962). Razzell and Trussell (1963) recovered 45 per cent of copper from chalcopyrite using cultures in a stationary state. The yields of metal was further improved by implementing agitation in reaction tanks to 35 per cent in 33 days, 72 per cent in 12 days and 100 per cent in 26 days (Duncan et al, 1964). Other workers achieved a yield of 50 per cent in 5 days, 60 per cent in 4 days, 79 per cent in 6 days and 55 per cent in 12 days (Duncan and Trussel, 1964; Sato, 1960).

Chalcocite: Copper sulphide mineral can be leached by *T. ferrooxidans* and can be oxidised to copper sulphate with digenite (Cu_9S_5) and covellite (CuS) as intermediate products as indicated by the following reactions:

$$10\ Cu_2S + 2\ H_2SO_4 + O2 \rightarrow 2\ Cu_9S_5 + 2\ CuSO_4 + 2H_2O$$

$$2\ Cu_9S_5 + 8\ H_2SO_4 + 40_2 \rightarrow 10\ CuS + 8\ CuSO_4 + 8\ H_2O$$

$$10\ CuS + 20\ O_2 \rightarrow 10\ CuSO_4$$

$$10\ Cu_2S + 10\ H_2SO_4 + 20\ O_2 \rightarrow 20\ CuSO_4 + 10\ H_2O$$

Copper was solubilized from chalcocite and covellite both in the absence and presence of iron. Ferric iron can oxidise these minerals to $CuSO_4$:

$$CuS + 2\ Fe\ (SO_4)3 \rightarrow 2\ CuSO_4 + 4\ FeSO_4 + S^\circ$$
$$Cu_9S_5 + 9\ Fe_2\ (SO_4)_3 \rightarrow 9\ CuSO_4 + 18\ FeSO_4 + 5\ S^\circ$$

Further, selenite (CuS.Se) can also be oxidised to copper sulphate and elemental selenium:

$$2\ CuS.Se + 2\ H_2SO_4 + O_2 \rightarrow 2\ CuSO_4 + 2\ Se^\circ + 2\ H_2O$$

Uranium

The standard free energy change (20.6 kCal/Mol) and the oxidation-reduction potential for the tetravalent hexavalent uranium system (446 mv) are similar to those of ferrous-ferric system (17.8 kCl/Mol and 747 mv respectively). Uranium oxidation followed by dissolution of tetravalent uranium proceeds by ferric ion as reaction indicates:

$$UO_2 + Fe\ (SO_4)_3 \rightarrow UO_2SO_4 + 2\ FeSO_4$$

The tetravalent uranium is oxidised by *T. ferrooxidans* (Fishner, 1966; Macgregor, 1966).

$$2\ UO_2 + O_2 + 2H_2SO_4 \rightarrow 2UO_2SO_4 + 2\ H_2O$$

Biometallurgy was first used in 1970 in Canada for the extraction of uranium (Murr et al, 1978; Brierley, 1978), and the United States (Fishner, 1966; Macgregor, 1966). The ferric ion required for oxidation of the tetravalent uranium is supplied through the biological oxidation of either soluble ferrous or iron containing sulphide minerals present in the uranium ores. Using percolation column, recoveries of uranium upto 90 per cent in 20 weeks were achieved from low grade ores with particle sizes of 4 mm (Cow et al, 1988). Other micro-organisms—*Sulfolobus* and *Thermophilic thiobacilli* have been found suitable for uranium extraction (Matic and Mrost, 1964).

Gold and Silver

It has been shown that some bacteria may play an important role in the formation of gold and silver deposits in sediments laid down by river-alluvial deposits. The negatively charged polymer on the outside of the bacteria attracts the positively charged gold/silver particles in the soil, so that they clump together as grains and eventually as gold/silver nuggets. It has been demonstrated that metals like gold/silver continue to accumulate on negatively charged polymers after the bacteria have died, so that the process can continue indefinitely giving nuggets.

It has been argued that genes responsible for mineralization in a variety of micro-organisms can be isolated, cloned and transferred into *E. coli,* which may then help in metal deposition more efficiently. Improvement of strains of bacteria with increased rate of leaching for metal extraction have already been attempted. For example, plasmids have been constructed, which when present with *T. ferrooxidans* increases its resistance to arsenite and arsenate (these may not allow growth of bacteria), and enhances recovery of gold from arsenopyrite-pyrite ores. It is hoped that genetically engineered bacteria with enhanced leaching or nucleating capabilities will soon be available to companies in mining and metal extraction (Gupta, 1995).

Chapter 10

Thuringiensis Toxin as a Natural Pesticide

Insect damage to crop plants results in 15-20% yield reduction worldwide. Introduction of pesticides is a common measure in plant protection programmes. Most of the pesticides are chemically synthesized, but it has also increased the cost of inputs and became detrimental to the environment. Discovery of the insecticidal proteins in a soil bacterium *Bacillus thuringiensis* (Bt) spurred tremendous interest in the development of what are known as the biopesticides and their application in agriculture. The spore preparation of this bacterium has been used as a natural pesticide during the last 20 years. Insecticidal activity of this species depends on a protein (delta endotoxins) synthesized during sporulation (Gupta, 1994). Bt insecticidal protein, also called as crystal protein, dissolve in the insect midgut and are processed into an active, toxic form. The toxin bind to cell membrane receptors along with the gastrointestinal tract, break down the membrane and kill the insect larvae. The toxins kill only insects and are harmless to animals, especially to mammals. Strains of Bt produce toxins which differ in their spectrum of activity against different classes of insects, including the larvae/caterpillar stages of Lepidoptera (moths and butterflies), Diptera (flies and Coleoptera (beetles). Most of the *B. thuringiensis* strains commercially released are of high potency against pests (Jayaraman, 1993), and thus, are like environment-friendly pesticides. Since these toxins are very specific in their action, they are safe insecticides, but their use is limited due to high production cost and due to instability of crystal protein when exposed in the field. It is sold in several countries the EPA has registered it for use on field crops, trees, ornamentals, home vegetable gardens and stored products (primarily grains and grain products) to control Lepidopteran larvae.

The primary insecticidal activity of *B. thuringiensis* resides in a parasporal glycoprotein crystal that is synthesized and crystallized within the parent cell (Bechtel and Bulla, 1976; Bulla et al, 1977). Most studies of *B. thuringiensis* have focussed on the field application and efficacy of the organism. However, there has been a recent upsurge of interest in the molecular biology of *B. thuringiensis*, particularly as it relates to toxin synthesis. This organism provides an ideal model for studying the formulation and regulation of a protoxin and its subsequent conversion to a toxin.

The Toxin of *B. Thuringiensis* Subspecies *Kurstaki*

The insecticidal activity of *B. thuringiensis* subspecies *kurstaki* as well as that of other subspecies, is associated with a parasporal glycoprotein crystal that is synthesized within the organism during its sporulation cycle. In its native state, the glycoprotein is a protoxin that is solubilized and activated after ingestion by an insect that is susceptible to the toxic product (Bulla et al, 1980). The protoxin can be activated in-vitro by exposure to alkaline buffer. Maximum solubility of the crystal protoxin occurs within several hours after the crystal is titrated with alkali. The protoxin is stable for several hours thereafter, but then begins to degrade into small fragments with a concomitant low in insecticidal activity. Reaggregation also occurs, especially after the pH is lowered to near neutrality or near the experimentally determined pH of the protoxin pH 7 (Bulla et al, 1981). The crystal protein may be an intact product of translation since methionine is the NH_2-terminal residue.

Figure 61 shows a schematic diagram showing the behaviour of the protoxin and toxic solution. The crystal (An) is made up of many subunits that are dissociated in native conformation (nA) by mild alkali titration (reaction 1).

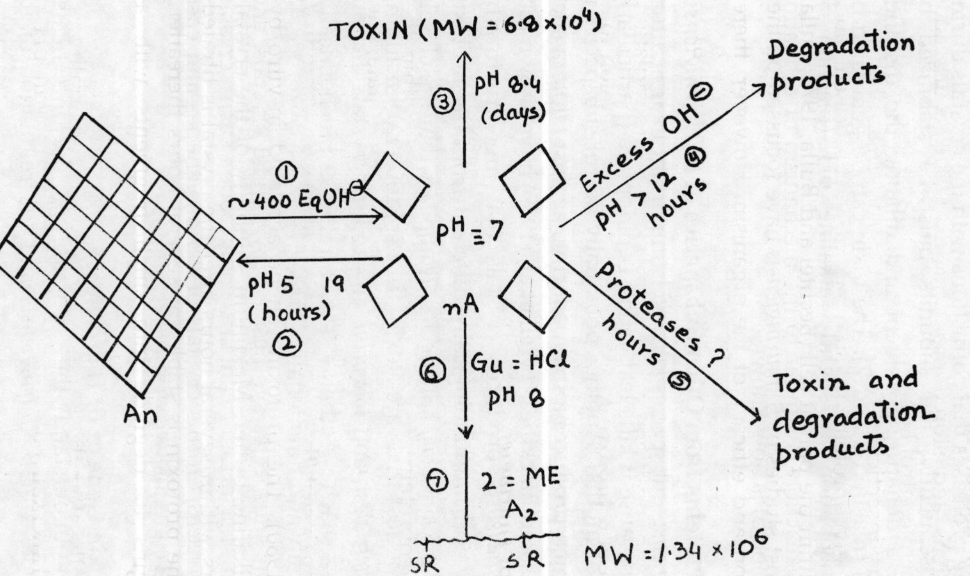

Figure 61

Schematic diagram of the behaviour of the protein from *B. thuringiensis* in solution. (EqOH = equivalent base; GuHCl = guanidine hudrochloride; 2 = ME ≠ 2-mercaptoethanol; R = alkylation reaction; SR = alkalated disulphide linkages)

Reaggregation occurs slowly (reaction 2), especially at pH 5, 12 hours; also the subunit may break down into smaller fragments (reaction 4). Degradation may be further stimulated by contaminating proteases that cofractions with the crystals during isolation subsequently, the subunits must be stabilized by placing it in a denaturing solvent and dilating the disulfide linkages (SR) (reaction 6 and 7). The toxin can be generated by prolonged incubation at slightly alkaline pH (reaction 3). Whether this reaction is similar to the mechanism of activation in-vivo (reaction 5) is now known.

Insecticidal Proteins from Other Subspecies of *B. Thuringiensis*

B. thuringiensis subspecies *kurstaki* is important in agriculture and forestry because it kills leaf-eating Lepidoptera (particularly larval forms of moths). Another potentially useful subspecies is *israelensis,* which produces a toxin effective against Diptera such as mosquitoes and blackflies. Tyrell et al, (1979) have shown that subspecies *israelensis* produces crystals that are structurally, biochemically and immunologically different from the protoxin produced by the *kurustaki* subspecies. It appears that the closely related subspecies that are toxic to Lepidoptera, including *kurstaki, berliner, alesti* and *tolworthi,* all produce crystals that are similar to each other structurally, biochemically, immunologically and functionally (Table 61).

These four subspecies synthesize bipyramidal crystals, and usually each bacterium contains only one such crystal. Subspecies *israelensis* forms crystals of various shapes and each bacterium cell contains two or three crystals.

The Insect Toxin Gene of *B. Thuringiensis*

To better understand the structure of the insecticidal crystalline protein of *B. thuringiensis* and the regulation of its synthesis, investigations in several laboratories have cloned the gene that codes for the protein in subspecies *kurstaki.* In particular, a recombinant Charon 4 A phase, C4K6C, was isolated by Held et al, (1982) and found to produce, when grown in *Escherichia coli,* a protein that reacts with antibodies specific for the crystalline protoxin protein of *B. thuringiensis* subspecies

Table 61
Characteristics of the subspecies of *B. thuringiensis* (Miller et al, 1983).

israelensis	*kurstaki, berliner, alesti, loborthi*
1. Crystal size and shape varies.	1. Crystal are about the same size and have a bipyramidal shape.
2. Protoxin subunit has apparent molecular weight of 1.34×10^5.	2. Protoxin subunit has apparent molecular weight of 1.34×10^4.
3. Major crystal breakdown product has apparent molecular weight of 2.6×10^4.	3. Major crystal breakdown product has apparent molecular weight of 6.8×10^4.
4. More lysine, threonine, proline, valine methionine, and isoleucine in crystals.	4. More arginine, serine, glutamine and glycine in crystals; four strains ver similar.
5. Ineffective against Lepidoptera; toxic to mosquitoes.	5. Ineffective against mosquitoes, toxic to Lepidoptera.
6. Crystals are immunologically different from other strains.	6. Crystals are immunologically related.
7. Crystals contain glucose, mannose, fucose, rhamnose, xylose and galactosamine.	7. Crystals contain glucose and mannose.
8. Crystal peptide map has spots not found in maps of other strains.	8. Crystal peptide maps are similar in the four strains.
9. Crystal protein contained in spore coat along with low molecular weight proteins; the size of the protein differs from that in the other strains.	9. Crystal protein contained in spore coats along with low molecular weight protein; similar sizes in the four strains.

kurstaki. Alkali extracts of C4K6c lysates are toxic to insect larvae, and the degree of toxicity is consistent with the quantity of antigen found in lysates of phage-infected cells. Cells of *E. coli* infected with C4K6c produce a polypeptide antigen of the same size as the protoxin molecule obtained by alkali solubilization of *kurustaki* protoxin. Held et al (1982) identified a 4.6 kilobase-pair (kbp) Eco RI fragment from C4K6c that contains all or part of the toxin gene and subcloned it into two plasmids, pBR328 and pHV33. Both *E. coli* and *Bacillus subtilis* containing these recombinant plasmids produced antigen that cross reacted with antibody directed against the protoxin. The 4.6 kbp Eco RI fragment is of chromosomal origin, although a homologous protoxin gene is also present on a 45 kbp plasmid carried by the *kurustaki* subspecies. Several acrystalliferous nontoxic mutants that have been isolated lack the 45 kbp plasmid and, in some mutants, all plasmids are absent. All of the mutants contain the chromosomal gene but do not produce protoxin antigen. The function of the plasmid gene in the expression of insecticidal activity is not known (Miller et al, 1983).

A more efficient way of using Bt toxin is make plants synthesize the chemical instead of spraying it on them. The above toxin gene (Bt 2) from *B. thuringiensis* has been isolated and used for *Agrobacterium* Ti plasmid mediated transformation of many crops like maize, rice, cotton, potato, tobacco, cotton and tomato plants. Bt transgenic crop plants endowed with complete insect resistance are expected to be released for commercial cultivation in the near future (1995-1996).

The advantage of using transgenic plants to control insect damage include reduced chemical usage and the pollution from them, decreased input costs to the farmers, season long protection, effective control of burrowing insects which chemical sprays cannot reach, control at all stages of insect development and confinement of the protection only to the crop plant. But transfer of Bt-toxin is not the only strategy available to counter insect pests. There are other approaches as well. They are the use of genes of proteinase inhibitors, lectins and amylase inhibitors which affect the insect digestive system. Insect feeding on plants having these genes cannot survive for long because of the damage to their digestive system (Kumar, 1994).

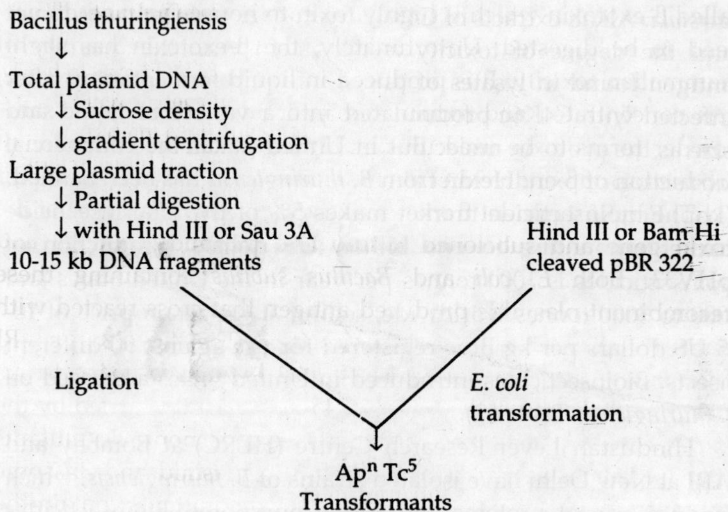

Figure 62
A methodological approach for isolation of the crystal protein gene from *B. thuringiensis*

When inheritance of insect resistance was studied using crosses with normal control plants, F_1 generation showed resistance and F_2 generation exhibited expected segregation.

The concern that transgenic plants may have deleterious effect on the environment has been carefully ruled out as the introduced gene products forms only a small percentage of the total plant product and do not have toxic effects on the animals etc. Moreover, as these gene products are extremely specific in their target action, their introduction into plants will not have any serious implications (Jayaraman, 1993).

Production of *B. Thuringiensis*

Bt was discovered in silkworms in Japan in the early part of the twentieth century. It first became a commercial product in France in 1938 and has been on sale ever since. The bacterium produces a protenaceous crystal which becomes insecticidal after digestion in the alkaline environment found in the mid-gut of susceptible species. Some strains of Bt produce a soluble toxin called β-exotoxin which is highly toxin to house-flies and do not need to be digested. Unfortunately, the β-exotoxin has slight mammalian toxicity. It is produced in liquid fermentations after it is concentrated and formulated into a variety of liquid and powder forms to be used. But in United States the commercial production of β-endotoxin from *B. thuringiensis* has been banned.

The bioinsecticide market makes 5% of the total insecticide market amounting to one billion US dollars. Production of bioinsecticides is 1000 tonnes per year in the United States and 2000 to 3000 tonnes worldwide, selling at the retain price of 10-25 US dollars per kg Bt is registered for use against 90 different insects. Bioinsecticides introduced in United States are based on *B. thuringiensis* Table 62).

Hindustan Lever Research Centre (HLRC) at Bombay and IARI at New Delhi have isolated strains of *B. thuringiensis,* which produce crystal proteins and forms no spores. Bioinsecticides based on these mutant strains will be ecologically safe, and have been successfully tested on insects affecting cotton, maize, cabbage, sunflower, pigeaonpea and safflower.

Problems of Bt Production

Although the fermentation of Bt presents few conceptual problems, its cost will determine the competitiveness of the company making it. Costs may be reduced atleast 30% over by the use of inexpensive raw materials, by faster fermentations, and by the simplification of the concentration steps used after the fermentation. At Microbial Resources Bt H-14 (Bt *israelensis)* is made into five separate products:

(1) A wettable powder for formulation at the point of use;
(2) A flowable liquid for general use;
(3) An emulsion for rapid dispersion over large areas of water;
(4) A sand-granule for aerial application through foliage; and
(5) An ultra-low volume liquid for efficient application by aircraft when there is no canopy.

Table 62

Commercially available Bacillus products (Burgess, 1984; Kurtsak, 1982; and Miller et al, 1983)

Species	*Commercial name*	*Target*
Bacillus moritai	Lavillus M	Diptera
Bacillus popilliae	Doom, Japidemic	Coleopteran larvae
B. thuringiensis	Dipel	Lepidoptera
(- endotoxin + spores)	Thuricide	"
	Certan	"
	Bactur	"
	Bactospeine, Blantibae	"
	Bugline, Sporeine	"
	Biospor, Bathurin	"
	Bantukal, Dendrobacillin	"
	Entobacterin - 3, Insektin	"
B. thuringiensis (exotoxin)	Biotoksylacillin, Eksotokin	"
	Toxobakterin	"
B. thuringiensis	Bactinos, Vectobac	Culex, Aedes
	Teknar	Anopheles

Stability of Bt

It is essential to know about the stability of the microbial biocide. With no additive Bt H-14 loses half of its activity in 0.05 months at 42°C.

Chapter 11

Biological Control of Insects Swarming Agricultural Fields

The agro-ecosystem are very suitable habitats offering ample opportunities for pest invasion. They are universal, proliferating in varying number each year on every farm, and because of their universal presence, there is a tendency to regard pests as unavoidable problems in farming. In fact, as agriculturists worst enemy, pests have recently come under world-wide attention as man has realised the problem.

The term 'pest' has been derived from the Greek word 'pestis' which means 'to annoy'. The precise definition of pest may be put in a few words as: any organism that causes annoyance to human beings is a pest. A pest is a destructive organism present at a place, where it is not desired or "a organism out of place". Ironically, because of their almost universal presence, pests tend to be regarded in the same way as death and taxes, not pleasant but unavoidable. It includes within its fold the fungi, bacteria, viruses, insects, nematodes, rodents and a multitude of other animals which transmit disease, destroy agricultural crops, stored products, and directly or indirectly affect human health and welfare.

Insects have been classified as "key pests", occasional pests, potential pests, migratory pests, sporadic pests and minor pests.

Key pests are those which are generally abundant or the kind of damage caused by the pest and the damage potential of a single individual is great.

Occasional pests or sporadic pests are those that cause economic damage only in certain places or at certain times.

The potential pests in general considered as minor pests, cause no apparent injury to the crop under the prevailing situation. It has been termed as potential pest because at certain situations it may appear as problem and may be elevated to the position of key pest.

The migratory pests are normally non-resident of any agro-ecosystem. This type of pest usually appears all of a sudden in crop area from places of their breeding, causes heavy damage to the crop and again leaves the area. Their association with agro-ecosystem is of transitory nature. The appearance of locust, army worms and similar highly mobile insects spells disaster, since they devour the leaf and even the bark of the plants, ultimately resulting in the death of the adult plant (Agarwal, 1971).

The aforesaid classification of pets are based either on taxonomic category of the organism involved, or on the hosts they utilize as food or from the point of view of their capacity to cause damage to objects of human interest. But according to Conway (1976) the more illuminating classification would be that with respect to the biological strategies of insects (Table 63); as discussed by Southwood (1976) who adopted the terms like *r-pests* and *k-pests.*

Table 63
Biological characteristics of some well known pests

Pests	*Reproducttive capacity*	*Duration Food plants of generation*
r-pests:		
Schistocerca gregaria	400	1-2 month Polyphagus
Lipaphis erysimi	120	6-12 days Cruciferous
Heliothis armigera	1500	21-45 days Polyphagus
k-pests		
Orycies rhinoceres	50	3-4 month Coconut
Lachnosterna consanguinea		12 month Polyphagus

r-pests include features like high potential rates of increase, strong dispersal and host finding ability and small size. This group includes the classical pests and they fit in most closely with the sense of plague. It includes locusts, aphids and other pests.

k-pests are characterised by low rates of potential increase, greater interspecific competitive capacity, more specialized food habit and bigger size.

Southwood (1976) observed that there is a continuum between the two extremes of r- and k-pests, and most of the crop pests lie between these two extremes and are termed as *'intermediate'* pests. Some of the important pests included in this category are *Nilaparvata lugens* on rice, *Labioproctus polei* on citrus, *Euproctis* spp on castor etc.

Biological Control

Biological control of pests may be defined as the eradication or suppression of insects by encouragement, artificial introduction or increase of their natural enemies such as parasites, predators and disease causing organisms. Biological pests control agents are gaining prominence for the control of insect pests in agriculture and forestry. The growth of interest in biological pest control agents occurred in part because of public awareness of the dangers of indiscriminate use of chemical pesticides. It has also been proved that there exists a relationship between chemical pesticides and mutation, cancer, human disease, damage to wildlife and toxicity to plants. Developments in biotechnology are generating interest and the means necessary for displacement of synthetic chemical insecticides.

The success in chemical control of phytopathogens, weeds and insects was accomplished by environmental pollution and ecological imbalance. Extreme large scale application of various pesticides resulted in their accumulation in soil, water, plants and eventually in food products and fodder.

The ecological problem became urgent, with restriction and strict regulation in the use of pesticides. Scientific endeavour was directed towards the elaboration and wide application of biological pest control which, as a rule, are not only harmless but in many cases have proved to be more effective. Classical biological control includes the introduction of a new organism into the environment as a natural enemy to the unfavourable parasite. The elucidation of the chemical structures of natural parasites seems to be an outstanding achievement in biological pest control. Thus, the discovery of phytoalexins responsible for resistance against insects and diseases provided new horizons for pest control using biological chemicals and their derivatives. Natural attractants, pheromones and their chemical modifications are another example of evolution in the field of biological pest control. Genetic alterations in insect heredity appeared to be a new perspective in pest elimination. Genetic engineering opened a new era in the creation of unprecedented efficient approaches and organisms for biological pest control.

It has been observed that insect poses can be controlled by various naturally occurring organisms including viruses, bacteria, fungi, protozoa and nematodes. These organisms utilize insects as their substrata for subsistence. Biological control is, therefore, essentially an application of knowledge of ecology to pest population control. The potential for widespread use of biological pest control agents is timely because of their relative host specificity and ecologically non-disruptive nature.

Microbial Pesticides

Approximately 1500 naturally occurring micro-organisms or microbial products have been identified as potentially useful pesticidal agents. This interest in microbial pesticides largely a result of the many problems associated with the extensive use of chemical pesticides. Not only do chemical pesticides generally affect beneficial insects as well as pest species, but insects tend to acquire resistance to the chemicals so that new pest problems develop. Furthermore, chemical pesticides pose environmental hazards and health concerns. Thus, microbial pesticides are seen as an alternative area of pest control that can play an important role in integrated pest management systems and reduce our dependence on chemical pesticides. Micro-organisms that are pathogenic to insects have relatively narrow host range, which makes it possible to reduce specific pest population while useful predators and beneficial insects are preserved or given the opportunity to become re-established. Insect resistance to microbial pesticides tend to be less common or to develop more slowly than chance to chemical pesticides. As far as known, entomopathogens now produced commercially do not affect humans or animals other than their target species, and their by-products are bio-degradable.

A microbial pesticide is expected to be (1) convenient to apply (as a dust, spray, or bait, alone or in conjunction with chemicals, predators, parasites or other pathogens), (3) storable, (3) economical, (4) easy to produce, (5) virulent, and (6) safe, and (7) aesthetically acceptable (Figure 63). Because they must be ingested to cause infection, viruses and bacteria need to be applied at the time the target insect begins feeding, that is, before it has caused any major economic damage.

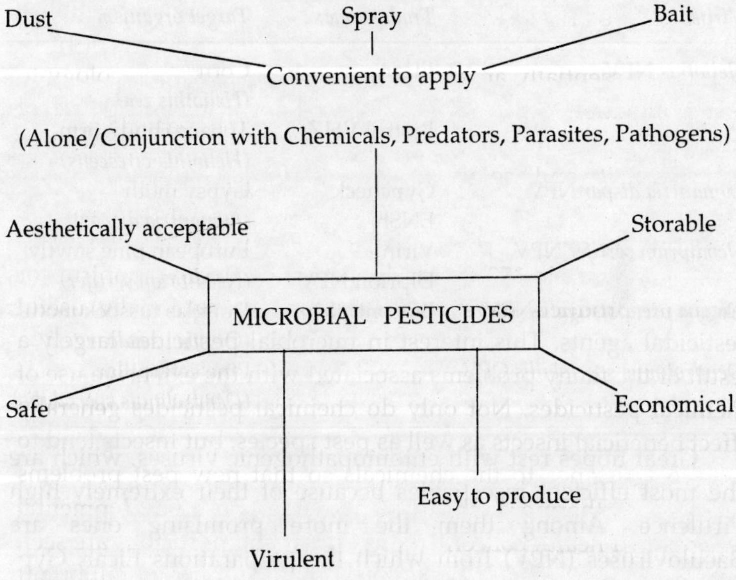

Figure 63
Characteristics of microbial pesticides

Viral Insecticides

Viruses have an interesting ecological specificity and unique location within the biological world. For example, the insect viruses which are members of the family Baculoviride (BV) are responsible for diseases of insects and mites and do not contain structural or biochemical similarities to viruses of the vertibrates or higher plants.

There are about 650 viruses that have been isolated from insects. About 540 were obtained from Lepidopteran, 90 from Hymenopteran and 20 from Orthopteran, Coleopteran and Dipteran insects. The insecticidal potential of viruses has been realized (Table 64) for control of insects populations. Insect viruses infections of susceptible larvae result in pathological symptoms.

Table 64
Commercial viral insecticides (Burgess, 1984 and Payne, 1982)

Virus	*Trade name*	*Target organism*
Heliothis NPV	Elcar	Corn *(Heliothis zea)*
	Biotrol VHZ	Tobacco budworm *(Heliothis virescens)*
Lymantria dispar NPV	Gypcheck ENSh	Gypsy moth *(Lymantria dispar)*
Neodiprion sertifer NPV	Virin - Diprion NPV	European pine sawfly *(Neodiprion sertifer)*
Orgyia pseudotsugata NPV	Biocontrol-1	Douglas tussock moth *(Orgyia pseudosugata)*
Dendrolimus spectabillis CPV		Pine caterpillar *(Dendrolimus spectabilis)*

Great hopes rest with entomopathogenic viruses, which are the most efficient insecticides because of their extremely high virulence. Among them the more promising ones are Baculoviruses (NPV) from which the preparations Eicar, Gyp-Check and Biocontrol-1 for the control of cotton and forest pests, have been registered in United States. Viral preparations of Virin are now produced in Russia.

Mass production of many of the insect viruses involves incubation of infected insects, and harvesting of diseased larvae or adult insects. The subsequent purification of the viral preparation is time consuming and labour intensive, the expenses of which is added to the cost of viral insecticide production increasing costs 4-5 fold. The obstacles to commercial production and use are numerous and include: (1) the development of efficient processes for large scale production; (2) marketability of the products especially where the selective host ranges are required; (3) development of procedures for scheduling and monitoring of filed applications; (4) stabilization of the product under field conditions e.g., sunlight and desiccation; (5) operational hazards, especially possible mutations during large-scale use of viruses; and (6) complicated growth conditions for cultivation of viruses.

Bacterial Insecticides

A number of facultative pathogenic bacteria are known to attack insects, by a number of mechanisms such as invasion of tissues and/or production of toxins. The bacteria can be divided into three ecological types: (1) those which produce within the susceptible host system, (2) those which do not reproduce, necessitating repeated application; and (3) those of either type (1) or (2), depending on the circumstances. *Bacillus thuringiensis* has been tried against various Lepidopterous larvae in Europe, United States and many other countries and success has been achieved. *B. papiliae* causes milky disease to the larvae of Japanese beetle *(Popilia japonica)* which is a serious pest of fruit trees in United States.

Most bacteria pathogenic to insects are members of the families Pseudomonadaceae, Enterobacteriaceae, Lactobacillaceae, Micrococeaeceae, and Bacillaceae. Except for the Bacillaceae, these families contain non-sporulating micro-organisms. Most spore-forming bacteria pathogenic to insects belong to the family Bacillaceae. about 100 bacteria have been reported as entomopathogens, but only four *(Bacillus thuringiensis, B. papillae, B. tentimorbus* and *B. sphaericus)* have been examined as insect control agents. These species are spore formers, the first two produce, in addition to the spore, discrete crystalline inclusions within the sporulating cells; the last two normally do not. Of these four organisms, *B. thuringiensis* is the only bacterium that has been developed successfully as a commercial insecticide on a very large scale and now sold in several countries, for use on field crops, trees, ornamentals, home vegetable gardens, and stored products.

Fungal Insecticides

Entomopathogenous fungi are widely distributed and represent the largest and most diverse group of insect pathogens. For practical purposes, the commonly used ones are the representatives of facultative and obligate pathogens. Among the fungal pathogens well known for many years are *Beauveria bassiana* affecting more than 500 species of insects, and *Metarrhizium anisopilae* which are pathogenic to approximately 200 insect species. These two fungal pathogens have been used with success against certain beetles and bugs in United States and Europe. Obligate insect pathogens belonging to the genera *Entomophthora, Coelomomyces* and some others have a specific host range and growth requirements.

Biological pest control, of late, has become the focal point in plant protection owing to various risk factors, such as development of pest resistance and secondary pest outbreaks and environmental hazards, including human poisoning. Among the biological pest control agents, fungal organisms are predominant and potent against the pest specially those of rice as the required humidities and temperatures are very much available there-in.

More than 500 species of fungi are capable of infecting insects and there are susceptible host species in all the major orders of the Insecta. Some fungi can maintain themselves in susceptible populations through many insect generations without requiring repeated introduction. Since most species do not have to be ingested to cause infection, they can be used to control insect populations in non-feeding stages, often before any economic damage has been done.

Ten mycoinsecticdes have been developed and brought into market so far (Table 65). Though a few products, as Vertalec, Mycar and Collego have been made available by the developed countries, large scale use of these products has not been forthcoming in this part of the world. By contrast, the use of mycoinsecticides in developing countries is reported to be extensive. The demand for cheaper and safer alternatives to chemical pesticides is bound to increase as concern over environmental change mounts. Mycoinsecticides have many advantages:

1. They can be applied using the conventional applicators.
2. These biopesticides comply with the existing law on pesticides registration and commercial production.
3. They pose negligible or no risk of environmental pollution.
4. A much reduced threat to natural enemies compared with conventional pesticides is evident.

Table 65
Commercial mycoinsecticides

Fungus	*Trade name*	*Target organism*
Beauveria bassiana	Boverin	Colorado potato beetle European corn borer Pine caterpillar
Culicinomyces clavisporus	EAO	Mosquito larvae
Hirsutella thompsonii	Mycar	Citrus rust mite
Metarhizium anisopliae	Metaquino	Spittle bug Sugarcane frog hopper
Normurae rileyi	EAO	Lepidopteran larvae
Verticillium lecanii	Vertalac	Aphids, Coffae green bug
	Mycotal	Green house whitefly
	Thriptal	Thrips
Colletotrichum gloeosporioides	—	Northern joint-vetch
Cercospora rodmanii	EAO	Water hyacinth
Paniophora gigantea	—	H. annosum of conifer tree roots
Trichoderma virdea	—	Apple canker disease

Inspite of their potential as agents of insect control, however, worldwide use of fungi on a commercial basis is limited. Comparatively little is known about the mode of pathogenesis of many entomopathogenic fungi and many are inhibited by micro-and-macro-climate. Technology for the large scale production of these fungi is difficult and in many case not yet developed, and the safety of the agents in regard to non-target insects and other animals has not been adequately established, in part because of a lack of reliable bioassay procedures. The few fungal insecticides that are used on a large scale are as follows: Boverin, prepared from the conidia of *Beauveria bassiana* is used for control of the Colorado potato beetle *(Leptinotarsa decemlineata)* in the Soviet Union. It is usually applied in combination with reduced dosages of trichlorphon or malathion. A preparation of *Metarhizium anisopline* called Metaquino is produced in Brazil and is used primarily for controlling a spittle bug *(Mahanarva posticata)* in sugarcane. Considerable use of *B. bassiana* for control of European corn borer *(Ostrinia aulilalis)* have been reported in China. In the United States, *Hirsutella thompsonii* is commercially produced and used for control of the citrus rust mite. *Nomurea rileyi* is being used in large scale field trials as a control for the cabbage looper and the velvet bean caterpillar. *Verticillium lecanii* is produced commercially in Great Britain for use against aphids and other pests of glass-house crops. In India, *Pandora delphacis,* an entomophthoralean fungus had showed its potency to kill Brown Plant Hopper *(Nilaparvata lugens)* pest of rice. *P. delphacis* was utilized in the preparation of four forms of Mycoinsecticides as Dust (10%) and Wettable powder (70%) and they were tested for efficiency against the BPH pest under field conditions (Table 66).

Most fungal entomopathogens progress through consistent steps in the infection process. Initially the conidium or zoospore attaches to the insect cuticle. If conditions in the microclimate are appropriate the attached spore germinates and penetrates through the cuticle by means of a germ tube or by infection pegs arising from a knob-like structure on the end of a hypha called an appressorium. Once entrance to the haemocoel is gained, yeast-like cells are produced and the host is killed as a result of mechanical disruption or the production of toxic metabolites. After death of the host, a mycelial growth phase usually follows with eventual representation of the cuticle of the insect to the exterior where new reproductive units can form.

Table 66

Efficacy of *Pandora delphacis* myco-insecticides against brown plant hopper *(Nilaparvata lugens)* pest of rice (Narayanasamy, 1995)

Treatment		*% Mortality and BPH at*	
	Dose kg/ac	*Mariyappa nagar*	*Keera palayam*
1. Mycoinsecticide 10% Dust	15	46.55	46.50
2. Mycoinsecticide 70% WP	1	66.15	64.11
3. Mycoinsecticide 70% WP	1	61.83	61.70
4. Mycoinsecticide 70% WP	1	55.97	58.34
5. BHC 10% dust (Control 1)	12	33.75	43.89
6. Plain water spray (Control II)	2001	48.97	32.10

Beauveria bassiana and *B. brongniartii* cause the disease commonly known as white muscardine. Although their host range is broad, these fungi occur most commonly in populations of hypogean pests or in pests occurring in locations with temperature and moisture and other physical conditions similar to those in soil. Their pathogenicity and ability to overcome host defence mechanism is in part due to production of toxins. *Beauveria bassiana* produces the cyclic depsipeptide beauvericin, which is reported to be toxic to mosquito larvae and adult housefly. However, beauvericin was not found to be toxic to Lepidopteran larvae. Not all isolates of *B. bassiana* produce beauvericin. Other cyclodepsipeptides, including the beauverolides, bassionalids, and isarolids are produced by *B. bassiana* and *B. brongniarii* but, as with beauvericin, broad toxicity screening has not been conducted and conducive evidence for the role of these matabolites in the pathogenicity of *Beauveria* has not been obtained.

The efficacy of *Beauveria* is affected in part by temperature and humidity in the macroclimate; microclimate conditions may also be important. Sunlight and possibly biological activity of other organisms affect the ability of *B. bassiana* to survive and to initiate infection. *B. bassiana* and *B. brongniartii* applied with low doses of chemical insecticide can act synergistically to cause mortality in host insect populations. These species are important candidates in integrated pest management.

Hirsutella thompsonii fungus causes natural epizootics among populations of the Citrus rust mite *(Phyllocoptruta oleivora)* commonly during the summer in United States. The fungus initially causes large reductions in the mite populations, then maintains the populations at low levels for relatively long periods (6 months). *H. thompsonii* can be applied as a conidial formulation or as mycelial fragments. Growth and spoilation can occur after application on leaf surface if the agent is applied with a carrier that can act as a substrate for fungal growth. The fungus may not be particularly suited for incorporation into integrated pest management programs because it is sensitive to many fungicides and moderately sensitive to many insecticides.

Two varieties of *Metarhizium anisopliae* var *anisopliae* and var *major* cause the syndrome known as green muscardine. *M. anisopliae* is able to infect the Old World desert locust *(Schistocerca gregaria)* and a weevil *(Hylobius pales)* by ingestion.

Common hosts of *Vertillium lecanni* are the Hemoptera, but also infects Collembola, Diptera and various mites and spiders. Although seemingly widespread in nature, it only produces epizootics in tropical climates or in greenhouse environments that approximate tropical conditions. Virulence of the organism in the absence of a host seems to be remarkably stable. Use of *V. lecanii* may be restricted by the inability of its conidia to survive dry conditions, making it difficult to produce and store the organism. There is evidence that *V. lecanii* must be applied directly to aphids in wet conditions to be effective. It can be grown and distributed on a local basis for immediate use, and is being commercially produced in Britain.

Nomuraea rileyi produces natural infection in Lepidoptera. Infection of the host insect is typically through the integument, and toxin may be involved in death of the host. Conidia produced on semi-solid media, are usually applied, but it may be possible to use blastospores or mycelium. Nutrients are required for germination of the conidia and invasion of the host may be through the integument or via the alimentary tract.

Entomophorales species produce disease in a wide variety of insects, although individual strains or species are often host specific. Inspite of their widespread occurrence in nature and ability to produce epizootics, use of these fungi in pest control are influenced a great deal by climate. Conidia of many species are extremely short-lived in air at a relative humidity of less than 80 per cent. Even culture maintenance was difficult until liquid nitrogen storage system were developed. Resting spores produced by some species are more resistant to temperature, heat and chemicals than conidia and could be used in control programs, but the spores do not readily germinate, and although they can survive for several years, they may be less infective after storage. The growth of Entomophorales species in artificial media is difficult, but some species can be produced in the form of hyphal bodies and media suitable for production of resting spores have been developed. Mycoinsecticides when applied with low doses of chemical insecticide can act synergistically to cause mortality in host insect populations (Miller et al, 1983).

Preparations from entomopathogenic fungus have proved promising in pest control, but their large-scale industrial production has not yet been organised. This is mainly due to the difficulties in cultivation of these micro-organisms as well as in maintaining activity for longer than a year. There exists also strong consumer requirements for their high efficiency and safety, which they do not meet as yet. Solutions to these problems are characterized by the lack of appropriate knowledge of the physiological and biochemical properties of entomopathogenic fungi, of their specific action mechanism as well as of reliable standardization methods to determine the activity of fungal insecticides.

Protozoa

Very few Protozoa have been mass produced industrially or used as biopesticides in large scale field trials (Tale 67). *Nosema locustae* is an obligate parasite and is cultured in the host insect e.g., grasshoppers, necessitating extraction from ground whole insects by filtration and centrifugation, to arrive at a pre preparation.

Table 67
Entomopathogenic protozoa in experimental production (Burgess, 1984)

Protozoa	*Target organism*
Nosema locustae	Grasshoppers
Nosema pyrausta	European corn borer
Nosema fumiferanae	Spruce budworm
Vairimorpha necatrix	Cabbage looper
	Corn earworm
	Tobacco budworm

Parasitic Namatodes

Lepidopterous and Coleopterous larvae have quite often been found to be infected by namatode parasites. In certain experiments with nematode parasites of insect pest—*Neoaplectena glasseri* against Japanese beetle encouraging results have been obtained.

Parasitic and Predatory Insects

Insects have been widely used for biological control of insect pests. Such natural enemies of insects include members mostly from Diptera, Hymenoptera and Coleoptera. They are either parasitic or predators.

Predatory Invertibrates other than Insects and Nematodes

This group of organisms includes various types of invertibrates which can utilize insects as their food. There exists quite a few species of arachnids that have been observed to maintain the field population of pest at a low level.

Predatory Vertibrates

Vertibrates like fishes, amphibians, reptiles and birds are well known to utilize insects as food.

Attributes of Effective Natural Enemy

It is rather impossible to find a natural enemy which is endowed with all the desirable characters which would make it qualify as the most suitable one for all the purpose of biological control of pests. The attributes of an ideally effective natural enemy are:

(1) High Searching Capacity

This is a prime requisite particularly in a situation of low host density.

(2) Host Specificity

A fairly host specific rather than a highly polyphagous habit is desirable. Most success have been obtained from the introduction of host specific entomophagous species. A high degree of host specificity implies good biophysiological adaptation to the host and a fairly direct dependence on changes in the hosts population. However, polyphagous species may also have certain obvious advantages.

(3) Potential Rate of Increase

This includes a short development period and relatively high fecundity. Thus, several generations of natural enemy can overtake its host quickly when the host begins to increase in numbers.

(4) Survival Capacity

This includes the ability of the natural enemy to occupy all host inhabiting niches. Ideally, the hosts and parasites should have absolutely equivalent distribution in time and space.

(5) Amenability to Culture

The natural enemy should be amenable to culture under artificial conditions. This facilitates the raring of natural enemies required for colonization and distribution for early control of pest.

Feasibility of Biological Control

The feasibility of biological control depends on many factors:

(1) Tolerance Limit of Injury

It is probably the most fundamental consideration because if little or no injury is tolerable then the adoption of biological control is probably entirely unsuitable. The damage by the pest on the vital parts of the plants particularly the marketable fruits or fresh vegetables is an important factor deserving consideration. The time required by the biological controlling agents for exerting its effects on pest population is not affordable in such situations.

(2) Value of the Crops

The economics of the situation regulates the adoption of control measures. Once the role of parasites and predators of a particular situation in maintaining the balance of pest population to the desired level is known and pray/predator ratio established, the maintenance of such operative forces in nature may be inexpensive.

(3) Crop Duration

Crops having longer duration—perennial crops appear to be more amenable to biological control. This however, does not exclude the possible of adoption of biological control in animal crops.

(4) Indigenous or Exotic Pests

Biological control of both these types of pests can be achieved with equal success. For exotic pests, the natural enemies may be searched in their homelands and for indigenous pests the natural enemy complex can be searched in the same locality. However, exploitation of both indigenous and exotic natural enemies can play beneficial role in both types of pests.

(5) Level of Control Provided by Natural Enemies

Every pest species has its natural enemy complex. The regulatory effects of these natural enemies on the pest population if maintained at the desirable level under certain sets of conditions require to be pin pointed and manipulated to advantage by the pest controller.

(6) Pest Complex and Desirable Level of Complex

Crops are usually ravaged by multiple pests. When the problem is unitary this can be tackled effectively by natural enemies if certain amount of damage is tolerable. Existence of pest complex aggregates the situation. The applied biological control when affected by introduction and colonisation of the natural enemy is far from being effective against pest complex and most frequently is aimed at a target pest—the key pest. The potential pest in such a situation may build upto alarming proportion. In such cases, biological control, particularly if tolerance to damage is minimal, may not be efficient.

(7) Identical Situation of Successful Examples

It has been found that pests in isolated areas have been successfully controlled by natural enemies. Introduction of natural enemies from such areas to new areas having similar pest situation have been profitably exploited. So, gathering such information and evaluation of the experience of one locality in other identical situation may be useful.

Advantages of Biological Control

The biological control of insects swarming the agricultural fields has some common advantages: (1) This is usually harmless to non-target organisms and residue or pollution hazards are no problem. (2) It is usually highly specific, so beneficial insects escape injury. For pathogens, even if they are capable of attacking several species, these are often restricted to single taxonomic group. As for example *B. thuringiensis* affects the Lepidopteran larvae alone and is innocuous to parasitic/ predatory Hymenopteran, Dipteran etc. (3) Pathogenic agents are usually compatible with chemical insecticides and thus advantageous in integrated control but they may be adversely acted upon by fungicides (Benz, 1971). (4) There is apparently slow development of resistance by pests to pathogens and the insect natural enemies. (5) Both microbial and insect agents for biocontrol have sustained effects on the pest population as the disease may be borne by the residual population of the pests and a good natural energy usually orients its activity with the fluctuating population of the pests. Therefore, expenditure on repetitive control measures as in chemical pest control is avoided.

As with all methods of pest control, biocontrol also has quite a few limitations and disadvantages:

(a) The initial cost in evolving biocontrol measures is quite expensive as it is basically an ecological control method and needs full evaluation of interacting phenomena under natural conditions. Along with these, exploration for effective natural enemy and the implementation of procedures also involves quite substantial cost.
(b) Evolving biological control in time consuming.
(c) During epidemic outbreak of pest, the time leg between the release of natural enemy and its effect on the suppression of pest population may be such that heavy economic loss would have to be incurred.
(d) In case of pest complex the specificity of pathogens and insects natural enemies may pose problems.
(e) A pathogen is often difficult to maintain in a virulent and viable state.
(f) Unpredictable changes in climate conditions may impair the efficacy of pathogens and natural enemies.
(g) Finally, the operation of this method depends much on the expert handling of the controlling agents and overall cooperation of population in general so that the codes of practice are followed piously with a view not to hamper the activities of natural controlling agents brought under the folds of artificial control of pests (Ghosh, 1989).

Biological control or the use of natural enemies against pests has permanently solved many of the pest problems throughout the world. For example, biological control of aphids on peach, cherry, sonchus, thistles and roses can be controlled by aphid midge—*Aphidoletes aphidimyza* which feeds predaceously on different species of aphids. This midge is also beneficial on aphids of tomato, cucumber, pepper etc. Another aphid midge—*Andpahis aphidimyza* is an internal parasitoid of aphid—*Uroleucon gobonis* which is a serious pest of safflower *(Carthamus tinctorious).* Gall midges have varied habits. They are phytophagous, mycetophagous, saprophagous and Zoophagous. Thus, they are reported to be good biocontrol agents (Kashyap, 1992).

The following are some success stories of biological control of insects swarming agricultural fields:

1. In the 1930's and 1940's the parasite *Aphelinus mali* was introduced in various parts of India to check the woolly aphid, a serious pest of apples. It was found to succeed in Kulu, Shimla, Punjab, Kodaikonal and Coonoor (Nilgiri Hills), in certain areas completely eliminating the need for pesticides.
2. In 1970 an area of 600 ha of coconut plantation was under a devastating attack by the Black-headed leaf eating caterpillar *(Nephantis serinopa)*, and was counter-attacked by biocontrol through the release of the bacterium—*Serratis macrcescens* and other parasites. From an infestation of 2-3 caterpillars per leaflet the incidence was brought down to 2-3 caterpillars per palm.
3. In Jammu and Kashmir, the parasites *Eucaisia pernicious* and *Aphytis diaspidia* are so well established against the dreaded apple pest, the SanJose scale, that a serious review of whether any more pesticide spraying is required or not, has been recommended.
4. Another success has been that of the exotic parasite—*Leptomastix dactylopii* against the mealy bug *Plonococcus citri,* a serious pest of citrus and coffee. Introduced in Coorg and other areas of South India in 1982-83, this parasite has almost completely eliminated the need to use pesticide on these plants.

Chapter 12

Biofertilizers

In recent past, there has been great increase in food production. This has primarily been due to crop improvement, use of fertilizers to maintain soil fertility and control of pests.

The chemical fertilizers, no doubt increased the production but also produced many harmful effects. Some of these are as follows:

(1) Chemical fertilizers are very costly.

(2) These are manufactured using coal and petroleum as sources of energy which themselves are decreasing very fast.

(3) Chemical fertilizers are often released in the environment causing air, water and soil pollution.

The need for the use of biofertilizers has arisen, primarily for three reasons: *First,* because increase in the use of fertilizers leads to increased crop productivity. *Second,* because increased use of chemical fertilizers leads to damage in soil texture and other environmental problems. *Third,* chemical fertilizers are manufactured from petroleum oil or coal. The increased cost of these fossil fuel result in the ever increasing cost of chemical fertilizers. Therefore in developing countries like India, the use of biofertilizers is both economical and environment friendly. Biologically fixed nitrogen can supply an adequate amount of nitrogen to plants and other nutrients to some extent. It is a non-hazardous way of fertilization of soil. Moreover, biologically fixed nitrogen consumes about 25-30 per cent less energy than the Haber-Bosch chemical process which consumes high energy (about 13500 K Cal/Kg N fixed).

Biofertilizers are defined as biologically active products or microbial inoculants of bacteria, algae and fungi (separately or in combination), which may help biological nitrogen fixation for the benefit of plants. Biofertilizers also include organic fertilizers (manure etc), which are rendered in an available form due to the interaction of micro-organisms or due to their association with plants. The term 'biofertilizers' denotes all the "nutrient inputs of biological origin for plant growth" (Subba Rao, 1982). Here biological origin should be referred to as microbiological process synthesizing complex compounds and their further release into outer medium, to the close vicinity of plant roots which are again taken up by plants.

An alternative was, therefore, very badly needed for overcoming the harmful effects of such chemicals. Certain plants have been found useful as fertilizers. These do not have any bad effect on the environment and the organisms. Use of biofertilizers is economical as well as environmental friendly.

Biofertilizers include the following: (i) symbiotic nitrogen fixers—*Rhizobium* spp., (ii) asymbiotic free nitrogen fixers—*Azotobacter, Azospirillum* etc., (iii) algal biofertilizers (blue green algae in association with *Azolla;* (iv) phosphate solubilising bacteria; (v) mycorrhizae; (vi) green manure.

Bacteria

Many free living and symbiotic bacteria fix atmospheric nitrogen. Increase of such bacteria in soil may increase the gross yield of nitrogen. The two methods—*bacterization* and *green-manuring* techniques are commonly employed in this connection.

Bacterization is a technique of seed-dressing with bacteria (as water suspension) for example *Azotobacter, Bacillus, Rhizobium* etc. It has been proved that bacteria can successfully be established in root region of plants which in turn improve the growth of hosts. Bacterial fertilizers named 'azotobakterin' (containing cells of *Azotobacter, Chroococcum)* and 'phosphobacterin' (containing cells of *Baccilus megaterium* var *phosphaticum)* have been used in Russia and East European countries. These fertilizers increased the crop yield about 10-20 per cent (Cooper, 1959). Subsequently bacterization of seeds in

Russia, Czechoslovakia, Rumania, Poland, Bulgaria, Hungary, England and India has clearly demonstrated the increase in crop yield of what, barley, maize, sugarbeet, carrot, cabbage and potato. In rhizosphere, bacteria secrete growth substances and antibiotic secondary metabolites which contribute to seed germination and plant growth (Subba Rao, 1982; Dwivedi et al, 1989). Free living bacteria *(Azotobacter)*, associate *(Azospirillum)*, symbiotic *(Rhizobium)*, phosphate solubilizing *(Bacillus megaterium, B. polymyxa* and *Pseudomonas striata)* are gaining popularity for bacterization.

Rhizobium are able to enter into symbiotic relationship with legumes. They fix atmospheric nitrogen and thus not only increase the production of the inoculated crops, but also leave a fair amount of nitrogen in the soil, which benefit the subsequent crop. Following seven groups of *Rhizobium* have been recognised for inoculating legumes in India: *R. leguminosarum, R. melilotti, R. trifoli, R. phaseoli, R. lupinli, R. japonicum,* and *Rhizobium* spp. The nitrogen fixing ability of legumes inoculated with these Rhizobia ranges from 50 kg—150 kg per hectare. Therefore, to increase this ability to fix atmospheric nitrogen, following efforts are being made:

(a) Efficient Rhizobia for different crops and locations which are tolerant to various stresses (drought, temperature, high or low pH) are being isolated, maintained and used.

(b) A hup^+ gene (uptake hydrogenase) is being used, which helps in recycling of H_2 produced during nitrogen fixation. This saves energy required for reduction. Rhizobium strains with hup^+ gene have been shown to fix more nitrogen.

$$N_2 \xrightarrow[\text{ATP, reduc tan t}]{\text{nitrogenase (nif)}} 2\ NH_3 + 3\ H_2$$

(c) Strains deficient in nitrate reductase have improved nitrogen fixing ability. Nitrate reductase catalyses the reaction $N\overline{O}_3 — N\overline{O}_2$ and $N\overline{O}_2$ inhibits nitrogen fixation.

Azotobacter and *Azospirillum* when applied to rhizosphere, fix atmospheric nitrogen and make it available to crop plants. They also synthesize growth promoting antibiotic substance, helpful to the plant. Most efficient strains of *Azotobacter* fix 30 kg of nitrogen from 1000 kg of organic matter. When applied to fields, positive responses by field crops were observed leading to a saving of 10-25 kg/ha of nitrogenous fertilizers. Similarly, *Azospirillum* with farmyard manure, leads to a saving of 15-25 kg/ha of nitrogen in crops like sorghum or other millets.

Pseudomonas fluorescens and *P. putida* are important phosphate solubilising bacterial fertilizers. They convert non-available inorganic phosphates into soluble phosphates, which can be utilized by crop plants.

Algae

The role of blue-green algae (*Aulosira fertilissima, Anabaena ambigua, Cylindrospermum gorakhporense, Nostoc sphaericum, Plectonema nostocorum, Tolypothrix tenuis* etc in the paddy fields was realised by Singh (1961). In water-logged conditions, they multiply, fix atmospheric nitrogen and release it, into the surroundings in the form of amino acids, proteins and other growth promoting substances (Sewart, 1970). Composite culture of blue-green algal genera have been found to be more effective than single generic cultures. Application of dried blue-green algae flakes at the rate of 10 kg/ha is recommended ten days after transplantation of rice. Besides being a source of nitrogen, they provide the following other advantages: (i) algal biomass accumulates as organic matter, (ii) growth promoting substances are produced which stimulate growth of rice seedlings, (iii) they promote partial tolerance to pesticides and fungicides, and (iv) they also help in reclamation of saline and alkaline soils.

Tyagi (1991) observed that inoculation of algae increased biomass and yield of rice (Table 68). Though *Spirogyra* is a non-nitrogen fixing algae, its inoculation provided organic biomass for the growing rice plants. Further, the mucilaginous sheath of the filaments of *Spirogyra* is an abode of many micro-organisms, some of which may fix nitrogen. *Nostoc* and *Aulosira* are strong nitrogen fixing blue green algae and therefore their inoculation together gave the highest biomass and yield in rice plants. Effect of *Nostoc,* when inoculated alone, was intermediate, that is, its effect was higher than that of *Spirogyra* but obviously lower than that of *Aulosira.* The beneficial effect of *Nostoc* and *Aulosira* was due to the transfer of fixed nitrogen to the growing rice plants, as well as the growth promoting substances liberated into the medium.

Table 68

Effect of algae on biomass and number of grains in rice (Tyagi, 1991)

Treatment	*Root weight (gm/plant)*	*Straw weight (gm/plant)*	*Number of grains/ plant*
Control	11.66	15.30	505
Spirogyra	13.30	21.96	716
Nostoc	14.33	21.40	829
Nostoc + Aulosira	15.76	28.83	1111

Algalization is the process of application of blue-green algal culture in field as biofertilizer (Venkataraman, 1972). The scum of blue-green algae growing on water surface is scrapped, dried, powdered and stored in bags for supply to farmers. They are applied at 10 kg/ha in rice fields after transplantation of rice seedlings. The benefit from algalization is about 25-30 kg N/ha per cropping season (Subba Rao, 1982).

Azolla

Azolla is the only genus of Pteridophytes which exists in a symbiotic association with a heterosystous blue-green algae *Anabaena azolae.* Although *Azolla* has long been recognised as a green manure in paddy cultivation in Vietnam and China, but its potential as a biofertilizer is known recently.

The genus *Azolla* belongs to a group of heterosporous, leptosporangiate ferns of the order Salviniales with seven living species namely—*A. caroliniana, A. filiculoides, A. pinnata, A. mexicana, A. microphylla, A. rubra* and *A. nilotica. A. pinnata* is commonly found in India in ponds ditches throughout the country, which multiplies vegetatively and whose growth is prolific (Sharma, 1991).

Azolla consists of a dichotomously branched free floating plant with deeply bilobed leaves and true roots, forming a thick mat on the surface of water. Alternately arranged small leaves cover the entire stem and branches. Each leaf is composed of a dorsal and a ventral lobe. The fleshy dorsal lobe is aerial, chlorophyllous and contains symbiont within the special cavities. The thinner ventral lobe is partially submerged and achlorophyllous. Both the host and the symbiont are capable of photosynthesis, but the nitrogen requirements of the host are fulfilled by the symbiont through nitrogen fixation.

An interesting feature of *Azolla-Anabaena* association is that the symbiosis is maintained during the sporophytic and gametophytic cycles of the fern. Akinetes of *Anabaena azollae* get enclosed within the megasporocarp of the fern and germinate to produce undifferentiated filaments during the development of female gametophyte and later the young sporophyte after fertilization. These filaments become associated with the apical meristem of the developing sporophyte. Thus, providing continuity of the symbiosis and precluding the necessity for a free living stage of the endophyte—*Anabaena* (Peters et al, 1982). The symbiotic *Anabaena* undergoes a pattern of development and differentiation in parallel to that of the fern. Filaments of *A. azollae* associated with the stem apex and meristematic leaves of *Azolla* actively divide, lack heterocysts and do not exhibit nitrogenase activity (Hill, 1977). Heterocyst differentiation begins only after the apical *Anabaena* colony becomes partitioned into the developing leaf cavities of the fern. As the leaves mature, cell division of the symbiont appears to decrease, individual algal cell enlarges with an increased rate of differentiation into heterocusts. The *Anabaena azollae* has very high heterocyst to vegetative cell ratio ranging from 15-18 per cent of the number of cells (Peters, 1979). Nitrogenase activity which is negligible in leaves of the apical region of *Azolla*, rapidly increases to a maximum level in mature leaves, which again decreases in aging leaves (Kaplan et al, 1986).

Azolla mat is harvested and dried to use as green manure. There are two methods for its application in the field: (a) incorporation of *Azolla* in soil prior to rice plantation, and (b) transplantation of rice followed by water draining and incorporation of Azolla (Singh, 1977, 1979).

Azolla multiplies very fast, winter season being more suitable for its multiplication. *Azolla* doubles its biomass in 2 days. The daily nitrogen fixing rate is as high as 7 kg N/ha in optimum conditions for *A. filiculoides* and 3 kg N/ha for *A. pinnata* (Singh and Singh, 1989).

Singh (1979) reported that a layer of *Azolla* covering one hectare of rice field conditions about 10 ton of green mature and ensures about 25-30 kg nitrogen. *Azolla* is capable of producing about 900 kg N/ha/year. One crop of *Azolla* increased the grain yield and nitrogen uptake as obtained with 25-30 kg N/ha whereas 2-3 crops of *Azolla* could increase the yield and nitrogen uptake as obtained with 60-90 kg N/ha.

Application of 0.5-2 ton/ha of fresh *Azolla* in the paddy fields, 20 days in advance of transplantation has been found to be more efficient over the basal application of 20 kg mineral N/ha. *Azolla* application increased the yield by 12.7 per cent over control (Table 69). When used in combination with the chemical fertilizers, application of similar rate of *Azolla* with 20 kg N/ha gave a yield higher to that obtained by using 40 kg N/ha alone. Incorporation of 1 ton *Azolla/ha,* 7 days after transplantation, which covered the entire field in 20-25 days produce an average of 10-15 ton of fresh biomass containing approximately 30 kg N/ha. Grain yield was found to be increased to the tune of 10-12 per cent over control and was higher to the yield obtained by using 20 kg N/ha. Inter-cropping, of *Azolla* with different levels of nitrogen has also provided encouraging results. Use of one layer of *Azolla* with 40 kg N/ha gave a yield higher than that of yield obtained by using 60 kg of mineral nitrogen.

Incorporation of one layer of *Azolla* in the soil after 20 days of incubation just before transplantation followed by addition of fresh *Azolla* at the rate of 1 ton per hectare as intercrop, gave an increase in yield comparable to that obtained by using 60 kg N/ha. These observations clearly indicates that the application of *Azolla* both as basal manure plus intercrop manure can replace 60 kg N/ha and this gives way to initiate the cultivation of paddy without using chemical nitrogen.

Table 69
Effect of Azolla green manuring on grain yield of paddy

S. No.	*Treatments*	*Grain yield in Quintals per hactare*	*Percentage increase over control*
1.	No N, No *Azolla* (control)	14.5	—
2.	Only nitrogen		
	(a) 20 Kg/hac.	16.29	12.34
	(b) 40 Kg/hac.	17.98	24.00
	(c) 60 Kg/hac.	20.22	39.44
3.	*Azolla* as basal application plus		
	(a) 0 Kg/hac.	16.34	12.68
	(b) 20 Kg/hac.	18.23	25.72
	(c) 40 Kg/hac.	21.55	48.62
	(d) 60 Kg/hac.	29.85	105.86
4.	*Azolla* as inter crop plus		
	(a) 0 Kg/hac.	15.98	10.20
	(b) 20 Kg/hac.	18.50	24.48
	(c) 40 Kg/hac.	19.75	36.20
	(d) 60 Kg/hac.	27.22	87.72
5.	*Azolla* as basal plus intercrop (No nitrogen)	19.85	36.89

Azolla in comparison to other nitrogen fixing systems, has many advantages when used as a biofertilizer for rice crop. These include its high nitrogen-fixing activity, rapid availability of the fixed nitrogen to the standing rice crop, quick growth in water-logged rice fields, suppression of the growth of weeds in dual culture with rice plants, and its non-interference with normal cultivation practices and crop development. *Azolla* adds large quantities of organic matter to the soil, thus enriching and improving its texture (Satapathy and Singh, 1985).

The utility of *Azolla* as green manure for paddy has largely been demonstrated throughout the country, but the most important limiting factor is the adoption of this technology is the non availability of fresh *Azolla* inoculum to the farmers. Government of India initiated a project on Use and Development of Biofertilizers with a National Centre at Ghaziabad and six regional centres at Hissar, Jabalpur, Pune, Bangalore, Bhubaneswar and Imphal. These centres maintain mother stocks of all biofertilizers including *Azolla,* which is disseminated to block level production units supported by State Departments of Agriculture.

Azolla pinnata absorbs heavy metals into cell walls and vacuoles through evolution of specific metal resistant enzymes. Therefore, heavy metal resistant strains of *A. pinnata* can be incorporated as green manure in rice fields near the polluted areas where heavy metal concentration is between 0.01 and 1.5 mg/litre.

There are certain limitations in using *Azolla* as a biofertilizer:

1. *Azolla* as a green manure crop is labour intensive.
2. Raising of *Azolla* needs assured and adequate water supply.
3. Damage of *Azolla* is caused by pest diseases, heat etc.
4. Optimum temperature is required for *Azolla* multiplication.

Mycorrhiza

Mycorrhiza is a symbiotic association of fungi with roots of plants, so that the nutrients absorbed from soil by the fungus are released to the host cells and in turn, the fungus takes its food requirements from the host. Mycorrhiza are of two types: (i) Ectomycorrhiza, and (ii) Endomycorrhiza.

Ectomycorrhiza are found on the roots of forest trees (e.g., Pina, Oak, Beech, Eucalyptus etc). They absorb nitrogen, phosphorus, potassium and calcium. They also covert complex organic molecules into simpler available forms, protect the roots from the pathogens, and produce growth promoting substances (cytokinins).

Endomycorrhiza are found in the roots of most fruit and other horticultural crops (e.g., coffee, pepper, cardamom, and betelvine). They particularly help in phosphorus nutrition. They also produce growth promoting substances and offer resistance against pathogens. The seedlings in nurseries are inoculated with endomycorrhizae and then transplanted in the field as done by citrus growers. Fungus strains of *Azotobacter, Aspergillus, Azospirillum, Beijerineleia* and *Glomus* are used as endomycorrhizae.

Green Manure

A crop grown with the objective of burying in its green matter in the soil to decay and improve soil fertility is called a 'green manure' crop. Many legumes are grown as green manure because of their ability to fix significant amounts of atmospheric nitrogen biologically, though some non-legumes are also known to be used as green manures. Documented literature suggests the prevalence of the practice of green manuring in India and China from early times. In addition to supplying biologically fixed nitrogen, green manures also supply significant amounts of organic matter which is essential for the health of the soil and hence for the following crop.

There are about 18000 species of legumes, but only a few hundred are cultivated. Of the ones which have been studied for nitrogen fixation include annual legumes, perennial legumes and wild annual legumes (Table 70).

Table 70

Crops which can be used as green manure (Ghai and Thomas, 1988)

I. **Cultivated annual legumes:**
Astralagus sinicus, Crotalaria juncea, Crotalaria striata, Crotalaria anagyroides, Cassia mimosoides, Cyamopsis tetragonoloba, Desmodium conum, Desmodium hetrophylum,' Desmodium ovalifolium, Desmodium virgatus, Glycine wightii, Indigofera anil, Indigofera linifolia, Indigofera teysmaii, Phaseolus lunatus, Psophocarpus ttetragonolobus, Sesbaria aculeata, Sesbania rostrata, Sesbania speciosa, Vigna radiata, Vigna unquiculata etc.

II. **Perinnial legumes:**
Acacia nilotica, Cassia hirsuta, Desmodium gyroides, Gliricidia maculata, Sesbania aegyptica, Sesbania grandiflora, Tephrosia candida, Tephorisisa vogelle, Leucaena leucocephala, Prosopis juliflora etc.

III. **Wild annual legume:**
Aeschynomene americana, Calopogonium caerulum, Calopogonium muconoides, Cassia cobanensis, Cassa occodentalis, Cassia ttorra Centrosema pubescens, Erythrina lithosperma, Lathyrus sativus, Macrotyoma axillaire, Macroptilium atropurpureum, Macroptilium lathyroides, Mimosa invisa, Mucuna bracteata, Mucuna cohinensis, Pueraria phaseoloides, Stylosanthes guinensis, Styulosanthes hamata, Stylosanthes humilis, Tephrosia purpurea etc.

The information about the exact nitrogen contribution by many of these is not known. *Sesbania aculeata* contributes 70-120 kg N/ha. Mungbeen straw contributes about 70 kg N/ha. *Sesbania rostrata* contributes 200-280 kg N/ha when grown as green manure at 52 days growth stage. *Leucaenia leucocephala* contributes 500-600 kg N/ha from green foliage. Nitrogen contribution of other legumes range from 34-300 kg N/ha when used as green manure. Besides legumes nodulated non-legumes, blue-green algae and symbiotic nitrogen fixation are also known to fix atmospheric nitrogen and hence used as green manure (Table 71).

Table 71

Contribution of N by some nodulated legumes when used as green manure (Ghai and Thomas, 1988).

N-Fixing system	*Amount of N contributed ($Kg\ ha^{-1}$)*
I. Green manuring type legumes:	
Sesbania aculeata-Rhizobium	70-120
Sesbania sp. PLSe 17-*Rhizobium*	70
Vigna radiata (straw)-Rhizobium	60
Sesbania rostrata Rhizobium (root)	
Rhizobium (shoot)	280
Leucaena leucocephala - Rhizobium	500-600
Beans (broad bean/lupines/soybean/ lentil etc.)-*Rhizobium*	60-210
Fodders (*Trifolium/Medicago/ Melilotus etc.) - Rhizobium*	100-300
II. Cover crop type legumes:	
Pueraria phaseoloides/Calopogonium mucunoides/Centrosema pubescens Rhizobium	226-350
Lablab purpureus-Rhizobium	240
Glycine jawanica - Rhizobium	210

Contd.

N-Fixing system	*Amount of N contributed ($Kg\ ha^{-1}$)*
III. Non legumes:	
Casuarina equistifolia - Frankia	100
Alnus - Frankia	30-300
Parasponia andersonil - Rhizobium	850
IV. Others:	
Azolla - Anabaena	25-150
Grasses - Azpospirillum	15-100
Azotobacter/Kleibsiella/Clostridial Beijernkia etc.	20-80

Cassia auriculata was collected and dried, chopped into small pieces upto 5 cm and mixed in soil @ 20 Qt/ha before monsoon, it increased the yield of mustard from 1.0 Qt/ha in control to 6.7 Qt/ha in the treated plots (Suwalka et al, 1990). Incorporation of dry chopped wild leguminous herb *Tephrosia purpurea* at 20 Qt/ha increased the yield of mustard by 3.1 Qt/ha (Suwalka and Karan, 1990). Narrow C/N ratio and fast decomposing nature of *Crotalaria burhia* improved physico-chemical properties of the saline-sodic soils after mixing it at the rate of 2 t/ha. It increased mustard yield from 1.0 Qt/ha in control to 4.7 Qt/ha in treated plota (Suwalka et al, 1991).

Aquatic Plants

Aquatic plants can be used as green manure. They are applied to the soil as a surface mulch or are ploughed under the soil. As a mulch, they suppress weeds, reduce evaporation and reduce erosion. They are also used to make compost after decomposition. Compost contains 1.5 to 4 per cent nitrogen, 0.5—1.5 per cent phosphorus and 1-2 per cent potassium (Nag and Singh, 1985). Aquatic plants can form valuable source of nutrient rich compost (Table 72).

Table 72
N and P Contents of Certain Aquatic Weeds on Dry Weight Basis (After Steward, 1970)

	Weed Species	*N (%)*	*P (%)*	*N (kg ha^{-1})*	*P (kg ha^{-1})*
1.	*Cyperus papyrus*	1.5	0.15	1250	125
2.	*Eichhornia crassipes*	4.0	0.4	1400	140
3.	*Hydrilla verticillata*	4.0	0.4	100	10
4.	*Myriophyllum exalbescens*	3.9	0.4	98	10
5.	*Phragmites communis*	1.5	0.15	490	50
6.	*Sagittaria subulata*	4.0	0.4	916	91
7.	*Thalassia testudinum*	4.0	0.4	1324	132
8.	*Typha latifolia*	1.5	0.18	760	90

The fresh water hyacinth contains about 3.5% organic matter, 0.04% nitrogen, 0.06% P_2O_5 and 0.02% K_2O. Its high water content of over 90% is the main hurdle in its composting, because it leads to anaerobic decomposition of the material. To activate aerobic digestion, frequent stirring of the pile is essential, or the freshly collected material should be left on shore for a few days to wilt. Then it should be mixed with rotten compost, dung etc and then put into ditches to compost. Yet another method of preparing compost is to chop the waterhyacinth into about 15 cm long pieces and left to wilt for sometime. Long bamboos are erected in pits at convenient distance and the wilted hyacinth are filled in the pits. After 1 m filling the pits are inoculated with dung. Periodically water is sprinkled on the pits. After 10 days additional wilted hyacinth are added in pits to fill the space made by the sinking of initial material. Then the pits are packed with soil and bamboos are slowly pulled out to leave aeration holes behind. The compost is ready in about 3-4 months.

Bacterial concentrates have been used to expedite aerobic decomposition of chopped, woody, raw material. The bacteria are largely species of *Bacillus*, beside some species of Azotobacter (Gupta, 1987).

Chapter 13

Food Biotechnology

All human and animal food is a product of bioprocesses. The ability to manipulate these processes towards specific needs—is obviously a worthy objective. Food processors may know what they want, but they do not know what is technically feasible, how much it could cost, and what is feasible in short-term as well as in long-term (Clausi, 1986). The food processing industry has a constant demand for acceptable additives such as colourants, sweetners, bulk protein, novel catalysts and preservatives. The incorporation of novel fungi into milk also increases the monetary value of products such as 'yoghurt' and cheese (Ignacimuthu, 1995).

Biotechnologists have given a boost to increased food production. Single cepp proteins are regarded as wholesome food in many aspects. *Sun-bean* a hybrid of sunflower and French-bean has been developed using biotechnological methods.

Biotechnology has helped us to provide animals which could yield more milk to meet the demands of large population; it has also succeeded in increasing the meat yield. It has evolved ways and means to efficiently increase the conversion potential of food and feed efficiency (Ignacimuthu, 1995).

The benefit of genetic engineering in food-front are both in pre-and post-harvest technologies. The potential application of genetic engineering in this area are in the improvements by gene-cloning, hybrids, somaclonal variations and cell fusion (Table 73).

Table 73
Genetic engineering and allied techniques and their application in food biotechnology (Ravishankar and Venkataraman, 1991)

	Techniques	*Application realized*
1.	*Agrobacterium tumefaciens*	Production of food additives, insect resistance plants, herbicide tolerant plant
2.	Somaclonal variation	Herbicide droughts pesticide tolerant plants. High yielding and easily maturing plants. Removal of anti-nutritional factors and enrichment of nutritional value. Improved qualities for efficient food processing
3.	Haploids culture	Increased productivity in crop plants and mutagenesis
4.	Cybrid from protoplast fusion methods	Transfer of herbicide/pesticide tolerance of male sterile plants
5.	Hybrids from protoplast fusion methods	Disease resistance and high yield of crop plants
6.	Microbial gene manipulation	Increased production of vitamins, proteins, oil, enzymes of food industrial applications for the manufacture of food additives, biocatalyst for biotransformation of substrate of food value products
7.	Cell culture for metabolites	Production of food colours, flavours, sweetners, enzymes
8.	DNA/RNA probes	Virus detection

Biotechnology offers promise for increasing quality by elevating functional attributes such as nutrition, flavour, processibility, texture and reduction of antinutritive substances. Moreover, elimination of diseases and rapid propagation offers scope for increased food production. The qualitative improvements in developing pesticide and herbicide tolerant plants and for cultivation in adverse environment such as salinity, alkalinity, frost, drought etc.

Commonly occurring anti-nutritive substances (Table 74) show that oxalates have been eliminated from *Amaranthus* (Tautonico and Knorr, 1986). The genetic changes which take place in cultured tissues during callus formation is explored to select for variants. Larkin et al (1984) observed genetic changes such as chromosomal rearrangement, translocation, reciprocal and non-reciprocal arrangements, single nucleotide substitution etc in wheat.

Table 74

Nutritional stress factors occurring naturally in foods (Ravishankar and Venkataraman, 1991)

Food	*Inhibitor*	*Nature*	*Action*
Cottonseed meal	Gossypol	Polyphenol	Decreased food intake
Castor bean	Toxalbumin	Protein	Paralyzes/respiratory/vasomotor
Cereals	Phytin	Inositol hexaphosphate	Makes calcium unavailable
Spinach, Knolkol, tomato	Oxalic acid	Organic acid	Makes calcium unavailable
Lima beans	Antitrypsin	Protein	Inactivates trypsin
Corn	—	—	Makes nicotinic acid unavailable
Lathyrus sativus	Toxic amino acid	Amino acid	Neurotoxin

Gene Cloning

There has been significant improvements in genetic engineering techniques in the last couple of years. Gene cloning has been fairly well understood. We can now put any identified gene into recipient cell by use of vectors or microinjection method. The exploitation of socalled natural genetic engineering of *Agrobacterium tumefaciens* and *A. rhizogenes* has become very popular method for dicotyledonous plants. Infectivity of these bacteria have been reported in monocotyledonous plants—*Asparagus officinalis* (Hernalsteens et al, 1984) and *Amarillus* (Slogteren et al, 1984). In this method the desired is tailored on Ti plasmid of the *Agrobacterium* and infected on the recipient host.

The tumour tissue formed is regenerated into plants. The plants are later tested for expression of inserted gene. This procedure has been applied for expression of *Bacillus thuringensis* (Bt) gene for the insecticidal toxic protein in tobacco plants (Adang et al, 1986). The transformed tobacco plants have been found to have resistance from form larvae. There is yet another example of genetic engineering technique used to produce Thaumatic (a plant protein) in microbial cells (Bhaduri, 1987). Thaumatin is a protein produced in the berry of *Thaumatococcus danielli* and is about 3000 times sweeter than cane sugar. This non-nutritive sweetner from genetic engineering has found application in ice-cream, sweets etc. This would provide an opportunity to produce Thaumatin without dependence on environmental and geographic factors. These techniques can be used for development of food biotechnology. Similarly, micro-organisms could be adopted as an agent for production of many plant derived additives such as colours and flavours. The advantage of using micro-organisms is that they can grow faster and productive rate is high. In contrast plant cells have doubling time of 16-24 hours and have limited capability to produce metabolites (Ravishankar and Venkataraman, 1991).

There are a number of examples in crop species where a particular gene function has to be removed in order to fit the crop product to market requirements. Notable examples are the removal of erucic acid and glucosinolate from rapeseeds and neurotoxic aminoacids in *Lathyrus sativus.*

Haploid Plants, Protoplasts and Somatic Hybrids

Since the first report of production of haploid plants from immature pollen in *Datura innoxia* by Guha and Maheshwari (1964), there are over 237 sps from 83 genera and 37 families wherein haploids have been reported (Dunwell, 1985). This technique has been extremely useful in developing homozygous diploid line for plant breeding. Seven improved varieties of rice has been developed in China using haploid breeding methods in 1983, China had 100,000 ha of land under rice production using haploid varieties (Ying, 1983).

Protoplast fusion method has proved to be less important for somatic hybrid development. However Cybrid (i.e. protoplast and cytoplasmic hybrid) have proven to be of immense use in transfer of cytoplasmic inheritance such as herbicide/pesticide tolerance, male sterility, toxin resistance etc.

Somaclonal Variation

This is commonly observed phenomenon in plants regenerated from callus in-vitro. The genetic variability is induced in such regenerants due to chromosomal changes occurring during in-vitro culture of cells and tissues or due to aneuploid cells i.e. monosomics (2n-1) and trisomics (2n+1).

Somaclonal variation can be induced type, that is by introduction of stress factors such as sodium chloride, drought, pesticide, herbicide etc. It is also found to be heritable.

Few examples of application of somaclonal variation are: (i) increased seedling vigour of lettuce, (ii) improved rice protein content, (iii) *Fusarium* resistance is alfaalfa, (iv) increased solid content in tomato, and (v) *Pseudomonas* and *Alternaria* resistance on tobacco. These examples show enormous potentialities of this technique in improvement of crop quality and quantity, which would increase the food output.

Somatic Cell Genetics

This technique permits selection of variant cells at cellular level in presence of the agent for which resistance is to be tested. This has been applied to amino acid analogues (to increase particular amino acid), toxin producing pathogens (to induce disease resistance), sodium chloride (salt tolerance), and polyethylene glycol (drought tolerance). The notable examples is of increased lysine content in rice by using S-aminoethylcysteine and hydroxylysine (Schaeffer and Sharpe, 1981).

Plant resistant to herbicide glypphosate has been developed by 20 fold amplification of the gene from glyphosate tolerant cell lines of Petunia hybrid producing enzyme 5-enolpyruvyl Shikimate 3 phosphate synthase which is responsible for tolerance (Horsch et al, 1986; Comai et al, 1983).

Disease Elimination/Resistance

The viral disease elimination has been achieved by meristem culture methods. Resistance to bacterial and fungal pathogens has been induced by use of toxins secreted by the invading organisms (Helgeson, 1983). This is applicable to all food crops for yield improvement In-vitro tubinization of potato has tremendous application in obtaining disease resistant plants. The advantage of microtubers are easy transport and efficient transplantation. Another interesting possibility is a somatic hybrid of *D. alata* (edible underground tuber), and *D. bulbizea* (aerial tubers) which yield aerial delicious tubers that can easily be harvested instead of digging out from soil (Ammirato, 1984).

Tasty and Long Lived Fruits and Vegetables

More than half of the fruits and vegetables produced annually are lost due to spoilage caused mainly by ethylene formation that triggers fruit ripening. Ethylene is believed to regulate fruit ripening by the expression of genes responsible for enhancing a rise in the rate of respiration, autocatalytic ethylene production, chlorophyll degradation, carotenoid synthesis, conversion of starch to sugars, and increased activity of cell wall degrading enzymes.

Tomatoes become mushy due to the production of a softening enzyme called Polygalaturonase (PG). The expression of existing genes can be inhibited by inserting a gene constructed to generate anti-sense RNA. Initial attempts to inhibit tomato fruit softening by anti-sense polygalacturonase (PG) RNA, a gene thought to be responsible for cell was hydrolysis during ripening, failed to give a desirable effect. PG antisense RNA dramatically inhibited PG mRNA accumulation and enzyme activity, but still fruit ripening was not completely arrested (Ignacimuthu, 1996).

Another approach used to prevent fruit ripening was to inhibit ethylene production. The inhibition of 1-aminocyclopropane-1-carxylic acid (ACC) synthetase with anti-sense RNA led to severe inhibition of ethylene production which is required for the ripening of fruits. The transgenic tomato is more resistant to mechanical stress associated with handling, packaging and transport without losing compressibility. This tomato is the first genetically engineered food crop released in the US market under the name "Flavr Savor" (Kaul and Nirmala, 1995).

Food Processing Improvements

Genetic engineering for manipulation of carbohydrate metabolism (Shannon and Garwood, 1984) can improve texture and cooking properties of rice, enhanced sweetness and mouthfeel of corns, improved staling characteristics of wheat flour for backed goods (Lawrance,1987).

Food Additives

Plant products such as colours, flavours and sweetners have been sought for as food additives (Table 75). These food value phytochemicals can be obtained by plant cell culture. However, the present technology of plant cell culture does not allow the production of low cost compounds. It has been estimated that any compound costing over US $ 500 per kg can be considered for production by plant cell culture. Though many synthetic food additives can be produced at much lower costs, many are being phased out due to proven toxic effects.

Capsaicin, is one of the widely investigated food additives for production by cell culture. It is produced in fruits of *Capsicum annuum* and *C. fruitescens*. Immobilised *Capsicum* cells are known to produce several fold higher amounts of capsaicin than freely suspended cells.

Cynara cardunculus, known for its production of phenol and flavonoid derivatives used widely for milk clotting and cheese production has been obtained by cell culture.

Embryos of *Theobroma cacao* produce aromatic butter Embryogenesis in this species has been investigated as a source for production of *Theobroma* butter.

Production of pigments has been reported in a wide range of plants, such as *Beta vulgaris, Catharanthus roseus, Crocus sativus, Daucus carota, Euphorbia milli, Haplopappus gracilis, Lithospermum erythrorhizon, Prunus persica, Rubia tinctorium, Vitis vinifera*. Among the food colours anthocyain has been widely worked out. Anthocyanin from conventional sources are expensive (US $ 1250 per kg). Stable anthocyanin has been produced from cell cultures of *Daucus carota,* which is 10 times the colour intensity, non-hygroscopic and stable for over 6 months at pH 2 (Vunsh et al, 1986). Crocin has been obtained from saffron stigma callus. In-vitro production of stigma-like structures which produce pigment crocin, has been achieved by culture of immature ovaries and stigma. Shikonin, a red pigment, from *Lithospermum erythrorhizon* cells has been commercially produced by Mitsui Petrochemicals Ltd, Japan. The advantage of this technology is high production capacity at 12-20 per cent of dry weight of cells as compared to intact plants containing 1-2 per cent Shikonin in roots (Fujita et al, 1981).

Table 75
Metabolites of food value that can be produced by plant cell cultures (Ravishankar and Venkataraman, 1991)

Product	*Plant*	*Source*
Food & agriculture industry		
Colourants		
Anthocyanin carotenoids	*Dacus carota*	Roots
Betaninee	*Beta vulgaris*	
Flavour		
Angelica	*Angelica sylvestris*	Flower
Capsicum	*Capsicum annum*	Fruit
Celery	*Apium graveolens*	Leaves
Garlic	*Alium sativum*	Bulbs
Ginger oil	*Zingiber officinale*	Roots
Hops	*Humulus lupulus*	Flowers
Onion	*Alium cepa*	Bulbs
Non-nutritive sweetners		
Thaumatin	*Thamatococcus danielli*	Fruits
Steroiside	*Stevia rebaudiana*	—
Insecticide		
Pyrethrins (for storage of field grains)	*Chrysanthemum cinerariaefolium*	Flowers

Flavours in a plant species is generally a mixture of compounds in definite proportion. Nearly 4300 different flavour compounds have been identified in foods. However, certain flavours consist of one or a few compounds only: 2-isobutyl thiazole (tomato), methyl ethyl cinnamates (straeberry), methyl anthranilate (grape), benzaldehyde (cherry), methanol (mint) and safranal (saffron). It is generally believed that essential oil cannot be produced by cell cultures since they lack oil glands. Atleast a certain level of differentiation is necessary for the biosynthesis of flavour compounds in tissue culture. Rudementary differentiation of flavour compounds in cell suspension culture of *Ruta graveolens*, perilla, coriander and peppermint show synthesis of flavour compounds. Capsaicinoids of green pepper can be produced in immobilised cell cultures of *Capsicum fruitescens* and *C. annuum*. The flavour of Onion and Garlic are also produced in tissue culture. Callus and cell culture of saffron (dry stigmas of *Crocus sativus)* showed production of the high flavour precursor, picrocrosin.

The demand of non-nutritive sweetners is increasing rapidly for use as food additives and for pharmaceutical applications. The sweetners which have been worked out for cell culture production include glycirrhizin *(Glycirrhiza glabra)*, steviosida *(Stevia rebaudiana)* and thaumatin *(Thaumatacoccus daniellii)*. Thaumatin—encoding mRNA species purified by extraction from a polyacrylamide gel, converted into cDNA and double stranded DNA and then cloned to *Saccharomyces cervisiae* and *Kluyveromyces lactis* (yeasts) can process the plant protein precisely and completely to yield Thaumatin. Intense sweetners and flavour enhancing properties of thaumatin make it a possible sugar substitute and a competitor with other low-calorie sweetners like saccharine and espartame.

Bioinsecticides are being increasingly applied in the storage of food grains. The well known biopesticide studied in tissue cultures are pyrethrins, nicotine and phytocidysones (Ignacimuthu, 1996).

Single Cell Proteins (SCP)

Single cell proteins are the dried cells of micro-organisms such as algae, bacteria, yeast, moulds and higher fungi that are grown in large-scale culture systems as protein for human or animal consumption. The products also contain other nutrients including carbohydrates, fats, vitamins and minerals.

Advances in scientific knowledge regarding physiology, nutrition and genetics of micro-organisms have led to significant improvement in SCP production from a wide range of micro-organisms (Table 76). Both photosynthetic and non-photosynthetic micro-organisms have been used in SCP production. At a minimum, these organisms require a carbon and energy source, a nitrogen source, and supplies of other nutrients including phosphorous, sulphur, iron, calcium, magnesium, manganese, sodium, potassium and trace elements for growth in aquatic environment. Some organisms cannot synthesize amino acids, vitamins and other cellular constituents from simple carbon and nitrogen source and in that event these substances must be supplied for the organism to grow.

Among algae *Chlorella* sp, *Scenedesmus acutus* and *Spirulina maxima* are cultivated for SCP. Large pond areas are needed because the algal growth occurs mainly in the top 20-30 cm where the light intensity is the greatest. Pond waters are agitated continuously or intermittently to prevent the algae from settling. This ensures that the cells are uniformly exposed to light and nutreints.

Among bacteria *Candida utilis, Geotrichum candidum, Candida guillienmondis, Kluyeromyces fragilis and Rhodopseudomonas* sp have been used for SCP production Bacteria multiply very fast and are capable of growing on a variety of substrates from carbohydrates such as starch and sugars, to gaseous and liquid hydrocarbons such as methane and petroelum fractions.

Yeast *(Saccharomyces cervisiae)* is used for the production of SCP. Yeasts have certain advantages over bacteria for the production of SCP. Yeasts tolerate a more acid environment, in the range of 3.5—4.5 instead of near neutral pH preferred by bacteria.

Table 76

Single cell protein (SCP) and mycoprotein produced on the selected substrates (Dubey, 1993)

Microgial groups	*Microorganisms*	*Protein (% /100g., on dry weight basis)*	*Substrates*
Algae	*Chlorella pyrenoidosa* (36)[a]	-	CO_2 (10%), light
	Scenedesmus acutus (20)[b]	-	CO_2 sunlight
	Spirulina maxima (15)[b]	53	CO_2 (5%), combustion gases, bicarbonate, sunlight (in pond)
Bacteria	*Achromobacter delvacvate*	-	Diesel oil in fermenter
	Bacillus megaterium	-	Collagen meat packing waste in fermenter
	Cellumonas sp. (0.45)[e]	87	Bagasse
	Methylomonas clara (0.5)[e]	13	Methanol
	Pseudomonas sp. (1.0)[e]	-	n-alkanes fuel oil
Actinomycetes	*Nocardia* sp. (0.98)[c]	-	n-alkanes
	Thermomonospora fusca (0.4)[c]	5.6	Cellulose pulp
Fungi			
I. Yeasts	*Candida lipolytica* (0.88)[c]	65-69	n-alkanes
	C. utilis (0.39)[c]	-	Potato starch waste
	B. utilis	54	Sulphite liquor
	Saccharomyces cerevisae (0.5)[c]	53	Molasses
	Saccharomyces cerevisae (0.5)[c]	45	Beer
	S. fragilis	54	Milk whey
	Rhodotorula gluinis	-	Domestic sewage
	Torulopsis sp.	-	Methanol

Contd.

Microgial groups	*Microorganisms*	*Protein (% /100g., on dry weight basis)*	*Substrates*
II. Moulds			
	Aspergillus niger	50	Molasses
	Trichoderma viride	64	Straw, starch
	Paecilomyces varioti	55	Sulphite waste liquor
III. Mushrooms			
	Agaricus campestris	36-45	Glucose
	Morchella crassipes	31	Glucose, cheese whey, sulphite, liquor.

(a) yield g/day (on dry weight basis); *(b)* yield g/m^2/day (on dry weight basis); *(c)* yield dry weight basis (g/g substrate used). Values in parantheses denote yield of biomass.

Among higher fungi *Fusarium graminearum, Penicillium cyclipium, Trichoderma harzianum, Paecilomyces varioti, Chaetomium cellulolyticum* are used for the production of SCP. Most of them are cultivated at a pH below 5.

The protein percentage for the various SCP products are crude value based on measurements of the total nitrogen contents of the material. A significant fraction is contributed by nucleic acids. In addition, SCP may be deficient in certain aminoacids that animals cannot synthesize for themselves. SCP will most probably be used in human foods primarily as protein supplements.

Table 77

Composition of dried powder of *Spirulina fusiformis* (constituents are in %) (Dubey, 1993)

A.	**Major constituents (%)**		**C.**	**Minerals (mg)**	
	Total protein	64.6		Calcium	6.58
	Fat	6.7		Phosphorus	977
	Crude fibre	9.3		Iron	44.7
	Carbohydrates	16.1		Sodium	796
	Calories	346		Potassium	1.28
B.	**Vitamins**		**D.**	**Essential aminoacids (%)**	
	Beta - carotene	320,000 IU		Lysin	2.99
	Biotin	0.22 mg		Cystine	0.474
	Cyanocobalamin (B_{12})	65.7 µg		Methionine	1.38
	Folic acid	17.6 µg		Phenylalanine	2.87
	Riboflavin	1.78 mg		Threonine	3.04
	Thiamin	0.118 mg			
	Tocopherol 0.773 IU				

* Analysed by Michelson Labortories Inc., California, USA (1988).

Nutritional value of SCP

Now-adays, considerable information is available on the composition—protein, amino acids, vitamins and minerals of microbial cells (Table 77). Commercial value of SCP depends on their nutritional performance. Average crude protein in dry matter of algae and yeasts ranges from 50-60 per cent, for alkane yeast between 55-65 per cent, and for bacteria about 80 per cent. A content of nucleic acid free protein is extremely important for the economic efficiency of the method in SCP production.

The most important measure of nutritional value is the actual performance of SCP products. The determinants of the utility of SCP products as food for human beings and as feed for animals differ. For human beings, protein digestibility and protein efficiency ratio (PER), biological value or net protein utilization (NPU) are the parameters. Whereas for aimals, metabolizable energy, protein digestibility and feed conversion ratio are the measures of performance in brilers, chickens, swine and calves.

Digestibility (D) is the percentage of total nitrogen consumed, which is absorbed through the alimentary canal. It is calculated as below:

$$D = \frac{Ni - Fn}{Ni} \times 100$$

Where,
Ni = nitrogen ingested from SCP
Fn = nitrogen content in faeces after feeding SCP

Biological value (B) is the percentage of total nitrogen assimilated which is retained by the body, taking into account the simultaneous loss of endogenous nitrogen through excretion in urine. This is expressed by the following formuale:

$$B = \frac{Ni - (Fn + Un)}{(Ni - Fn)} \times 100$$

Where,
Un = nitrogen content in urine after feeding SCP

Nutritional value of SCP products (Table 78) indicate that protein digestibility range from good to very good and is true for bacteria and yeasts growth on non-conventional substrates (Roth, 1982).

However, there are certain problems which warrant the use of SCP products as human foods—(i) high content of nucleic acid leading to development of kidney stone and gout if consumed in high quantity, (ii) possibility for the presence of toxic secondary metabolites, and (iii) poor digestibility and stimulation of gastrointestinal and skin reaction (litchfield, 1979).

Table 78
Nutritional value of food protein (Dubey, 1993)

Food protein	Analytical composition (%)[A]		Essential amino acids (g/100g crude protein)[B]		Biological coefficient (%)[A]		Digestibility of crude protein (%)[B]	Metabolisable energy for poultry (k cal/Kg)[A]
	Total nitrogen	Crude protein (N × 6.25)	Lysine	Threonine	NPU	NPV		
Algae	8.0	45-71	5.7	5.2	—	—	82	—
Dried skimmed milk	5.7	35.9	8.0	—	87	31.2	—	2510
Soybean meal	7.0	44	6.4	4.0	64	65	—	2240
Alkane yeast (toprina - LBP)	11.2	7.0	7.4	4.2	91	96	92	2540
Bacteria from methanol	11.5	72	6.2	4.6	84	88.4	91	3468

Source: A, Senez (1986); B, Roth (1982); *Digestibility determined by pig feeding; NPU = Net protein utilization; NPV = Net protein value.

Advantages of producing SCP

Roth (1982) described a number of advantages in the production of microbial protein:

(1) Rapid succession of generations (algae 2-6 h, yeast 1-3 h, bacteria 0.5-2 h).
(2) Easily modifiable genetically.
(3) Broad spectrum of original raw material used for the production, which also include waste products.
(4) Production in continuous cultures, consistent quality not dependent on climate in determinate amount, low land requirements, ecologically beneficial.

Other advantages are: (a) high solar energy conversion efficiency per unit area, (b) easy regulation of environmental factors, (c) cellular, molecular and genetic alterations, and (d) algal culture in space, which is normally unused instead of competing for land (Vijayan, 1988).

References

ABE, 1985. Towards a perspective on energy demand and supply in India in 2004/05. Advisory Board on energy. Goverment of India, New Delhi.

ACC, 1965. *Air conservation.* Amer. Assoc. Adv. Sci. Washington DC.

Adang, M.J., Firoojabady, E., Debov, D.L., Klein, J., Merlo, D.J., Murray, E., Rachelean, T., Raksha, K., Staffeld, G., and Stock, C., 1986. In *Plant tissue and cell culture.* Eds. D.A. Somers, B.G. Gengenbach, D.D. Bioesboer, W.P. Hackett and C.E. Green. University of Minnesota, p. 471.

Agarwal, S.K., 1971. Ph.D. Thesis, Udaipur University, Udaipur.

Agarwal, S.K., 1980. *Acta Ecologica,* 2:1-6.

Agarwal, S.K., 1991. *Pollution ecology.* Himanshu Publications, Udaipur.

Agarwal, S.K., 1993. *Resource ecology.* Himanshu Publications, Udaipur.

Agarwal, S.K., 1996. *Industrial environment: Assessment and strategy.* APH Publishing Corporation, New Delhi.

Ahuja, D.R., 1985. *Energy Environment Monitor,* 112: 3-20.

Alexander, M., 1983. Testimony before US House Committee on science and technology. Hearing on environmental implications of genetic engineering. June 22. Committee print No. 26.

Allen, M., 1989. *Ecoforum,* 14:2: 1-10.

Altenbach, S., Pearson, K., Mecker, G., Staraci, L., and Sun, S., 1990. *Plant Mol. Biol.* 13: 513.

Ambasht, R.S., 1981. *Cent. Bd. Prev. Contr. Water Pollut.* Osmania University, Hyderabad. pp. 61-65.

Ammirato, P.V., 1984. In *Handbook of plant cell culture.* Vol. 3. *Crop Science.* Eds., P.V. Ammirato, D.A. Evens, W.R. Sharp, and Y. Yamada. Macmillan, New York. pp. 327.

Annonymous, 1977. Report on methane generation from human, animal and agricultural wastes. National Academy of Science, USA., Washington DC.

Annonymous, 1980. *Nat. Acad. Sci.* Washington DC.

Arbridge, M.V., and Pitcher, W.H. 1989. *Trends in Biotech.* 7: 330-335.
Bachtel, D.B., and Bulla, L.A. Jr. 1976. *J. Bacteriol.* 127: 1472.
Bailey, J.A., 1987. In *Genetics and plant pathogenesis.* Ed. P. Day and G. Ellis. Blackwell Sci. Publ. Scotland.
Bailey, J.E., Hijortso, M., Lee, S.B., and Sriene, F., 1983. In *Biochemical engineering.* eds. K. Venkatasubramanian, A. Constantinides, W.R. Vieth. Academy of Science, New York. pp. 3: 71-87.
Bajaj, Y.P.S., and Gill, M.S., 1986. In *Biotechnology of Plants and micro-organisms.* Eds. O.J. Crocomo et al. Ohio State University Press, Columbus. pp. 118-151.
Banz, G., 1971. In *Microbial control of insets and mites.* Eds. Burgesss and Husser. Academic Press, London. pp. 327-355.
Barcelo, P., Hagel, C., Becker, D., Martin, A., and Lorz, H., 1984. *Plant J.* 5: 583-592.
Barton, K.A., and Brill, W.J., 1983. *Science,* 219; 671-675.
Barton, K.A., Whiteley, H., and Yang, N., 1987. *Plant Physiol.* 85:1103.
Batch, W.E., Fox, G.E., Mangrum, L.J., Woese, C.R., and Walfe, R.S., 1979. *Microbiol. Rev.* 43:260.
Bates, G.W., Piatuch, W., Riggs, C.D., and Rubussay, D., 1988. *Plant cell, tissue and organ culture,* 12: 213-218.
Beachy, R., Loesch-Fries, S., and Tumer, N., 1990. *Ann. Rev. Phytopathol.* 28:451.
Beachy, R.N., Stark, D.M., Deom, C.M., and Fraley, R.T., 1987. In *Tailoring genes for crop improvement.* Eds. G. Bruening, J. Harada and T. Kosuge, Plenum Press, New York. pp. 169-180.
Becker, D., Brettschneider, R., and Lorz, H., 1994. *Plant J.* 5: 299-307.
Benin, A., Navrot, J., Noi, U., and Yoles, D., 1981. *J. Environ. Qual.* 10: 4: 536.
Berrow, M.L., and Webber, J., 1972. *J. Sci. Fd. Agric.* 23: 93-100.
Beversdorf, W.D., Weiss-Lerman, J., Erickson, L.R., and Souza Machado, V., 1990. *Can. J. Genet. Cytol.* 22: 73.
Bhaduri, S., 1987. In *Food biotechnology.* Ed. D. Knorr. Marcel Dekker, New York. pp. 139-161.
Bhattacharya, P., and Mishra, U.C., 1994. *Sym. Workshop Environ. Biotechnol.* NEERI, Nagpur, pp. 72-83.
Bhawalkar, V., and Bhawalkar, U., 1992. *Vermiculture biotechnology.* Bhawalkar Earthworm Research Institute, Poone.
Bhoota, B.V., 1975. *Commerce,* 131: 157-162.
Bhur, H.O., and Andrews, J.E., 1977. *Water Res.* 11: 2: 129.
Black, J.L., 1987. In *Merino improvement programs in Australia.* Australian Wool Corporation. Ed. B.J. McGuirk. pp. 457-480.

Black, J.L., and Reiz, P.J., 1979. In *Physiological and environmental limitations to wool growth.* Eds. J.L. Black and P.J. Reiz. University of New England, NSW. Australia. pp. 270-294.

Bostock, R.M., and Stermer, B.A., 1989. *Ann. Rev. phytopath.* 27: 343.

Botterman, J., and Leemans, J., 1988. *Trends Genet.* 4: 219.

Bowles, D.J., 1990. *Ann. Rev. Biochem.* 59: 873.

Bowman, H.G., and Hultmarb, D., 1987. *Ann. Rev. Microbiol.* 41: 103.

Braithwaite, J.W., and Wadsworth, M.E., 1976. *Extractive metallurgy of copper.* Vol. 2. The American Institute of Mining, Metallurgical and Petroleum Engineers, New York.

Branco, J.R.C., and Costa Rebeiro, C., 1979. *Anaerobic fermentation of sillage.* Ist International Symposium on anaerobic digestion. Cardiff. 17-21 September.

Brierley, C.L., 1977. *Dev. Indus. Microbiol.* 18: 273-275.

Brierley, C.L., 1978 *Cirt. Rev. Microbiol.* 6: 207-215.

Brierley, C.L., and Brierley, J.A., 1973. *Canadian J. Microbiol.* 19: 183-186.

Brierley, C.L., and Brierley, J.A., 1975. Microbial leaching of metals. Paper presented at conference on heavy metals in the environment. Toronto. Oct. 27-31.

Brierley, J.A., 1976. *Appl. Environ. Microbiol.* 36: 523-525.

Brierley, J.A., and Lockwood, S.J., 1977. *FEMS Microbiol. Lett.* 2: 163-165.

Brierley, J.A., Norris, P.R., Kelley, D.P., and Leroux, N.W., 1978. *En. J. Appl. Microbiol. Biotechnol.* 5: 291-295.

Brinster, R.L., Chen, H.Y., Trumbarer, M.E., Yagle, M.K., and Palmitzer, R.D., 1985. *Proc. Natl. Acad. Sci. USA.* 82: 4438.

Broglie, K., Chet, I., Holliday, M., Cressman, R., Biddley, P., Knowlton, S., Mauvais, C.J., and Broglie, R., 1991. *Science,* 254: 1194.

Brown, C.M., Campbell, I., and Priest, F.G., 1987. *Introduction to biotechnology.* Blackwell Scientific Publications, Oxford.

Brunke, K., and Meeusen, R., 1991. *Trends in Biotechnology,* 9: 197.

Bryner, L.C., and Anderson, R., 1957. *Indus. Engng. Chem.* 49: 1721-1724.

Bryner, L.C., Black, J.E., Davis, L.B., and Wilson, D.G., 1954. *Industr. Engng. Chem.* 46: 2587-2592.

Bryngelsson, T., and Collinge, D.B., 1991. In *Barley: Genetics biochemistry, molecular biology and biotechnology.* Ed. P.R. Shewry. CAB Intl. Awllingford, Oxon, UK. pp. 452-473.

Bull, A.T., 1992. *In the treatment and handling of wastes.* Eds. A.D. Bradshaw, R. Southwood and F. Warner. Chapman and Hall, London. pp. 155-166.

Bull, A.T., 1996. *Biotechnology and Conservation.* 5: 1-25.

Bulla, L.A. Jr., Bechtel, D.B., Kramer, K.J., Shethna, Y.I., Aronson, A.I., and Filz-James, P.C., 1980. *CRC Crit. Rev. Microbiol.* 8: 147.

Bulla, L.A. Jr., Kramer, K.J., Cox, D.B., Jonnes, B.L., Davidson, I.I., and Lookhart, G.I., 1981. *J. Biol. Chem.* 256: 3000.

Bulla, L.A. Jr., Kramer, K.J., and Davidson, L.I., 1977. *J. Bacteriol.* 130): 375.

Burgess, H.D., (ed.), 1984. *Microbial control of pests and plant diseases.* Academic Press, NY. pp. 1970-1980.

Busse, H.J, El-Banna, T., Oyaizu, H., and Auling, G., 1992. *Int. J. Syst. Bateriol.* 42: 19-26.

Buston, M.M., Kalkan, F.A., Vandenbosch, K.A., and Hall, T.C., 1991. *Plant Mol. Biol.* 16: 381.

Calvin, M., 1979. *Bioscience,* 29: 533-538.

Chahal, D.C., and Overend, R.P., 1982. In *Advances in agricultural microbiology.* Ed. Subba Rao. Oxford and IBH Publication, New Delhi. pp. 585-641.

Chakrabarty, A.M., 1992. In *Harnessing biotechnology for 21st century.* Eds. R. Ladisch and A. Bose. American Chemical Society.

Chakrabarty, C., and Chakrabarty, T., 1988. *Environ. Pollut.* 52: 219-235.

Chakraborty, S., and Old, K.M., 1982. *Soil Biol. Biochem.* 14: 247-255.

Chakraborty, S., and Warcup, J.H., 1983 (a). *Soil Biol. Biochem.* 15: 181-185.

Chakraborty, S., Old, K.M., and Warcup, J.H., 1983 (b). *Soil Biol. Biochem.* 15: 17-24.

Chan, M.T., Chang, H.H., Ho, S.L., Tong, W.F., and Yu, S.M., 1993. *Plant Mol. Biol.* 22: 441-506.

Chaney, R.L., 1973. In *Recycling municipal sludge and effluents on land.* National Association of State Universities of Land Grant College, Washington DC. pp. 129-141.

Chang, S.L., 1968. *Bull. W.H.O.,* 14: 949-1006.

Chapman, H.D., 1966. In *Diagnostic criteria for plant* and *soils.* University of California, riverside, USA. pp. 484-499.

Chapman, R.E., and Ward, K.A., 1979. In *Physiological and environmental limitations for wool growth.* Eds. J.L. Block and P.J. Reiz. University of New England. NSW. Australia. pp. 193-208.

Chaudhary, A., Chowdhry, C.N., Maheshwari, N., Maheshwari, S.C., and Tyagi, A.K., 1994. *J. Plant Biochem. Biotech.* 3: 9-13.

Cheung, A., Bogorad, J., Van Montagu, M., and Schell, J., 1988. *Proc. Natl. Acad. Sci.* USA. 85: 391.

Chrispeels, M.J., and Raikhel, N.V., 1991. *Plant Cell.* 3: 1.

Christou, P., 1992. *Plant J.* 2: 275-281.

Christou, P., 1994. *Euphytica.* 74: 165-185.

Christou, P., Ford, T.L., and Kofron, A., 1991. *Biotechnology.* 9: 957-962.

Christou, P., Ford, T.L., and Kofron, A., 1992. *Tib. Tech.* 10: 239-246.

Clapham, W.B. Jr., 1973. *Natural ecosystems.* The macmillan, NY.

Clausi, A., 1986. In *Biotechnology of plants and micro-organisms.* Eds. O.J. Crocomo, W.R. Sharp, J.E. Bravo, F.C.A. Tavares, and E.F. Paddock. Ohio State University Press, Columbus. pp. 3-8.

Comai, L., Ficciotti, D., Hiat, W.R., Thompson, G., Rose, R.E., and Stalker, D.M., 1985. *Nature,* 317; 741.

Comai, L., Sen, L.C., and Stalker, D.M., 1983. *Science,* 221: 370.

Comai, L., and Stalker, D., 1986. *Oxford Surv. Plant Mol. Cell. Biol.* 3: 166.

Convey, G.R., *In Theoretical ecology: Principles and practices.* Ed. R.M. May. Saunders, Philadelphia. pp. 257-281.

Cook, E.F., 1976. *Man energy society.* W.H. Freeman and Co., Sanfrancisco.

Cooper, R., 1959. *Soil fertilizers.* 22: 227-233.

Costa Rebeiro, C., and Branco, J.R.C., 1980. The use of stillage in the ferti-irrigation of sugarcane cropland dedicated to full alcohol production. Paper presented at 2nd International symposium on Waste treatment and utilization. University of Waterloo, Ontario, Canada.

Cullis, C.A., 1987. *Ohio. J. Sci.* 87: 143-147.

Cuozzo, M.O., Connel, K., Kaniewski, W., Fong, R., Chua, N., and Tumer, N., 1988. *Biotechnology,* 6: 549.

Das, D., Sirdar, K., and Chatterjee, A.K., 1994. *Indian J. Environ. Hlth.* 36: 186-191.

Das, R.C., 1988. *Energy Environment Monitor,* 4: 41-49.

Dastefano-Behran, L., Nagpala, P.G., Kim, J., Dodds, J.H., and Jaynes, J.M., 1991. In *Molecular methods for potato improvement.* International Potato Center, Peru. pp. 49-64.

Da Silva, E.J., 1981. In *Fuel gas production from biomass.* Vol. I. Ed. D.L. Wise. CBC Press Inc. Florida. pp. 53-71.

Da Silva, E.J., and Sasson, A., 1989. *MIRCEN J.* 5: 115-118.

Da Silva, E.J., and Sasson, A., 1995. *Biotechnology and Development Review.* 4: 37-47.

Datta, S.K., Datta, K., and Potrykus, I., 1990. *Plant Cell Rep.* 9: 253-256.

Datta, S.K., Peterhaus, A., Datta, K., and Potrykus, I., 1990. *Biotechnology,* 8: 736-740.

De, A.K., 1990. *Environmental Chemistry.* Wiley Eastern Ltd., New Delhi.

De Block, M., Botterman, J., Vandewicle, M., Dock, X., Thoen, C., Gossele, V., Rao, M.N., Thompson, V., Motaga, M., and Leemans, J., 1987. *EMBO J.* 6: 2513.

De Clercq, A., Vandeweile, M., Van Damme, J., Guerche, P., Van Montagu, M., Vandekerchkhove, J., and Krebbers, E., 1990. *Plant Physiol.* 94: 970.

Deffardius, A.E., and Gardaner, H.B., 1989. *Mol. Plant Microb. Inter.* 2:26.

DeGreef, W., Delon, R., DeBlock, M., Leemans, J., and Botterman, J., 1989. *Biotechnology,* 7:61.

DeKeyser, R., Claes, B., Marchal, M., VanMontagu, M., and Captan, A., 1989. *Plant Physiol.* 90: 217-223.

Della Ciopp, G., Bauer, S., Taylor, M., Rochester, D., Klein, B., Shah, D., Fraley, R., and Kishore, G., 1987. *Biotechnology,* 5: 579.

DeLumen, B.O., 1990. *J. Agric Food Chem.* 38: 1779.

Deoras, P.J., 1981. *Indian J. Environ. Prot.* 1: 21-25.

Deshpande, C.M., 1985. *Indian J. Environ. Prot.* 2: 43-52.

Detroy, R.W., Cunningham, R.L., Bothast, R.J., Bagby, M.O., and Herman, A., 1982. *Biotechnol. Bioengr.* 24: 1105-1113.

D'Halluin, K., Bonne, E., Bossut, M., De Beuckeleer, M., and Leemans, J., 1992. *Plant Cell.* 4: 1495-1505.

Dickman, M.D., Podila, G.K., and Kolatlubudy, P.E., 1989. *Nature,* 342: 446.

Dill, N.H., 1986. In *environment and natural resources.* Eds. V.P. Agarwal and S.V.S. Rana. Society of Bioscience, Muzaffarnagar, pp. 145-149.

Dilworth, M., 1991. *Plant Cell.* 3: 213.

Dimarce, J.L., Zachary, D., Hoffman, J.A., Hoffman, D., and Riechart, J.M., 1990. *EMBO J.* 9: 2507.

Dixon, B., 1995. *Biotechnology,* 13: 308.

Donn, G., Tischer, E., Smith, J.A., and Goodman, H.M., 1984. *J. Mol. Appl. Genet.* 2: 621.

Dorfman, R., 1991. In *The global possible: Resource development and the new century.* Ed. R. Repetto. East West Press, New Delhi. pp. 67-96.

Dubey, P.S., Pawar, K., Shringi, S.K., and Tripathi, L., 1982. *Agroecosystems,* 8: 137-140.

Dubey, R.C., 1993. *A text book of biotechnology.* S. Chand and Company Ltd, New Delhi.

Duncan, D.W., and Bruynesteyn, A., 1971. *Canadian Min. Metal Bull.* 64: 32-35.

Duncan, D.W., and Trussel, P.C., 1964. *Canadian Mettal. Qual.* 3: 43-45.

Duncan, D.W., Trudsell, P.C., and Walden, C.C., 1964. *Appl. Microbiol.* 12: 122-126.

Duncan, D.W., and Walden, C.C., 1972. *Industr. Microbiol.* 13: 66-68.

Dunwell, J.M., 1985. In *Biotechnology in plant science.* Eds. M. Zaitlin, P. Day, and A. Hollaender. Academic Press, Orlando. pp. 49-76.

Dwivedi, R.S., Dubey, R.C., and Dwivedi, S.K., 1989. In *Plant-microbe interactions.* Ed. K.S. Bilgrami. Narendra Publ. House, New Delhi. pp. 217-238.

Eckes, P., Schmitt, P., Daub, W., and Wengenmayor, F., 1989. *Mol. Gen. Genet*. 217: 263.

Edwards, P., 1985. *Aquaculture: a component of low cost senitation technology*. The World Bank, Washington DC.

El-Halwagi, M.M., 1984. *Biogas technology transfer and diffusion*. Elsevier Applied Science Publishers, London.

Ellis, M.M., 1937. *Bull. U.S. Bur. Fish. Washington*. 48: 365-437.

Eswara Rao, M., and Nagendra, M., 1982. *IAWPC Tech Annul*. 9: 163-164.

Faizi, S., 1995. *Humanscape*, October, pp. 18-19.

Fennel, A., and Hauptmann, 1992. *Plant Cell Rep*. 11: 567-570.

Fiksel, J.R., and Covello, V.J., 1986. In *Biotechnology risk assessment*. Eds. J. Fiksel and V.J. Covello. Pergamon Press, New York. pp. 1-34.

Finnegan, J., and McElroy, D., 1994. *Biotechnology*. 12: 883.

Fischhoff, D.A., Bowdish, K.S., Perlak, F.J., Marrone, P.G., McCormick, S.M., Niedermeyer, J.G., Dean, D.A., Kusano-Kretzmer, K., Mayer, E.J., Rochester, D.E., Rogers, S.G., and Fraley, R.T., 1987. *Biotechnology*, 5: 807.

Fishner, J.R., 1966. *Trans. Canadian Inst. Min. Mettal*. 69: 167-169.

Forberg, C.H., and Tsao, G.T., 1983. *Am. Chem. Soc. Abstr*. No. 48.

Frear, D.S., and Swanson, H.R., 1970. *Phytochemistry*, 9: 2123.

Fromm, M.E., Morrish, F., Armstrong, C., Williams, R., Thomas, J., Klein, T.M., 1990. *Biotechnology*, 8: 833-839.

Fromm, M., Taylor, L.P., and Walbot, V., 1986. *Nature*. 319: 791-794.

Fujita, Y., Hara, Y., Suga, C., and Marimoto, T., 1987. *Plant Cell Reports*. i.

Garcia, M.J., and Page, A.L., 1976. *J. Soil Sci. Soc. Am*. 40: 658-663.

Garlach, W., Llewllyn, D., and Haschoff, J., 1987. *Nature*, 328: 799, 802.

George, A.D., Suzuki, I., and Less, H., 1967. *Canadian J. Microbiol*. 13: 1413-1418.

Ghai, S.K., and Thomas, G.V., 1989. In *Plant microbe interactions*. Ed. K.S. Bilgrami. Narendra Publ. House, New Delhi. pp. 47-60.

Ghosh, M.R., 1989. *Concepts of insect control*. Wiley Eastern Limited, New Delhi.

Golemboski, D., Lommonossoff, G., and Zaitlin, M., 1990. *Proc. Natl. Acad. Sci. USA*. 87: 6311.

Goswamy, K.P., and Choudhury, S., 1966. Fuel gas and manure from agricultural wastes. Get together on Science and Industry. CSIR, New Delhi (memeo).

Goswamy, K.P., and Choudhury, S., 1967. *Indian Farming*. April. 18-19.

Goswamy, K.P., and Rajor, A., 1983. *Urja*, May. 263-267.

Gordon-Kamm, W.J., Spenscer, T.M., Mangano, M.L., Adams, T.R., Davies, R.J., Start, W.J., O'brien, V.J., Chambers, S.A. Jr., Adams, W.R., Willets, N.G., Rice, T.B., Mackey, C.J., Krueger, R.W., Kausch, A.P., Lexaux, P.G. 1990. *Plant Cell*. 2: 603-618.

Gow, W.A., Mccreedy, H.H., Ritchey, G.M., Mcnamara, V.M., Harrison, V.I., and Lucas, B.H., 1988. *Recovery of Uranium.* Vienna Inst. St. Energy Agency.

Grierson, D., 1991. *Plant genetic engineering.* Vol. 1. *Plant biotechnology,* Blackie, Glasgow.

Guey, R., Torma, A.E., and Silver, M., 1975. *Ann. Microbiol.* 126 B: 209-213.

Guerche, P., DeAlmeida, E.R.P., Schwarztein, M.A., Gander, E., Krebbers, E., and Pelletier, G., 1990. *Mol. Gen. Genet.* 231: 306.

Guha, S., and Maheshwari, S.C., 1964. *Nature,* 204: 497-498.

Gupta, O.P., 1987. *Aquatic weed management.* Today and Tomorrow, New Delhi.

Gupta, P.K., 1994. *Elements of biotechnology.* Rastogi and Co., Meerut.

Gupta, P.K., 1995. *Elements of biotechnology.* Rastogi & Co., Meerut.

Hahne, H.C.H., and Kroonte, W., 1973. *J. Environ. Qual.* 2: 444-450.

Hain, R., Bieseeler, B., Kindi, H., Schroeder, G., and Stocker, R., 1990. *Plant Mol. Biol.* 15: 325.

Haitt, A., Cafferkey, R., and Bowdish, K., 1989. *Nature,* 342: 76-78.

Hall, D.O., Barnard, G.W., and Poss, P.A., 1982. *Biomass for energy in the developing countries.* Pergmon Press. Oxford.

Hammock, B.D., Bonning, B.C., Possee, R.D., Hanzlik, T.N., and Maeda, S., 1991. *Nature.* 344: 458-461.

Harrison, B., Mayo, M., and Baulcombe, D., 1987. *Nature.* 328: 799.

Haughn, G.W., Smith, J., Mazur, B., and Somerville, C., 1988. *Mol. Gen. Genet.* 211: 266.

Hauptmann, R.W., Vasil, V., Ozias-Akins, P., Tabacizadeh, Z., Rogers, S.G., Fraley, R.T., Horsch, R.M., and Vasil, I.K., 1988. *Plant Physiol.* 86: 602-606.

Hayashimoto, A., Li, Z., and Murai, N., 1990. *Plant Physiol.* 93: 857-863.

Held, G.A., Bulla, L.A. Jr., Ferrari, E., Bech, J., Aronson, A.I., and Minnich, S.A., 1982. *Proc. Natl. Acad. Sci.* USA. 79: 3000. Hel.

Helgeson, J.P., 1963. In *Use of tissue culture and prothology.* Eds. J.P. Helgesson and B.J. Deverall. Academic Press, Sydney. pp. 9-38.

Hernalsteens, J.P., Thia-Tong., Schell, J., and Montagu, M., 1984. *EMBO J.* 3: 3039-3041.

Hiei, Y., Ohta, S., Komari, T., and Kumashiro, T., 1994. *Plant J.* 6: 271-282.

Hilder, V., Gatehouse, A., and Boulter, D., 1990. In *Genetic engineering of crop plants.* Eds. G. Lycett and D. Grierson. Butterworths, UK. pp. 51-66.

Hilder, V., Gatehouse, A., Sherman, S., Barker, R., and Boulter, D., 1987. *Nature.* 330: 160.

Hill, D.D., 1977. *New Phytol.* 78: 611-616.

Hirschberg, J., and McIntosh, I., 1983. *Science.* 222: 1346.

Hofte, H., and Whiteley, H., 1989. *Microb. Rev.* 53: 242.

Holmes, P.A., 1985. *Phys. Technol.* 16: 32-36.

Horsch, R.B., Shah, D.S., Klea, J.H., Rodgers, S.G., and Fraley, R.T., 1986. In *Plant tissue and cell culture.* Eds. D.A. Somers, B.G. Gangenbach, D.D. Biesboer, W.P. Kackett and C.E. Green. University of Minnesota. p. 126.

Hoyle, D., 1995. *Biotechnology.* 10: 316.

Hubble, B.R., et al, 1982. In *Proceedings of a conference held in Portland.* Oregon. Eds. J.A. Cooper and D. Matek. Graduate Centre, Oregon. pp. 79-138.

Hullman, A., 1996. In *Biotechnology risk assessment—issues and method for environmental introductions.* Eds. J. Fiksel and V.J. Covello. Pergamon Press, New York. pp. 75-87.

Ignacimuthu, S., 1995. *Basic biotechnology.* Tata McGraw Hill, New Delhi.

Ignacimuthu, S., 1996. *Applied plant biotechnology.* Tata McGraw Hill Book Co Ltd, New Delhi.

Ingle, A., Charde, O., Chandekar, C.J., Bholey, A.D., Daginawala, H.F., and Parhad, N.M., 1994. *Symp. Workshop Environ. Biotechnology.* NEERI. Nagpur. pp. 158-167.

Jaenisch, R., 1977. *Cell.* 12: 691.

Jaenisch, R., and Mintz, B., 1974. *Proc. Natl. Acad. Sci.* USA. 71: 1250.

Jaenisch, R., et al, 1981. *Cell.* 24: 519.

Jagdish, K.S., 1994. In *Workshop on watershed development and bioenergy.* Choksi Hall, IISC. Bangalore. February 26-27.

Jahan, H., and Kaur, J., 1992. *Madhya Pradesh Paryavaran.* March-April. 50-52.

Jahner, D., and Jaenisch, R., 1980. *Nature.* 287: 456.

Jain, S.M., 1993. *Curr. Sci.* 64: 715-724.

Jain, S.M., Oker-Blom, C., Pehu, E., and Newton, R.J., 1992. *J. Agric. Sci. Finl.* 1: 323-338.

Jayaraman, K., 1993. *The Hindu Survey of the Environment.* pp. 40-41.

Jogdand, S.N., 1995. *Environmental biotechnology.* Himalaya Publishing House, New Delhi.

Kaeppler, H.F., Somers, D.A., Rines, H.W., and Cockburn, A.F., 1992. *Theor. Appl. Genet.* 84: 560-566.

Kaeppler, H.F., Gu, W., Somers, D.A., Rines, H.W., and Cockburn, A.F. 1990. *Plant Cell Rep.* 9: 415-418.

Kanekar, P., 1990. Microbiological approach to environmental protection against toxic substances and hazardous wastes. In *Environmental education and sustainable development.* Eds. Desh

Bandhu, Harjit Singh and A.K. Maitra. Indian Environment Society, New Delhi. pp. 353-359.

Kanekar, P., and Godbole, S.H., 1983. *Pseudomonas trinitro-toluenophila* sp nov. isolated from soil exposed to explosive wastes. *Biovigyanam.* 9: 5-8.

Kanekar, P., and Godbole, S.H., 1984. Microbial degradation of Trinitrotoluene (TNT). *Indian J. Environ. Hlth.* 26: 2: 89-101.

Kant, S., 1987. In *Environment and ecotoxicology*. Eds. R.C. Dalela, Y.N. Sahai and S. Gupta. The Academy of environmental biology, Muzaffarnagar. pp. 49-74.

Kaplan, D., Calvert, H.E., and Peters, G.A., 1986. *Plant Physiol.* 80: 884-890.

Karaivko, G.I., and Mnoshniakova, S.A., 1977. *Mikrobiologiya,* 43: 156-158.

Kashyap, V., 1992. In *Indian environment*. Ed. P. Singh. Ashish Publishing House, New Delhi. pp. 193-203.

Kaul, M.L.H., 1988. *Male sterility in higher plants.* Springer Verlag, Germany.

Kaul, M.L.H., and Nirmala, C., 1995. *J. Phytol Res.* 8: 1-26.

Kelling, K.A., Keeney, D.R., Walsh, L.M., and Ryan, J.A., 1977. *J. Environ. Qual.* 613: 525-528.

Khanna, P., 1994. *Symp. Workshop Env. Biotechnology*. NEERI, Nagpur. pp. 1-8.

Khoshoo, T.N., 1982. *Energy from plants: problems and prospects.* Presidential address, Botany Section, Indian Science Congress 69th Session, Mysore.

Kishore, G., and Shah, D., 1988. *Ann. Rev. Biochem.* 57: 627.

Klein, J.M., Wolf, E.D., Wu, R., Stanford, J.L., 1987. *Nature,* 327: 70-73.

Klein, L., 1957. *Aspects of river pollution.* Butterworth Publ. London.

Klein, L., 1973. *River pollution* II. *Causes and effects.* Butterworth Publ. London.

Konekamp, A.H., 1977. *Animal Research and Development*. 5: 29-35.

Kothari, S.L., and Chandra, N., 1995. *J. Indian Bot. Soc.* 74A: 323-342.

Krebbers, E., Van Rampaey, J., and Vandekerchov, J., 1993. In *Transgenic plants: fundamentals and applications.* Eds. A. Hiatt. Marcel Dekker Inc, New York. pp. 37-60.

Krebbers, E., and Vandekerchkhoeve, J., 1990. *Trends Biotechnol.* 8:1.

Krishnaswamy, K.V., Prasad, V.S., and Iyer, K.S., 1981. *Indian J. Environ. Hlth.* 23: 41-52.

Kriwisky, S., 1982. *Genetic alchemy: the social history of the recombinant DNA database.* MIT Press, Cambridge, M.A.

Kshirsagar, S.R., 1968. *Sewarage and sewage treatment*. Roorkee Publishing House, Roorkee.

Kumar, A., 1986. In *Pollution Control Handbook*. Ed. P.L. Diwaker Rao, Utility Publications, Secunderabad. pp. 57-62.

Kumar, H.D., 1977. *Modern concepts of ecology*. Vikas Publ. House Pvt Ltd, New Delhi.

Kumar, D., 1994. *Symp. Workshop Environ. Biotechnology*. NEERI, Nagpur. pp. 124-127.

Kumar, P.A., 1994. *Sci. Reportr*. 31: 20-23.

Kumar, S., and Upadhyay, S.N., 1983. *Proc. Symp. Air Pollution Control*. pp. 119-131.

Kunhi, A.A.M., 1991. In *Microbial gene technology*. Ed. H. Polasa South Asian Publishers, New Delhi. pp. 31-59.

Kurtsak, E. (d.), 1982. *Microbial and viral pesticides*. Marcel Dekker, new York.

Larkin, P.J., Ryan, S.A., Brettell, R.I.S., and Scowcroft, W.R., 1984. *Theoret. Appl. Genet*. 67: 443.

La Rossa, R.A., and Schloss, J.V., 1984. *J. Biol. Chem*. 259: 8753.

Lawrence, R.H., 1987. *Cereal Foods World*. pp. 758-765.

Leason, M., Cunliffe, D., Parkin, D., Lea, P.J., and Miflin, B.J., 1982. *Phytobiochemistry*. 21: 855.

Lee, K., Townsend, J., Tepperman, J., Black, M., and Chui, C., 1988. *EMBO J*. 7: 1241.

Leeper, G.W., 1978. *Managing the heavy metals on the land*. Marcel Dekker, New York.

Lehrer, R.I., Ganz, T., and Selsted, M.E., 1992. *Cell*. 64: 224.

Levy, S.B., 1986. In *Biogechnology risk assessment issues and methods for environmental introductions*. Eds. J. Fiksel and V.J. Covello. Pergamon Press, NY. pp. 56-74.

Litchfield, J.H., 1979. In *Microbial technology*. Vol. 1, 2nd ed. Eds. H.J. Peepler and D. Perlman. Acaemic Press, New York. pp. 93-155.

Lones, T., 1980. *New Scientist*. 87: 866-868.

Longmann, J., Jach, G., Tommercup, H., Mundy, J., and Schell, J., 1992. *Biotechnology*. 10: 305-308.

Lundgren, D.G., and Dean, W., 1979. In *Biogeochemical cyanide mineral forming elements*. Ed. P.A. Trudinger and P.J. Swain, Elsevier, New York.

Lyon, B.R., Llwellyn, D.J., Huppatz, J.L., Kennis, E.S., and Peacock, W.J., 1989. *Plant Mol. Biol*. 13: 533.

Macgregor, R.A., 1966. *Canadian Inst. Min. Metall. Bull*. 59: 583-586.

Mahon, K.A., Overbeck, P.A., and Westphal, H., 1988. *Proc. Natl. Acad. Sci*. USA. 85: 1165.

Mantell, S.H., Matthews, J.A., and McKee, R.A., 1985. *Principles of plant biotechnology*. Blackwell Scientific Publications, London.

Mann, H.S., 1981. In *Aridland ecosystem: structure functioning and management*. Vol. 2. Ed. W.D. Goodall and F.P. Perry. Canbridge Univ. Press. Great Britain. pp. 479-493.

Mariani, C., Benckeleer, M., Truettner, J., Leemans, J., and Goldberg, R.B., 1990. *Nature*, 347: 737.

Mariani, C., Cossela, D.V., De Beuckeleer, M., DeBlock, M., Goldberg, R.B., DeGreef, W., and Leemans, J.L., 1992. *Nature*, 357: 384.

Mark, W., Signorella, K., Lacy, E., 1985. *Cold spring harbour Symposium Quant. Biol.* 50: 453.

Massey, M.L., and Phland, F.G., 1978. *J. Water Pollut. Contr. Federation.* 50: 9: 2204.

Mathur, A.N., 1987. In *Bioenergy for arid and semiarid zones*. Ed. A.N. Mathur. Himanshu Publications, Udaipur. pp. 15-26.

Matic, M., and Mrost, M., 1964. *S. Afr. Industr. Chem.* 18: 127-130.

Mazur, B., and Falco, S., 1989. *Ann. Rev. Plant Physiol Plant Mol.* 40: 441.

McCarty, P.L., and Vath, C.A., 1963. *Int. J. Air Water Pollut.*6: 65.

McInerny, M.J., and Bryant, M.P., 1981. In *Fuel gas production from biomass*. Vol. 1. Ed. D.L. Wise, CBC Press Inc. Florida. pp. 19-46.

McGaughey, W., 1985. *Science*, 229: 193.

Mehta, R.S., 1968. *IAWPC Tech. Annual.* 4: 153-156.

Meshram, S., Saonali, U., Joshi, N., and Pande, S.S., 1994. *Symp. Workshop Env. Biotechnology*. NEERI, Nagpur. pp. 97-123.

Michell, R., 1979. *Imprint*. p. 180.

Milborrow, B.V., 1981. In L.G. Paley and D. Aspinall eds. pp. 347-383. Acad. Press, Sydney, Australia.

Miller, L.K., Lingg, A.J., and Bulla, L.A. Jr., 1983. *Science*, 219: 715-721.

Mishra, S.G., and Mani, D., 1992. *Metallic pollution*. Ashish Publishing House, New Delhi.

Mishra, S.G., and Shukla, P.K., 1992. In *Indian environment*. Ed. P. Singh. Ashish Publishing House, New Delhi. pp. 95-102.

Misra, R., 1971. In *Proceedings of the school of plant ecology*. Oxford & IBH Publ. Co, New Delhi. pp. 279-285.

Mizoguchi, T., Sato, T., and Okabe, T., 1976. *J. Fermentation Technol.* 54: 181-183.

Mock, J.J., and Eberhart, S.A., 1972. *Crop. Sic.* 12: 446.

Mookerjee, A., 1994. *Symp. Workshop Environ. Biotechnology*. NEERI, Nagpur. pp. 9-60.

Moore, L., Warren, G., and Strobel, G., 1979. *Plasmid*. 2: 617-626.

Mori, H., Yamane, T., Kobayashi, T., and Shimizu, S., 1983. *J. Fermentation Technol.* 61: 305-314.

Motwani, M.P., Banerjee, S., and Karamchandani, S.J., 1956. *Indian J. Fish.* 3: 334-367.

Murakami, T., Anzai, H., Imai, S., Satoh, A., Nagaoka, K., and Thompson, C.J., 1986. *Mol Gen. Genet.* 205: 42.

Murr, L.E., Torma, A.E., and Brierley, J.A., 1978. *Metallurgical application of bacterial leaching and related microbiological phenomenon.* Academic Press, New York.

Murray, L.E., Elliott, L.G., Capitant, S.A, West, J.A., Hanoson, K.K., Scarafia, L., Jhonston, S., DeLuca-Flacherty, C., Nichols, S., Cunanon, D., Dietrich, P.S., Mettles, E.J., Dewald, S., Warnick, D.A., Rhodes, C., Sinibaldi, R.M., and Brunke, K.J., 1993. *Biotechnology.* 11: 1559.

Muthana, K.D., 1977. *Aridzone Research in India.* CAZRI, Jodhpur. pp. 98-101.

Nair, K.V.K., Kannan, V., and Sebastian, S., 1982. *Indian J. Environ. Hlth.* 24: 277-284.

Nag, K.N., and Singh, P., 1985. In *Bioenergy for arid and semiarid zones.* Ed. A.N. Mathur. Himanshu Publications, Udaipur. pp. 87-96.

Narayansamy, P., 1995. *Biotechnology and Development Review,* 4: 15-20.

Neelakanton, L.S., 1975. *Agric Situation in India.* 30: 9.

Nehra, N.S., Chibbar, R.N., Leung, N., Caswell, K., Mallard, C., Steinhauer, L., Baga, M., and Kartha, K.K., 1994. *Plant J.* 5: 285-297.

Nelson, O.E., 1979. *Adv. Agron.* 21: 171.

Nemathy, E.K., Otvos, J.W., and Calvin, M., 1980. Hydrocarbons from Euphorbia lathyris. Reprint No. 11351. Lawrance Berkeley Laboratory. University of California.

Nirmala, C., and Kaul, M.L.H., 1975. *J. Phytol. Res.* 8: 1-26.

Novak, J.T., and Ramesh, M.S., 1975. *Water Res.* 9: 11: 963.

Oakes, J.V., Shewmaker, C.K., and Stalker, D.M., 1991. *Biotechnology,* 9: 982-986.

Ohad, N., and Hirchberg, J., 1990. *Photosynthesis Research,* 23; 73.

Ohtani, T., Galili, G., Wallace, J.C., Thompson, G.A., and Larkins, B.A., 1991. *Plant Mol. Biol.* 16: 117.

Omirulleh, S., Abraham, M., Golovkin, M., Stefanov, I., Karabaev, M.K., Mustardy, L., Monocz, S., Dudits, D., 1993. *Plant Mol. Biol.* 21: 415-428.

Orton, T.J., 1980. *Zeitschrift fur pflanzenphysiologie.* 105: 98.

Owen, O.S., 1971. *Natural resource conservation.* The Macmillan Co, New York.

Oza, G.M., 1984. Desertification—a global problem. *Environmental Awareness.* 7: 1-3.

Pandeya, J.D., 1985. *Energy Environment Monitor.* 1: 21-30.

Paszkowski, J., Shillito, R.D., Saul, M., Mendak, V., Hohn, T., and Potrykus, I., 1984. *EMBO J.* 3: 2717-2722.

Pathak, P.V., and Colah, J.J., 1976. *Science Today.* 11: 45-46.

Patterson, B.D., and Payne, L.A., 1983; *Hort. Sci.* 18: 340.

Payne, C.C., 1982. *Parantology.* 85: 35-77.

Pen. J., Molendik, L., Quax, W.J., Sijmons, P.C., VanOoyen, A.J.J., Vanden Elzen, P.J.M., Rietveld, K., and Hoekema, A., 1992. *Biotechnology.* 10: 292-296.

Peng, J., Wen, W., Lister, R.L., and Hodges, T.K., 1995. *Plant. Mol. Biol.* 27: 91-104.

Perlak, F., Fuchs, R., Dean, D., McPherson, S., and Fischhoff, D., 1991. *Proc. Natl. Acad. Sci.* USA. 88: 3324.

Perrin, D.R., and Bottomley, W., 1962. *J. Amer. Chem. Soc.* 84: 1919.

Peters, G.A., 1979. In *Genetic engineering for nitrogen fixation.* Eds. A. Hollaender et al. Plenum Press, London. pp. 231-258.

Peters, G.A., Calvery, H.E., Kaplan, D., Ito, O., and Tota, R.E., 1982. *Isreal J. Bot.* 31: 305-323.

Poehlman, J.M., and Sleper, D.A., 1995. *Preeding field crops.* Iowa State Univ. Press Ames. Iowa.

Pool, R., 1989. *Science,* 245: 1187-1189.

Potrykus, I., 1990. *Biotechnology.* 8: 535-542.

Potrykus, I., Saul, M., Petruska, J., Palzkowski, J., and Shillito, R.D., 1985. *Mol. Gen. Gent.* 199: 183-188.

Powel-Abel., Nelson, R., Hoffman, N., Rogers, S., and Fraley, R., 1986. *Science.* 232: 738.

Powell, P., Stark, D., Sanders, P., and Beachy, R.N., 1989. *Proc. Natl. Acad. Sci.* USA. 86: 6949.

Prakash, C., Prakash, P., and Rawat, D.C., 1980. *IAWPC tech. Annual.* 5: 140-148.

Prentis, S., 1985. *Biotechnology: New industrial revolution.* Orbis, London.

Raineri, D.M., Bottino, P., Gordon, M.P., and Nester, E.W., 1990. *Biotechnology.* 8: 33-38.

Raj, M., 1977. In *Current trends in Indian environment.* eds. Deshbandhu and E. Chauhan. Today and Tomorrow, New Delhi. pp. 99-106.

Rajor, A., and Goswamy, K.P., 1983. *Sci. & Cult.* 49: 357-359.

Randhawa, N.S., and Maheshwari, R.C., 1981. *Society and Science.* 5: 95-120.

Ramachandran, T.V., 1985. Ph.D. Thesis, Bombay University, Bombay.

Ramade, F., 1984. *Ecology of natural resources.* John Wiley & Sons. Chichester.

Rathore, K.S., Chowdhury, V.K., and Hodges, T.K., 1993. *Plant Mol. Biol.* 21: 871-884.

Ravishankar, G.A., and Vankataraman, LV., 1991. In *Microbial gene technology*. Ed. H. Polasa. South Asian Publishers, New Delhi. pp. 185-196.

Ray, T.B., 1984. *Plant Physiol.* 75: 827.

Razzet, W.E., 1962. *Canadian Inst. Min. Metall. Bull.* 65: 135-136.

Razzet, W.E., and Trussel, P.C., 1963. *Appl. Microbiol.* 11: 105-109.

Ream, W., 1989. *Ann. Rev. Phytopathol.* 27: 583-618.

Rehm, H.J., and Reed, G., 1986. *Biotechnology.* Vol. 8. VCH Weinhuim.

Reiz, P.J., 1989. In *The biology of wool and hair.* Eds. G.E. Rogers et al. Chapman and Hall. pp. 185-203.

Rezian, M., Skene, K., and Ellis, J., 1988. *Plant Mol. Biol.* 11: 463.

Rhodes, S.L., and Middleton, P., 1983. *Environmental Forum.* 2: 32-35.

Riggs, C.D., and Bates, G.W., 1986. *PNAS USA.* 83.

Roberts, C.H., Grihiey, J., and Willam, E.H., 1940. *Chemical methods for the study of river pollution.* London.

Roby, D., Broglie, K., Cressman, R., Biddle, P., Chet, I., and Broglie, R., 1990. *Plant Cell.* 2: 999.

Roth, F. X., 1962. In *Advances in agricultural microbiology.* Ed. N.S. Subba Rao. Oxford and IBH Publ. Co., New Delhi. pp. 663-676.

Ryder, M.L., and Stephenson, S.K., 1968. *Wool growth.* Academic Press, New York.

Sachs, M.M., and Ho, T.H.D., 1986. *Ann. Rev. Plant Physiol.* 37: 363.

Sanders, P.R., Sammons, B., Kaneewski, W., and Haley, L., 1992. *Phytopathology.* 82: 683-689.

Sanford, J.C., Klein, T.M., Wolf, E.D., and Allen, N., 1987. *Particle Sci. Technol.* 5: 27-37.

Sasson, A., 1984. *Biotechnology: Challenges and promises.* Oxford IBH Publishing Co, New Delhi.

Sassen, A., 1987. *Development and South-South Cooperation.* 3: 95-108.

Sastry, C.A., 1986. In *Pollution control handbook.* Ed. P.L. Diwakar Rao. Utility Publications, Secunderabad. pp. 39-48.

Sato, M., 1960. *Econ Geol.* 55: 1202-1205.

Satpathy, K.B., and Singh, P.K., 1985. In *Proc. Symp. on Newer Approaches to biological applications.* M.S. University, Baroda. pp. 270-276.

Saxena, M.M., 1990. *Applied environmental biology.* 1. *Resources and management.* Agrobotanical Publishers, Bikaner.

Schaeffer, G.W., and Sharpe, F.T., 1981. *In-vitro.* 17: 345-352.

Schlappi, M., and Hohn, B., 1992. *Plant Cell.* 21: 705-708.

Schroeder, H.A., 1965. *J. Chron. Dis.* 18: 647-656.

Seetharam, P.L., Thulajappa, Y., and Chavan, R.R., 1995. *Text book of biology.* Expert educational publishers, Bangalore.

Sen, B.P., and Bhaskaran, T.R., 1962. *J. Water Pollut. Contr. Fed.* 34: 10-105.

Sewart, W.D.P., 1970. *Nature,* 214: 603.

Shah, D., Gasser, C., Della-Cioppa, G., and Kishore, G., 1988. In *Plant gene research temporal and spetial regulation of plant genes.* eds. P.S. Verma. Vol. 5. Springer Verlag. pp. 297.

Shannon, J.C., and Garwood, D.L., 1984. In *Starch, Chemistry and Industry.* Eds. R.L. Whistler, E.F. Paschall, and J.N. BeMiller. Academic Press, New York: pp. 26-86.

Sharma, S., 1984. *Sci. Reptr.* 21: 426-427.

Sharma, V.K., 1991. In *Botanical researches in India.* Eds. N.C. Aery and B.L. Chaudhary. Himanshu Publications, Udaipur. pp. 213-216.

Sharples, F.E., 1983. Testimony before US House Committee on science and technology. Hearing on environmental implications of genetic engineering. June 22. Committee Print No. 26.

Shaver, D.L., and Anderson, P.C., 1985. In *Biotechnology in plant science: Relevance to agriculture in the eighties.* Eds. M. Zaitlin, P. Day and A. Hollaender. Academic Press, New York. p. 287.

Shekhawat, N.S., 1991. In *Botanical researches in India.* Eds. N.C. Aery and B.L. Chaudhary. Himanshu Publications, Udaipur. pp. 535-543.

Shivraj, D., and Seenayya, G., 1994. *Indian J. Environ. Hlth.* 36: 11-118.

Shotwel, M.A., and Larkins, B.A., 1989. *Biochem. Plant.* 15: 297.

Shrivastava, G.H., and Singh, V.P., 1990. *Oikoassay,* 7: 5-7.

Shuval, H.A., et al., 1986. *Waste water irrigation in developing countries: health effects and technical solutions.* The World Bank, Washington DC.

Silverman, M.P., and Ehrlich, H.K., 1964. *Adv. Appl. Microbiol.* 6: 153-206.

Sinha, C.S., and Kishore, V.V.N., 1991. *TERI Information Digest on Energy.* 1: 1-24.

Singh, A.J., and Bhullar, A.S., 1990. *Indian. J. Environ. Hlth.* 32: 407-412.

Singh, A.L., and Singh, P., 1989. *Nitrogen fraction in Indian rice fields (Azolla and blue green algae).* Agro Botanical Publishers, Bikaner.

Singh, D.P., and Singh, P.K., 1989. *Plant & Soil.* 114: 205-209.

Singh, J.S., Singh, S.P., and Jeet Ram., 1988. Report from a study sponsored by Planning Commission, Government of India for study Group on Fuel and Fodder Working Committee. p. 30.

Singh, P.K., 1977. *Curr. Sci.* 46: 642-644.

Singh, P.K., 1979 a. In *Nitrogen and rice.* IARI, Los Banos, Philippines.

Singh. P.K., 1979 b. *Ann. Rev. Plant Sci.* pp. 37-65.

Singh, P.K., 1980. *Curr. Sci.* 49: 155.

Singh, R.N., 1961. *The role of blue green algae in nitrogen economy of Indian agriculture.* ICAR, New Delhi.

Sleat, D.E., and Palukaitis, P., 1990. *Proc. Natl. Acad. Sci.* USA. 87: 2946.

Slogteren, G.M.S., Hooykaas, P.J.J., and Shilperoot, R.A., 1984. *Nature,* 311: 763-764.

Smith, K.R., 1985. *Traditional biomass fuels and air pollution: A global review.* Plenum Publishing Co., New York.

Smith, K.R., Aggarwal, A.L., and Dave, J.M., 1983. *Atmospheric environment.* 17: 2343-2362.

Smith, L.W., Goering, H.K., and Gorden, C.H., 1969. *Proceedings of conference on animal waste management.* Cornell University, Ithaca. New York. pp. 88-97.

Smith, S., and Walker, N., 1981. *New Phytol.* 89: 225-240.

Songstad, D.D., Halaka, F.G., Deboer, D.L., Armstrong, C.L., Hinchee, M.A.W., Ford-Sautino, C.G., Brown, S.M., Fromm, M.E., and Horsch, R.B., 1993. *Plant cell tiss. Org. Cult.* 33: 195-201.

Soper, S.R., 1985. In *The leaf hoppers and plant hoppers.* Eds. L.R. Kault and J.G. Radriguez. John Wiley & Sons, New York. 00.469-485.

Soriano, P., and Jaenisch, R., 1986. *Cell,* 46: 19.

Southwood, T.R.E., 1976. In *Theoretical ecology: Principles and practices.* Ed. R.M. May, Saunders, Philadelphia. pp. 26-48.

Spencer, T.M., O'Brien, J.V., Start, W.G., Adams, T.R., Gordon-Kamm, W.J., and Lemaux, P.G., 1992. *Plant Mol. Biol.* 18: 201-210.

Srivastava, P.K., and Maheswari, R.C., 1981. Position paper on utilization of agricultural wastes and byproducts in India. Paper presented in xviii Annual Convension of ISAE held at CSSRI, Karnal. Feb. 26-28.

Stainer, R.Y., Inagraham, J.L., Wheetis, M.L., and Painter, P.R., 1987. *General microbiology,* 5th Ed. Macmillan Educational Hall, London.

Stark, D., and Beachy, R., 1989. *Biotechnology.* 7: 1257.

Steinrucken, H., Schulz, A., Amrhein, N., Porter, C.A., and Fraley R.T., 1986. *Arch. Biochem. Biophys.* 244: 169.

Steponkus, P.L., Uemura, M., Balsamo, R.A., Arvinte, T., and Lynch, D.V., 1988. *Proc. Natl. Acad. Sci. USA.* 86: 9026.

Steward, K.K., 1970. *Hyacith. Contr. J.* 8: 34-35.

Stewart, L.M.D., Hirst, M., Ferber, M.L., Merryweather, A.T., Cayley, P.J., and Possee, R.D., 1991. *Nature.* 352: 85-88.

Stripe, F., Barbeiri, L., Baitelli, M.G., Soria, M., and Lappi, D.A., 1992. *Biotechnology.* 10: 405-412.

Strauss, M., 1990. *The World Hlth Forum.* 11: 46-59.

Streber, W.R., and Willmitzer, L., 1989. *Microb. Rev.* 53: 242.

Stuckey, D.C., 1986. In *Biogas technology transfer and diffusion.* Ed. M.M. El-Halwagi. Elsevier, London. pp. 18-44.

Subba Rao, N.S., 1982. In *Advances in agricultural microbiology*. Ed. N.S. Subba Rao. Oxford & IBH Publ. Co, New Delhi. pp. 219-242.

Sundarrarajan, R., Jayanthi, S., and Sadhasivan, P., 1996. *Indian J. Environ. Hlth.* 38: 7-12.

Suwalka, R.L., and Karan, F., 1990. *Acta Ecol.* 12: 60-63.

Suwalka, R.L., Karna, F., and Qureshi, F.M., 1990. *Acta Ecol.* 12: 56-58.

Suwalka, R.L., Qureshi, F.M., Chaudhary, S.R., and Sharma, N.N., 1991. *Acta Ecol.* 13: 63-65.

Swaminathan, M.S., 1991. *Biotechnology in agriculture: A dialague.* Macmillan, Madras.

Sylvester, R.O., 1961. *Tech. Rep. Tat. Sanit. Eng. Centre.* W. 61-3: 80-81.

Tal, M., 1990. In *Somaclonal variation in crop improvement.* Vol. 1. Eds. Y.P.S. Bajaj. Springer Verlag. pp. 236-257.

Tato, T., and Lundgren, D.G., 1978. *Appl. Environ. Microbiol.* 36: 1198-1201.

Tautonico, R.A., and Knorr, D., 1984. *Food Technology.* 120-127.

Titus, J.A., Luli, G.W., Dekleva, M.L., and Strohl, W.R., 1984. *Appl. Environ. Microbiol.* 47: 239-244.

Tomalski, M.D., and Miller, L.K., 1991. *Nature,* 352: 82-85.

Thomas, H., 1985. In *The global possible: Resource development and the new century.* Ed. R. Repetto. Yale University Press. pp. 33-46.

Thomashaw, M.F., 1990. *Ad. Genet.* 28: 99.

Thompson, C., Movva, N., Tizard, R., Grameri, R., Desies, J., Lauwerys, M., and Botterman, J., 1987. *EMBO J.* 6: 2519.

Torma, A.E., 1977. In *Adv. Biochem. Engng.* Eds. T.K. Ghosh, A. Freshter and N. Blackebrongh, Springer, NY.

Torma, A.E., and Sakaguchi, H., 1978. *J. Ferment. Technol.* 56: 173-176.

Trindade, S.C., 1980. Energy crops: The case of Brazil. Paper presented at International conference on energy from biomass. Brighton, 4-7 November.

Tripathi, A.K., Khare, S., and Ganguli, A., 1994. *Symp. Workshop Environ. Biotechnology.* NEERI, Nagpur. pp. 149-157.

Trivedi, R.K., Ashtekar, P.V., Patankar, S.Y., and Lokhande, S.V., 1982. In *Advances in environmental research.* Ed. S.K. Agarwal, IEO, Kota. pp. 53-60.

Trivedi, R.K., 1982. *Proc. Natl. Conf. Environ. Kolhapur.* Environmental Academy, Kolhapur. pp. 59-66.

Trudinger, P.A., 1971. *Mineral Sci. Engng.* 3: 13-15.

Tyagi, S., 1991. In *Botanical researches in India.* Eds. N.C. Aery and B.L. Chaudhary. Himanshu Publications, Udaipur. pp. 217-218.

Tyrell, B.J., Davidson, T.A., Bulla, L.A. Jr., and Ramoska, W.A. 1979. *Appl. Environ. Microbiol.* 38: 656.

Udvard, M.D.F., 1975. *A. classification of the biogeographical systems of the world.* IUCN, Occasional Paper. No. 18. Morges, Switzerland.

Ui, J., 1974. *Kogai,* 4: 14.

Ui, J., 1980. *Kogai,* No. 8: 1-24.

Underwood, E.J., 1977. *Trace elements in human and animal nutrition.* Academic Press, New York.

Vaeck, M., Raymaerto, A., Hofte, H., Jensens, S., De Beuckeller, M., Dean, C., Zabeau, M., Van Montagu, M., and Leemans, J., 1987. *Nature,* 328: 33.

Van Dekerckhove, J., et al., 1989. *Biotechnology.* 7: 929.

Varmus, H., 1982. *RNA Tumor viruses.* Eds. R. Weiss. N. Teich, H. Varmus, J. Collin. pp. 369-512.

Varshney, C.K., and Garg, K.K., 1978. *J. Air Pollut. Contr. Assoc.* 28: 1141-1142.

Vasil, I.K., 1991. *Curr. Opinion Biotechnol.* 2: 158-163.

Vasil, I.K., and Vasil, V. 1992. *Physiol. Plant.* 85: 279-283.

Vasil, V., Castillo, A.M., Fromm, M.E., and Vasil, I.K., 1992. *Biotechnology,* 10: 667-675.

Venkataraman, G.S., 1972. *Algal biofertilizers and rice cultivation.* Today and Tomorrow, New Delhi.

Vergara, W., et al, 1981. Biomass energy production system in Brazil. OLADE Biomass Seminar, Curitiba (Parana).

Verma, S.R., and Shukla, G.R., 1969. *Indian J. Environ. Hlth.* 11: 145-162.

Vigers, A., Roberts, W., and Selitrennikoff, C., 1991. *Mol. plant Microbe interaction.* 4: 315.

Vigers, A.J., Wiedemann, S., Roberts, W.K., Legrand, M., Schtrennikoff, C.P., and Fritig, B., 1992. *Plant Sci.* 83: 155-161.

Vijayan, M., 1988. *Sci. Reptr.* 25: 176-178.

Vimal, O.P., and Tyagi, P.D., 1988. *Bioenergy spectrum.* Bioenergy and wasteland development organisation, New Delhi.

Viswanathan, R., 1957. *Indian J. Med. Res.* 45: Suppl. No. 1.

Viraraghavan, T., 1980. *IAWPC Tech. Annual.* 6 & 7: 80-82.

Voelker, T., Herman, E., and Chrispeels, M., 1989. *Plant Cell.* 1: 95.

Von Ethen, H.D., Mathews, D.E., and Mathews, P.S., 1989. *Ann. Rev. Phytopathology.* 27: 143.

Vunsh, R., Ratilsky, M.B., Zur, K.M., and Robenfield, B., 1986. In *Plant tissue and cell culture.* eds. E.A. Somers et al. University of Minnesota. p. 119.

Vyas, L.N., 1991. In *Automobile pollution.* Ed. S.K. Agarwal, Ashish Publ. House, New Delhi. pp. 175-176.

Wagner, M., Amann, R., Lemmer, H., and Schleifer, K.H., 1993. *Appl. Environ. Microbiol.* 59: 1520-1525.

Warren, R.S., and Gould, A.R., 1982. *Zeitschrift fur Pflanzenphysiologie.* 107: 347.

Wate, S.R., Chakrabarti, T., and Subramanyam, P.V.R., 1983. *Indian J. Environ. Hlth.* 25: 179-190.

Waterworth, H.E., Kraper, J.M., and Tousignant, M.E., 1979. *Science,* 204: 845.

Webb, M., 1993. In *Molecular biology and bitechnology.* eds. J.M. Walkar and E.B. Gingold. Panima Educational Book Agency, New Delhi. pp. 389-411.

Weeks, J.T., Anderson, O.D., and Blechl, A., 1993. *Plant Physiol.* 102: 1077-1084.

W.H.O., 1969. *WHO Expert Committee on Amoebiasis Report.* Geneva. Tech. Rep. Ser. No. 421.

W.H.O., 1972. *Health Hazards of the Human Environment.* WHO. Geneva.

W.H.O., 1978. *W.H.O. Chronical.* 32: 302-310.

W.H.O., 1980. *Analysis and interpreting air monitoring data.* Offset Publications No. 51, Geneva.

Wilkins, T.A., Badnarrek, S.Y., and Raikhel, N.V., 1990. *Plant Cell.* 2: 301.

Willison, K., Rabinet, C., Bocdra, C., Kelley, P., 1983. In *Coldspring Harbor Cof. Cell Proliferation.* 10: 307.

Wolfe, R.S., 1979. In *Microbial biochemistry.* Vol. 21. Ed. J.R. Quayle. University Printing Press, Baltimore.

Woloshuk, C.P., Meulenhoff, J.S., Sela-Burriage, M., Van den Elzen, P.J.M., and Cornelissen, B.J.C., 1991. *Plant Cell.* 3: 619-628.

Yadav, A.K., Adkar, S.S., and Nigam, R., 1992. In *Perspectives in Pteridology: Present and future.* Part II. Eds. T.N. Bhardwaja and C.B. Gena. Today and Tomorrow, New Delhi. pp. 487-493.

Ying, C., 1983. In *Cell and Tissue culture techniques for cereal crop improvement.* Science Press, Beijing. pp. 11-26.

Yoder, J.I., and Goldbrought, A.P., 1994. *Biotechnology.* 12: 263-267.

Young, V., and Trindade, S.C., 1978. The Brazilian gashol programme. Symposium on energy from biomass and waste paper no. 42. p. 815. Institute of gas technology, Chicago.

Zachrisson, A., and Borman, C.H., 1986. *Physiol. Plant.* 67: 507-516.

Zhang, H.M., Yang, H, Rech, E.L., Golds, T.J., Davis, A.S., Mulligan, B.J., Cocking, E.C., and Davey, M.R., 1989. *Plant Cell Rep.* 7: 379-384.

Zhang-Yi, L., Yia, G.Y., Chen, H., and Guo, G.Q., 1992. *J. Plant Physiol.* 139: 414-418.

Zurawski, G., Bohnert, H.J., Whitefield, P.R., and Bottomlay, W., 1982. *Proc. Natl. Acad. Sci. USA.* 79: 7699.

Subject Index